Matroid Decomposition

1994

Matroid Decomposition

K. Truemper

University of Texas at Dallas
Richardson, Texas

ACADEMIC PRESS, INC.

Harcourt Brace Jovanovich, Publishers

Boston   San Diego   New York
London   Sydney   Tokyo   Toronto

Copyright © 1992 by Academic Press, Inc.
All rights reserved.
No part of this publication may be reproduced or
transmitted in any form or by any means, electronic
or mechanical, including photocopy, recording, or
any information storage and retrieval system, without
permission in writing from the publisher.

ACADEMIC PRESS, INC.
1250 Sixth Avenue, San Diego, CA 92101-4311

United Kingdom Edition published by
ACADEMIC PRESS LIMITED
24–28 Oval Road, London NW1 7DX

Library of Congress Cataloging-in-Publication Data

Truemper. K., 1942–
    Matroid decomposition / K. Truemper.
        p.   cm.
    Includes bibliographical references and indexes.
    ISBN 0-12-701225-7 (acid-free paper)
    1. Matroids. 2. Decomposition (Mathematics)   I. Title.
QA 166.6.T78   1992
511' .5—dc20   92-7074
                                                  CIP

Printed in the United States of America
92 93 94 95   9 8 7 6 5 4 3 2 1

Contents

v

Preface

Matroids were first defined in 1935 as an abstract generalization of graphs and matrices. In the subsequent two decades, comparatively few results were obtained. But starting in the mid-1950s, progress was made at an ever-increasing pace. As this book is being written, a large collection of deep matroid theorems already exists. These results have been used to solve difficult problems in diverse fields such as civil, electrical, and mechanical engineering, computer science, and mathematics.

There is now far too much matroid material to permit a comprehensive treatment in one book. Thus, we have confined ourselves to a part of particular interest to us, the one dealing with decomposition and composition of matroids. That part of matroid theory contains several profound theorems with numerous applications. At present, the literature for that material is quite difficult to read. One of our goals has been a clear and simple exposition that makes the main results readily accessible.

The book does not assume any prior knowledge of matroid theory. Indeed, for the reader unfamiliar with matroid theory, the book may serve as an introduction to that beautiful part of combinatorics. For the expert, we hope that the book will provide a pleasant tour over familiar terrain.

The help of many people and institutions has made this book possible. P. D. Seymour introduced me to matroids and to various decomposition notions during a sabbatical year supported by the University of Waterloo. The National Science Foundation funded the research and part of the writing of the book through several grants. Most of the the writing was made possible by the support of the Alexander von Humboldt-Foundation and of the University of Texas at Dallas, my home institution. The University of Bonn and Tel Aviv University assisted the search for and verification of reference material.

M. Grötschel of the University of Augsburg made the resources of the Institute of Applied Mathematics available for the editing, typesetting, and proofreading. He also supported the project in many other ways. P. Bauer, M. Jünger, A. Martin, G. Reinelt, M. Stoer, and G. Ziegler of the University of Augsburg were of much assistance.

T. Konnerth most ably typeset the manuscript in TeX. R. Karpelowitz and C.-S. Peng patiently prepared the numerous drawings.

A number of people helped with the collection of reference material, in particular S. Fujishige and M. Iri.

R. E. Bixby, A. Bouchet, T. J. Reid, G. Rinaldi, P. D. Seymour, M. Stoer, A. Tamir, and U. Truemper reviewed a first draft. Their critique helped considerably to clarify and simplify material.

To all who so generously gave of their time and who lent support in so many ways, I express my sincere thanks. Without their help, the book would not have been written.

Chapter 1

Introduction

1.1 Summary

A matroid may be specified by a finite set E and a nonempty set of so-called independent subsets of E. The independent subsets must observe two simple axioms. We will introduce them in Chapter 3. Matroids generalize the concept of linear independence of vectors, of determinants, and of the rank of matrices. In fact, any matrix over any field generates a matroid. But there are also matroids that cannot be produced this way.

With matroids, one may formulate rather compactly and solve a large number of interesting problems in diverse fields such as civil, electrical, and mechanical engineering, computer science, and mathematics. Frequently, the matroid formulation of a problem strips away many aspects that at the outset may have seemed important, but that in reality are quite irrelevant. Thus, the matroid formulation affords an uncluttered view of essential problem features. At the same time, the matroid formulation often permits solution of the entire problem, or at least of some subproblems, by powerful matroid techniques.

In this book, we are largely concerned with the binary matroids, which are produced by the matrices over the binary field GF(2). That field has just two elements, 0 and 1, and addition and multiplication obey very simple rules. Any undirected graph may be represented by a certain binary matrix. The graphic matroid produced by such a matrix is an abstraction of the related graph. Thus, binary matroids generalize undirected graphs.

During the past forty years or so, a large number of profound matroid theory results have been produced. A significant portion of these results concerns properties of binary matroids and related matroid decompositions and compositions. Unfortunately, much of the latter material is not easily accessible. This fact, and our own interest in combinatorial decomposition and composition, motivated us to assemble this book.

As we started the writing of the book, we faced a basic conflict. On one hand, we were tempted to prove all matroid results with as much generality as possible. On the other hand, we were also tempted to restrict ourselves to binary matroids, since the proofs would become less abstract. A major argument in favor of the second viewpoint was that the matroid classes analyzed here are binary anyway. Thus, that viewpoint won. Nevertheless, we have mentioned extensions of results to general matroids whenever such extensions are possible.

We proceed as follows. Chapter 2 contains basic definitions concerning graphs and matrices. In Chapter 3, we motivate and define binary matroids. We also prove a number of basic results. In particular, we classify whether the elements of a matroid are loosely or tightly bound together, using the idea of matroid separations and of matroid connectivity. In addition, we learn to shrink matroids to smaller ones by two operations called deletion and contraction. Any such reduced matroid is called a minor of the matroid producing it. Finally, we derive from any matroid another matroid by a certain dualizing operation. Appropriately, the latter matroid is called the dual matroid of the given one.

Chapters 4–6 contain fundamental matroid constructions, tools, and theorems. Chapter 4 is concerned with some elementary constructions of graphs and binary matroids. The constructions rely on replacement rules called series-parallel steps and delta-wye exchanges. In Chapter 5, we introduce a simple yet effective method called the path shortening technique. With its aid, we establish basic connectivity relationships and certain results about the intersection and partitioning of matroids. Chapter 6 contains another elementary matroid tool called the separation algorithm, which identifies certain matroid separations.

The techniques and results of Chapters 4–6 are put to a first use in Chapters 7 and 8. In Chapter 7, we prove the so-called splitter theorem, which links connectivity of a given matroid with the presence of certain minors. With that theorem, we show that a sufficiently connected matroid always contains minors that form a sequence with special properties. In Chapter 8, we establish fundamental notions and theorems about matroid decomposition and composition.

With Chapter 9, we begin the second half of the book. That chapter provides fundamental facts about a very important property of real matrices called total unimodularity. Several translations of the total unimodularity property into matroid language are possible. In one such translation, total unimodularity becomes a property of binary matroids called regularity. Establishing a real matrix to be totally unimodular then becomes equivalent to proving that a certain binary matroid is regular.

In Chapters 10–13, we prove a number of decomposition and composition results about the class of regular matroids and about other, closely related matroid classes. In Chapter 10, we begin with an analysis of the graphic matroids, which are regular. In Chapter 11, we examine the remaining regular matroids, i.e., the nongraphic ones. In Chapter 12, we explore nonregular matroids that, loosely speaking, have many regular minors. The matroids with that property are called almost regular. Finally in Chapter 13, we investigate flows in matroids by borrowing ideas from flows in graphs. A well-known result about the behavior of flows in graphs is the max-flow min-cut theorem. The matroids whose flows exhibit the nice behavior described in that theorem are called the max-flow min-cut matroids. The investigation of Chapter 13 focuses on these matroids.

For each of the classes of matroids mentioned so far (graphic, regular, almost regular, max-flow min-cut), Chapters 9–13 provide polynomial testing algorithms, representative applications, and, except for the almost-regular case, characterizations in terms of excluded minors. In addition, excluded minor characterizations of the binary matroids and of the ternary matroids are given in Chapters 3 and 9, respectively. The ternary matroids are the matroids produced by the matrices over GF(3).

The book may be read as follows. First, one should cover Chapters 2 through 9. During a first reading, one may skip the proofs of the chapters. Chapters 10–13 are largely independent. Thus, one may read them in any order, provided one is willing to occasionally interrupt the reading of a chapter for a quick glance at some auxiliary result of an earlier chapter. In the first section of each chapter, we list relevant earlier chapters, if any.

1.2 Historical Notes

In 1935, H. Whitney realized the mathematical importance of an abstraction of linear dependence. His pioneering paper (Whitney (1935)) contains a number of equivalent axiomatic systems for matroids, and thus laid the foundation for matroid theory. In the 1950s and 1960s, W. T. Tutte built upon H. Whitney's foundation a remarkable body of theory about the structural properties of matroids. In the 1960s, J. Edmonds connected matroids with combinatorial optimization. Within a few years, he produced several key results. In the process, he popularized matroid theory.

From 1965 on, an ever growing number of researchers became interested in matroids. In 1976, D. J. A. Welsh published a book (Welsh (1976)) that contained essentially all results known at that time. As these notes are written, a comprehensive treatment of matroid theory in one book is no longer possible. Selected topics are covered in Crapo and Rota (1970),

Lawler (1976), Aigner (1979), Lovász and Plummer (1986), White (1986), (1987), (1991), Kung (1986c), Schrijver (1986), Murota (1987), Recski (1989), Fujishige (1991), Oxley (1991), Bachem and Kern (1992), and Björner, Las Vergnas, Sturmfels, White, and Ziegler (1992). Kung (1986c) includes an excellent historical survey.

Central to this book is the work of P. D. Seymour, W. T. Tutte, and K. Wagner. In historical order, the key results are as follows: K. Wagner's decomposition of the graphs without minors isomorphic to the complete graph on five vertices (Wagner (1937a)); W. T. Tutte's characterization of the regular and graphic matroids (Tutte (1958)) and his efficient test of graphicness (Tutte (1960)); P. D. Seymour's characterization of the max-flow min-cut matroids (Seymour (1977a)), his decomposition of the regular matroids (Seymour (1980b)), and his results on matroid flows (Seymour (1981a)).

Some years ago, these results motivated us to start a systematic investigation using graph and matroid decomposition and composition as main tools (Truemper (1985a), (1985b), (1986), (1987a), (1987b), (1988), (1990), (1991), (1992), Tseng and Truemper (1986)). Except for some minor modifications and simplifications, this book comprises a large portion of that effort.

Due to space constraints, the book does not include details of several important matroid results that are related to the material covered here. In particular, we have omitted the principal partitioning results and related earlier material by M. Iri, N. Tomizawa, and others (Kishi and Kajitani (1967), Tsuchiya, Ohtsuki, Ishizaki, Watanabe, Kajitani, and Kishi (1967), Iri (1969), Bruno and Weinberg (1971), Ozawa (1971), Tomizawa (1976b), Iri (1979), Nakamura and Iri (1979), Narayanan and Vartak (1981), Tomizawa and Fujishige (1982), Murota and Iri (1985), and Murota, Iri, and Nakamura (1987)). Tomizawa and Fujishige (1982) provide a detailed historical survey of the work on principal partitions. We should also mention L. Lovász' matroid matching results (Lovász (1980), see also Lovász and Plummer (1986)). That work is not really related to the contents of this book. We are compelled to mention it here since it is one of the very profound achievements in matroid theory. We also have not included, but should mention here, work on oriented matroids. These matroids were independently defined by Bland and Las Vergnas (1978) and Folkman and Lawrence (1978). Two recent books cover most of the known results for oriented matroids (Bachem and Kern (1992), and Björner, Las Vergnas, Sturmfels, White and Ziegler (1992). Finally, many important matroid applications are described in Iri and Fujishige (1981), Iri (1983), Murota (1987), and Recski (1989).

Chapter 2

Basic Definitions

2.1 Overview and Notation

This chapter covers basic definitions about graphs and matrices, and the computational complexity of algorithms. For a first pass, the reader may just scan the material.

We first introduce notation and terminology connected with sets. An example of a set is $\{a, b, c\}$, the set with a, b, and c as elements. With two exceptions, all sets are assumed to be finite. The exceptions are the set of real numbers $\mathbb{R}$, and possibly the set of elements of an arbitrary field $\mathcal{F}$.

Let S and T be two sets. Then $S \cup T$ is $\{z \mid z \in S \text{ or } z \in T\}$, the *union* of S and T. The set $S \cap T$ is $\{z \mid z \in S \text{ and } z \in T\}$, the *intersection* of S and T. The set $S - T$ is $\{z \mid z \in S \text{ and } z \notin T\}$, the *difference* of S and T. The set $(S \cup T) - (S \cap T)$ is the *symmetric difference* of S and T.

Let T contain all elements of a set S. We denote this fact by $S \subseteq T$ and declare S to be a *subset* of T. We write $S \subset T$ if $S \subseteq T$ and $S \neq T$. The set S is then a *proper subset* of T. The set of all subsets of S is the *power set* of S. We denote by $|S|$ the *cardinality* of S. The set $\emptyset$ is the set without elements and is called the *empty set*.

The terms "maximal" and "minimal" are used frequently. The meaning depends on the context. When sets are involved, the interpretation is as follows. Let $\mathcal{I}$ be a collection, each of whose elements is a set. Then a set $Z \in \mathcal{I}$ is a *maximal set* of $\mathcal{I}$ if no set of $\mathcal{I}$ has Z as a proper subset. $Z \in \mathcal{I}$ is a *minimal set* of $\mathcal{I}$ if no proper subset of Z is in $\mathcal{I}$.

2.2 Graph Definitions

An undirected graph is customarily given by a set of *nodes* and a set of *edges*. For example, the graph

(2.2.1)

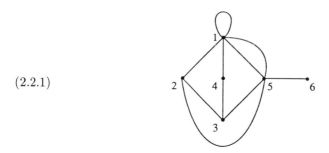

Graph with node labels

has nodes $1, 2, 3, 4, 5, 6$, and various edges connecting them. The nodes are sometimes called *vertices* or *points*. The edges are sometimes referred to as *arcs*.

Unless stated otherwise, we rely on a slightly different graph notation. We start with a nonempty set E of edges, say $E = \{e_1, e_2, \ldots, e_n\}$. Then we declare certain subsets of E to be the nodes. Each such subset specifies the edges incident at the respective node.

Let us apply this idea to the graph of (2.2.1). That graph has ten edges. Thus, we may choose $E = \{e_1, e_2, \ldots, e_{10}\}$. The graph of (2.2.1) becomes the following graph G.

(2.2.2)

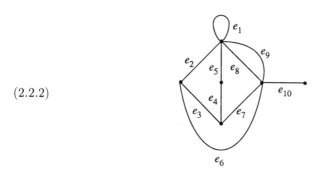

Graph G with edge labels

Node 1 of (2.2.1) is now the subset $\{e_1, e_2, e_5, e_8, e_9\}$ of E, and node 5 has become the subset $\{e_6, e_7, e_8, e_9, e_{10}\}$. Observe that each edge occurs in at most two nodes. The latter sets are the *endpoints* of the edge. An edge occurring in just one node is a *loop*. For example, e_1 occurs only in the node $\{e_1, e_2, e_5, e_8, e_9\}$ and thus is a loop. On the other hand, e_2 occurs in $\{e_1, e_2, e_5, e_8, e_9\}$, as well as in $\{e_2, e_3, e_6\}$. We also say that an edge is *incident* at a node, meaning that it is an element of that node.

There are two special cases where the set notation cannot properly represent the nodes. The first instance involves nodes that have no edges incident. We call such nodes *isolated*. According to our definition, each isolated node produces a copy of the empty set. Thus, if a graph has at least two isolated nodes, then we encounter multiple copies of the empty set. In the second case, the graph has two nodes that are connected with each other by any number of edges, but that have no other edges incident. Then the two nodes produce identical edge subsets. In principle, one may handle the two special cases with some auxiliary notation, say using labels on edge subsets. However, for almost all graphs discussed in this book, the special cases never arise. Indeed, in a moment we will see how isolated nodes may be avoided altogether. Thus, we use the above-defined set notation and implicitly assume that a more sophisticated version is employed if needed.

At first glance, our notation has little appeal even when one ignores the trouble caused by the above exceptional cases. But the utility of the idea will become apparent when we discuss graph minors and related reduction and extension operations. At any rate, we can avoid the cumbersome set notation by the introduction of additional symbols. For example, we may declare i to be the vertex $\{e_1, e_2, e_5, e_8, e_9\}$, and j to be $\{e_6, e_7, e_8, e_9, e_{10}\}$. We usually work with symbols such as i and j, and in graph drawings we may write them next to the nodes they reference. By this device, we approach the compactness of notation inherent in the customary notation. At times, we want to emphasize that a subset of edges defines a vertex. We then refer to that subset as a *star*, or more specifically, as a *k-star* if it contains k edges. The *degree* of a vertex is its cardinality.

We avoid isolated nodes as follows. We always start with graphs having no isolated nodes. Suppose we remove edges from a graph so that a node becomes isolated. Then we also remove that node as well. From now on, whenever we mention the *removal* of some edges from a graph, we implicitly assume the removal of isolated nodes. Note that the reduced graph is unique regardless of the order in which edges are removed. As an example, removal of the edges e_2, e_3, and e_6 from the graph G of (2.2.2) includes removal of the node $\{e_2, e_3, e_6\}$.

Subgraph

A *subgraph* is obtained from a given graph by the removal of some edges. A subgraph is *proper* if at least one edge is removed. Let J be a subset of the node set of a graph. Delete from the graph all edges that have at least one endpoint not in J. The resulting graph is the *subgraph induced by J* . For example, let i and j be the earlier defined nodes of G of (2.2.2), i.e., $i = \{e_1, e_2, e_5, e_8, e_9\}$ and $j = \{e_6, e_7, e_8, e_9, e_{10}\}$. The subgraph of G induced by $J = \{i, j\}$ is

(2.2.3)

Induced subgraph of graph G of (2.2.2)

The nodes of the induced subgraph may have fewer edges incident than the corresponding nodes of the original graph. This is so for the above example. As a consequence, the set J need not be the vertex set of the induced subgraph.

Path, Cycle, Tree, Cocycle, Cotree

Suppose we walk along the edges of a graph starting at some node s, never revisit any node, and stop at a node $t \neq s$. The set P of edges we have traversed is a *path from s to t*. The nodes of the path are the nodes of the graph we encountered during the walk. The nodes s and t are the *endpoints* of P. The *length* of the path P is $|P|$. For the graph G of (2.2.2), let i and j be the previously defined nodes. Then $P = \{e_2, e_3, e_7\}$ is a path from i to j. The length of P is 3. Two paths with equal endpoints are *internally node-disjoint* if they do not share any node except for the endpoints. Later in this section, the statement of Menger's Theorem relies on a particular fact about the number of internally node-disjoint paths connecting two nodes. If the two nodes, say i and j, are adjacent, then that number is unbounded since we may declare any number of paths to consist of just the edge connecting i and j. Evidently, these paths are internally node-disjoint.

Imagine another walk as described above, except that we return to s. The set C of edges we have traversed is a *cycle*. The *length* of the cycle is $|C|$. A set containing just a loop is a cycle of length 1. For the graph G of (2.2.2), $C = \{e_2, e_3, e_7, e_8\}$ is a cycle of length 4. Let H be a path or cycle of a graph G. An edge e of G is a *chord* for H if the endpoints of e have edges of H incident but are not connected by an edge of H.

In a slight abuse of language, we say at times that a graph G *is* a cycle or a path, meaning that the edge set of G is a cycle or path of G. We employ terms of later defined edge subsets such as trees and cotrees similarly. The reader may wonder why we introduce such inaccuracies. We must describe a number of diverse graph operations that are not easily expressed with one simple set of terms. So either we tolerate a slight abuse of language, or we are forced to introduce a number of different terms and sets. We have opted for the former solution in the interest of clarity.

Analogously to the use of "maximal" and "minimal" for sets, we use these terms in connection with graphs as follows. Suppose certain subgraphs of a graph G have a property $\mathcal{P}$, while others do not. Then a subgraph H of G is a *maximal subgraph* of G with respect to $\mathcal{P}$ if no other subgraph of G has $\mathcal{P}$ and has H as proper subgraph. A subgraph H of G is a *minimal subgraph* of G with respect to $\mathcal{P}$ if no proper subgraph of H has $\mathcal{P}$. We later use "maximal" and "minimal" in connection with matrices in a similar fashion. The above definitions become the appropriate ones when "graph" is replaced throughout by "matrix."

A graph is *connected* if for any two vertices s and t, there is a path from s to t. The *connected components* of a graph are the maximal connected subgraphs.

A *tree* T of a connected graph G is a maximal subset of edges not containing any cycle. Note that a tree is the empty set if and only if the connected G has just one node and all edges of G are loops. A *tip node* or *leaf node* of a tree is a node of G having just one edge of T incident. That edge is a *leaf edge* of the tree. For G of (2.2.2), $T = \{e_2, e_3, e_4, e_7, e_{10}\}$ is a tree. It is easy to show that the cardinality of a tree is equal to the number of nodes of G minus 1. A *cotree* of G is $E - T$ for some tree T of G. Suppose we select for each connected component of a graph G one tree. The union of these trees is a *forest* of G. An edge of a graph G that is not in any cycle is a *coloop*. In graph theory, such an edge is sometimes called a *bridge* or *isthmus*. It is easy to see that a coloop is in every forest of G. On the other hand, a loop cannot be part of any forest of G. The *rank* of a graph G is the cardinality of any one of its forests.

As a matter of convenience, we introduce the *empty graph*. That graph does not have any edges or nodes, and its rank is 0. We consider the empty graph to be connected.

As one removes edges from a graph, eventually the number of connected components must increase or nodes must disappear, or the empty graph must result. In each case, the rank is reduced. A minimal set of edges whose removal reduces the rank is a *cocycle* or *minimal cutset*. In the graph G of (2.2.2), the set $\{e_2, e_5, e_8, e_9\}$ is a cocycle for the following reason. Its removal reduces the rank from 5 to 4, while removal of any proper subset of $\{e_2, e_5, e_8, e_9\}$ maintains the rank at 5. Recall that a coloop is contained in every forest. Hence, removal of a coloop leads to a drop in rank. We conclude that a set containing just a coloop is a cocycle. The definitions of forest and cocycle imply that a cocycle is a minimal subset of edges that intersects every forest.

A subset of non-loop edges of a given graph G forms a *parallel class* if any two edges form a cycle and if the subset is maximal with respect to that property. We also say that the edges of the subset are *in parallel*. A subset of edges forms a *series class* (or *coparallel class*) if any two edges form a cocycle and if the subset is maximal with respect to that property. We also say that the edges of the subset are *in series* or *coparallel*. In the example graph G of (2.2.2), the edges e_8 and e_9 are in parallel, and e_4 and e_5 are in series. In the customary definition of "series," a series class of edges constitutes either a path in the graph all of whose intermediate vertices have degree 2, or a non-loop cycle all of whose vertices, save at most one, have degree 2. Our definition allows for these cases, but it also permits a slightly more general situation. For example, in the graph

(2.2.4)

Graph G

the edges e and f are in series since $\{e, f\}$ is a cocycle. A graph is *simple* if it has no loops and no parallel edges. It is *cosimple* if it has no coloops and no coparallel edges.

Deletion, Addition, Contraction, Expansion

We now introduce graph operations called *deletion, addition, contraction,* and *expansion*. It will become evident that these operations are easily accommodated by the graph notation where nodes are edge subsets. This is decidedly not so for the traditional notation displayed in (2.2.1). Before we provide details of the operations, let us examine their goals. Since additions and expansions are inverse to deletions and contractions, it suffices for us to state the goals for deletions and contractions. So let G be a connected graph, and e be an edge of G. Then the deletion (resp. contraction) of edge e is to result in a connected graph whose trees (resp. cotrees) are precisely the trees (resp. cotrees) of G that do not contain edge e. The reader should have no trouble verifying that the following definitions achieve these goals.

We start with the *deletion* of an edge e. If e is a coloop, then the deletion is carried out as a contraction, to be described in a moment. Otherwise, the deletion is the removal of the edge e from the graph. Accordingly, we remove e from the edge set E and from the one or two nodes containing it. *Addition* of an edge is the inverse of deletion. We consider the addition operation only if the corresponding deletion involves an edge that is not a coloop.

We define the *contraction* operation. If e is a loop, then the contraction of e is carried out as a deletion. If e is not a loop, then the contraction may be imagined to be a shrinking of the edge e until that edge disappears. In G, let the edge e have endpoints i and j. Accordingly, in the contraction of e we remove e from the edge set of G, and replace i and j by a new node $(i \cup j) - \{e\}$. For example, in a contraction of the edge e_8 of the graph G of (2.2.2), we replace the endpoints $i = \{e_1, e_2, e_5, e_8, e_9\}$ and $j = \{e_6, e_7, e_8, e_9, e_{10}\}$ by $(i \cup j) - \{e_8\} = \{e_1, e_2, e_5, e_9, e_6, e_7, e_{10}\}$. Note that the edge e_9 is an element of i and j. Thus, e_9 becomes a loop by the contraction. *Expansion* by an edge e is the inverse of contraction. We consider the expansion only if the corresponding contraction operation involves an edge that is not a loop.

We emphasize that the removal of an edge e may not be the same as its deletion. Indeed, the two operations produce different graphs if and only if e is a coloop both of whose endpoints have degree of at least 2.

A *reduction* is a deletion or a contraction. An *extension* is an addition or an expansion.

Uniqueness of Reductions

Suppose a given sequence of reductions for a given graph G produces a graph G'. One would wish that the same G' results if one changes the order in which the reductions are carried out. Unfortunately, this is not so. For example, suppose in the graph of (2.2.4) we first delete e and then delete f. After the deletion of e, the edge f is a coloop. Hence, the deletion of f becomes a contraction, and the resulting graph G' is given by (2.2.5) below.

(2.2.5)

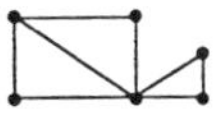

Graph G'

If we reverse the sequence of deletions, we get the graph G'' of (2.2.6) below. Clearly, G' and G'' are different graphs. The difference is produced by the fact that one of the deletions of e and f is actually carried out as a contraction.

(2.2.6)

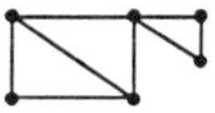

Graph G''

For obvious reasons, we want to avoid the nonuniqueness of reductions. A particularly simple method for achieving that goal relies on the following convention. Let G be a graph. When we consider reductions of G, we implicitly order the edges of G, and perform the reductions in the sequence that is compatible with that order. The same order of the edges of G is assumed for all reductions involving G. Indeed, that order is assumed to induce the related order of edges in all graphs producible from G by reductions. Trivially, this convention induces a unique outcome when a given subset U of edges is contracted and a given subset W of edges is deleted.

We denote deletion by "\\" and contraction by "/." Let G be a graph, and suppose U and W are disjoint edge subsets of G. Then $G' = G/U\backslash W$ denotes the graph produced from G by contraction of U and deletion of W. We implicitly assume that uniqueness of G' is achieved by the above convention, or possibly by some other device. G' is called a *minor* of G. As a matter of convenience, we consider G itself to be a minor of G. When U or W is empty, we may write $G\backslash W$ or G/U, respectively, instead of $G/U\backslash W$. Suppose U contains just one element u. We then write G/u instead of $G/\{u\}$ to unclutter the notation. Similarly, we write $G\backslash w$ and $G/u\backslash w$.

Cycle/Cocycle Condition

We should mention that our simple resolution of the nonuniqueness of reductions may be inappropriate in some settings. For such situations, a second method for achieving uniqueness of reductions may be a good choice. That method relies on a *cycle/cocycle condition*, which demands that the set U (resp. W) of edges to be contracted (resp. deleted) does not include a cycle (resp. cocycle). The cycle/cocycle condition guarantees uniqueness of reductions, since the order of any two successive reduction steps can be reversed without affecting the outcome of those two reductions. We leave it to the reader to fill in the elementary arguments. For the cycle/cocycle condition to be useful, we must still prove that any minor can be produced under that condition. The following arguments establish that fact. Let G' be any graph producible from a given graph G by some sequence of deletions and contractions. We claim that G' is $G/U\backslash W$ for some disjoint U and W that obey the cycle/cocycle condition. The following construction proves the claim. Start with $U = W = \emptyset$. One by one, perform the reductions that transform G to G'. Consider one such reduction, say involving edge z. Let G'' be the graph on hand at that time. Suppose z is to be contracted. If z is not (resp. is) a loop of G'', then add z to U (resp. W). Suppose z is to be deleted. If z is not (resp. is) a coloop of G'', then add z to W (resp. U). A simple inductive proof establishes that the final set U (resp. W) does not contain a cycle (resp. cocyle) of G. We omit the details. By the definition of U and W, $G/U\backslash W$ is the graph G' as desired. Thus, all minors of G producible under all possible implicit edge orderings are obtainable as minors under the cycle/cocycle condition.

The cycles and cocycles of a minor $G/U\backslash W$ can be readily deduced from those of G. We claim that the cycles of $G/U\backslash W$ are the minimal nonempty members of the collection $\{C - U \mid C \subseteq E - W;\ C =$

cycle of G}. The proof consists of two easy steps, the details of which we leave to the reader. First, one shows that every cycle of $G/U\backslash W$ occurs in the collection. Second, one establishes that each nonempty member of the collection contains a cycle of $G/U\backslash W$. An analogous proof procedure verifies that the cocycles of $G/U\backslash W$ are the minimal nonempty members of the collection $\{C^* - W \mid C^* \subseteq E - U;\ C^* = $ cocycle of $G\}$.

Addition is denoted by "+" and expansion by "&." Recall that addition of an edge is carried out only if that edge does not become a coloop. Similarly, expansion of an edge is done only if that edge does not become a loop. The latter operation is thus accomplished as follows. We split a node into two nodes and connect them by the new edge. Suppose we add the edges of a set U and expand by the edges of a set W. In the resulting graph, the sets U and W obey the cycle/cocycle condition. Thus, the earlier arguments about that condition imply that the same graph results, regardless of the order in which the additions and expansions are performed. That graph is denoted by $G\&U+W$. Analogously to $G/U\backslash W$, we simplify that notation at times. That is, we may write $G\&U$, $G+W$, $G\&u$ when $U = \{u\}$, etc.

Subdivision, Isomorphism, Homeomorphism

In a special case of expansion, we replace an edge e by a path P that contains e plus at least one more edge. We say that the edge e has been *subdivided*. The substitution process by the path is a *subdivision* of edge e.

Two graphs are *isomorphic* if they become identical upon a suitable relabeling of the edges. They are *homeomorphic* if they can be made isomorphic by repeated subdivision of certain edges in both graphs.

At times, a certain graph, say $\overline{G}$, may be a minor of a graph G, or may only be isomorphic to a minor of G. In the first case, we say, as expected, that $\overline{G}$ is a minor of G, or that G has $\overline{G}$ as a minor. For the second, rather frequently occurring case, the terminology "G has a minor isomorphic to $\overline{G}$" is technically correct but cumbersome. So instead, we say that *G has a $\overline{G}$ minor.*

Planar Graph

A graph is *planar* if it can be drawn in the plane such that the edges do not cross. The drawing need not be unique. Thus, we define a *plane graph* to be a drawing of a planar graph in the plane. Consider the following example.

(2.2.7)

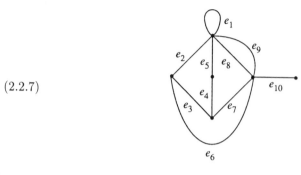

Graph G

Suppose one deletes from the plane all points lying on the edges or vertices of the plane graph. This step reduces the plane to one or more (topologically) open and connected regions, which are the *faces* of the plane graph. For example, the edges e_2, e_3, e_4, and e_5 and their endpoints border one face of the graph of (2.2.7). Note that each plane graph has exactly one unbounded face.

A connected plane graph has a *dual* produced as follows. Into the interior of each face, we place a new point. We connect two such points by an edge labeled e if the corresponding two faces contain the edge e in their boundaries. We use the asterisk to denote the dualizing operation. Thus, G^* denotes the dual of a plane graph G.

As an example, we derive G^* from the connected graph G of (2.2.7). Below, the dashed edges are those of G^*. We place each edge label near the intersection of the edge of G and the corresponding edge of G^*.

(2.2.8)

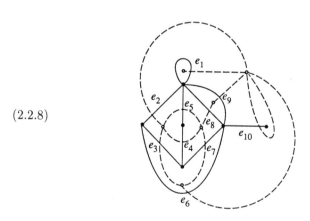

Graph G of (2.2.7) and its dual G^*

Note that according to our definition, the drawing of the dual graph G^* may not be unique. For example, in the drawing of G^* of (2.2.8) we could reroute the edge e_2 to change the unbounded face. One can avoid the defect of nonuniqueness of G^* by drawing the original graph G on the sphere instead of in the plane. Then each face is bounded, and the drawing of the dual graph G^* on the sphere is unique. Furthermore, one can show that the dual of G^* is G again, i.e., $(G^*)^* = G$. Using either type of drawing, one may verify the following relationships. Coloops of G become loops of G^*. Indeed, every cocycle of G is a cycle of G^*. Any cotree of G is a tree of G^*. Parallel edges of G are series (= coparallel) edges of G^*.

Vertex, Cycle, and Tutte Connectivity

There are several ways to specify the *connectivity* of graphs. Two commonly used concepts of graph theory are *vertex connectivity* and *cycle connectivity*. But here we employ *Tutte connectivity*. We define all three types, then justify our choice.

We need an auxiliary process called *node identification* of two nodes. Informally speaking, node identification amounts to a merging of two given nodes into just one node. For the precise definition, let G_1 and G_2 be two connected graphs. Then we *identify* a node i of G_1 with a node j of G_2 by redefining the nodes i and j to become just one node $i \cup j$. One extends this definition in the obvious way for the pairwise identification of $k \geq 1$ nodes of G_1 with k nodes of G_2.

Let (E_1, E_2) be a pair of nonempty sets that partition the edge set E of a connected graph G. Let G_1 (resp. G_2) be obtained from G by removal of the edges of E_2 (resp. E_1). Assume G_1 and G_2 to be connected. Suppose pairwise identification of k nodes of G_1 with k nodes of G_2 produces G. These k nodes of G_1 and G_2, as well as the k nodes of G they create, we call *connecting nodes*. Since G is connected, and since both G_1 and G_2 are nonempty, we have $k \geq 1$. If $k = 1$, the single connecting node of G is an *articulation point* of G. For general $k \geq 1$, (E_1, E_2) is a *vertex k-separation* of G if both G_1 and G_2 have at least $k + 1$ nodes. The pair (E_1, E_2) is a *cycle k-separation* if both E_1 and E_2 contain cycles of G. Finally, (E_1, E_2) is a *Tutte k-separation* if E_1 and E_2 have at least k edges each. Correspondingly, we call G *vertex k-separable, cycle k-separable,* or *Tutte k-separable.* For $k \geq 2$, the graph G is *vertex k-connected* (resp. *cycle k-connected, Tutte k-connected*) if G does not have any vertex l-separation (resp. cycle l-separation, Tutte l-separation) for $1 \leq l < k$. Note that the empty graph is vertex, cycle, and Tutte k-connected for every $k \geq 2$. The same conclusion holds for the connected graph with just one edge.

It is easy to see that any vertex l-separation or cycle l-separation is a Tutte l-separation. Thus, Tutte k-connectivity implies vertex k-connectivity and cycle k-connectivity. The converse does not hold, i.e., in general, vertex k-connectivity plus cycle k-connectivity do not imply Tutte k-connectivity. A counterexample is the simple graph on four nodes where any two nodes are connected by an edge. For any $k \geq 1$, that graph is readily checked to be both vertex k-connected and cycle k-connected. But it has a Tutte 3-separation (E_1, E_2) where E_1 is one of the 3-stars, and where E_2 contains the remaining three edges. In (2.2.11) below, we declare the graph of the counterexample to be the wheel W_3. There are not many other counterexamples. Indeed, it is not difficult to show that the wheels W_1 and W_2 of (2.2.11) constitute the only other counterexamples.

To summarize: Tutte connectivity implies vertex and cycle connectivity, while vertex and cycle connectivity imply Tutte connectivity except for the graphs W_1, W_2, and W_3.

Suppose G is a plane graph. We claim that (E_1, E_2) is a vertex k-separation of G if and only if it is a cycle k-separation of G^*. The proof follows by duality once one realizes the following: Each of the graphs G_1 and G_2 defined earlier from E_1 and E_2 has at least $k + 1$ vertices if and only if each one of E_1 and E_2 contains a cocycle of G.

Each one of vertex, cycle, and Tutte connectivity has its advantages and disadvantages. Thus, one should select the connectivity type depending on the situation at hand. In our case, we prefer a connectivity concept that is invariant under dualizing. That is, a plane graph should be k-connected if and only if its dual is k-connected. Tutte connectivity satisfies this requirement, while vertex and cycle connectivity do not. This feature of Tutte connectivity is one reason for our choice. A second, much more profound reason is the fact that Tutte k-connectivity for graphs is in pleasant agreement with Tutte k-connectivity for matroids, as we shall see in Chapter 3.

By Menger's Theorem (Menger (1927)), a connected graph G is vertex k-connected if and only if every two nodes are connected by k internally node-disjoint paths. Equivalent is the following statement. G is vertex k-connected if and only if any $m \leq k$ nodes are joined to any $n \leq k$ nodes by k internally node-disjoint paths. One may demand that the m nodes are disjoint from the n nodes, but need not do so. Also, the k paths can be so chosen that each of the specified nodes is an endpoint of at least one of the paths. By the above observations, the "only if" part remains valid when we assume G to be Tutte k-connected instead of vertex k-connected. Menger's Theorem also implies the following result. A graph is 2-connected if and only if any two edges lie on some cycle.

From now on we abbreviate the terms "Tutte k-connected," "Tutte k-separation," and "Tutte k-separable" to k-connected, k-separation, and k-separable.

The maximal 2-connected subgraphs of a connected graph G are the *2-connected components* of G. Consider the following process. At a node of one of the components, attach a second component. At a node of the resulting graph, attach a third component, and so on. Then the components and nodes of attachment can be so selected that this process creates G.

Complete Graph

The simple graph with $n \geq 2$ vertices, and with every two vertices connected by an edge, is denoted by K_n. It is the *complete graph* on n vertices. Small cases of K_n are as follows.

(2.2.9)

$K_2 \qquad K_3 \qquad K_4 \qquad K_5$
Small complete graphs

Bipartite Graph

A graph G is *bipartite* if its vertex set can be partitioned into two nonempty sets such that every edge has one endpoint in each of them. Note that a bipartite graph cannot have loops. The *complete bipartite graph* $K_{m,n}$ is simple, has m nodes on one side and n on the other one, and has all possible edges. Small cases are as follows.

(2.2.10)

$K_{1,1} \qquad K_{2,1} \qquad K_{2,2} \qquad K_{3,1} \qquad K_{3,2} \qquad K_{3,3}$
Small complete bipartite graphs

Evidently, $K_{1,1}$ is the complete graph K_2.

Wheel Graph

A *wheel* consists of a *rim* and *spokes*. The rim edges define a cycle, and the spokes are edges connecting an additional node with each node of the rim. The wheel with n spokes is denoted by W_n. Small wheels are as follows.

(2.2.11)

$W_1 \qquad W_2 \qquad W_3 \qquad W_4$
Small wheels

Evidently, W_3 is the complete graph K_4.

In the next section, we introduce definitions for matrices over fields.

2.3 Matrix Definitions

In this section, we define a few elementary concepts and tools of matrix theory. We make much use of the binary field GF(2). The ternary field GF(3) and the field $\mathbb{R}$ of real numbers are also employed. Occasionally, we refer to a general field $\mathcal{F}$.

The binary field GF(2) has only the elements 0 and 1. Addition is given by: $0 + 0 = 0$, $0 + 1 = 1$, and $1 + 1 = 0$. Multiplication is specified by: $0 \cdot 0 = 0$, $0 \cdot 1 = 0$, and $1 \cdot 1 = 1$. Note that the element 1 is also the additive inverse of 1, i.e., -1. Thus, we may view a $\{0, \pm 1\}$ matrix to be over GF(2). Each -1 then stands for the 1 of the field.

The ternary field GF(3) has 0, 1, and -1. Instead of the -1, we could also employ some other symbol, say 2, but will never do so. Addition is as follows: $0 + 0 = 0$, $0 + 1 = 1$, $0 + (-1) = -1$, $1 + 1 = -1$, $1 + (-1) = 0$, and $(-1) + (-1) = 1$. Multiplication is given by: $0 \cdot 0 = 0$, $0 \cdot 1 = 0$, $0 \cdot (-1) = 0$, $1 \cdot 1 = 1$, $1 \cdot (-1) = -1$, and $(-1) \cdot (-1) = 1$.

Submatrix, Trivial/Empty Matrix, Length, Order

A *row* (resp. *column*) *submatrix* is obtained from a given matrix by the deletion of some rows (resp. columns). A *submatrix* is produced by row or column deletions. A submatrix is *proper* if at least one row or column has been deleted from the given matrix. *Subvectors* are similarly defined.

We allow a matrix to have no rows or columns. Thus, for some $k \geq 1$, a matrix A may have size $k \times 0$ or $0 \times k$. Such a matrix is *trivial*. We even permit the case 0×0, in which case A is *empty*. The *length* of an $m \times n$ matrix is $m + n$. The *order* of a square matrix A is the number of rows of A. We denote any column vector containing only 1s by $\underline{1}$. Suppose a matrix A has been partitioned into two row submatrices B and C, say $A = [\frac{B}{C}]$. For typesetting reasons, we may denote this situation by $A = [B/C]$. In the special case where A, B, and C are column vectors, say a, b, and c, respectively, we may correspondingly write $a = [b/c]$.

Frequently, we index the rows and columns of a matrix. We write the row indices or index subsets to the left of a given matrix, and the column indices or index subsets above the matrix. For example, we might have

(2.3.1)
$$B = \begin{array}{c} \\ a \\ b \\ c \end{array} \begin{array}{c} d\ e\ f\ g \\ \left[\begin{array}{cccc} 1 & 0 & 1 & 1 \\ 1 & 1 & 0 & 1 \\ 0 & 1 & 1 & 1 \end{array}\right] \end{array}$$

Example matrix B

Matrix Isomorphism

We consider two matrices to be equal if up to permutation of rows and columns they are identical. Two matrices with row and column indices are *isomorphic* if they become equal upon a suitable change of the indices.

We may refer to a column directly, or by its index. For example, in a given matrix B let b be a column vector with column index y. We may refer to b as "the column vector b of B." We may also refer to b by saying "the column y of B." In the latter case, we should say more precisely "the column of B indexed by y." We have opted for the abbreviated expression "the column y of B" since references of that type occur very often in this book. We treat references to rows in an analogous manner.

Characteristic Vector, Support Matrix

Suppose a set E indexes the rows (resp. columns) of a column (resp. row) vector with $\{0, 1\}$ entries. Let E' be the subset of E corresponding to the 1s of the vector. Then that vector is the *characteristic column* (resp. *row*) *vector* of E'. We abbreviate this to *characteristic vector* when it is clear from the context whether it is a row or column vector. The *support* of a matrix A is a $\{0, 1\}$ matrix B of same size as A such that the 1s of B occur in the positions of the nonzeros of A. Sometimes, we view B to be a matrix over GF(2) or over some other field $\mathcal{F}$.

Occasionally, we append an identity to a given matrix. In that case the index of the ith column of the identity is taken to be that of the ith row of the given matrix. From the matrix B of (2.3.1), we thus may derive the following matrix A.

(2.3.2)
$$
A = \begin{array}{c} \\ a \\ b \\ c \end{array}
\begin{array}{ccc} a\ b\ c & d\ e\ f\ g \end{array}
\left[\begin{array}{ccc|cccc}
1 & 0 & 0 & 1 & 0 & 1 & 1 \\
0 & 1 & 0 & 1 & 1 & 0 & 1 \\
0 & 0 & 1 & 0 & 1 & 1 & 1
\end{array}\right]
$$

Matrix A produced from B of (2.3.1)

We often view a given $\{0, 1\}$ or $\{0, \pm 1\}$ matrix at one time to be over GF(2), at some other time to be over GF(3), and later still to be over $\mathbb{R}$ or over some other field $\mathcal{F}$. Thus, a terminology is in order that indicates the underlying field. For example, consider the rank of a matrix, i.e., the order of any maximal nonsingular submatrix. If the field is $\mathcal{F}$, we refer to the $\mathcal{F}$-*rank* of the matrix. For determinants we use "$\det_{\mathcal{F}}$," but in the case of GF(2) and GF(3) we simplify that notation to "$\det_2$" and "$\det_3$,", respectively. There is another good reason for emphasizing the underlying field. In Chapter 3, we introduce abstract matrices, which have abstract determinants, abstract rank, etc. In connection with these matrices, we just use "determinant" or "det," "rank," etc.

Pivot

Customarily, a pivot consists of the following row operations, to be performed on a given matrix A over a field $\mathcal{F}$. First, a specified row a is scaled so that a 1 is produced in a specified column d. Second, scalar multiples of the new row a are added to all other rows so that column d becomes a unit vector. In this book, the term $\mathcal{F}$-*pivot* refers to a closely related process. We explain the pivot operation using the GF(2) case. Let B be a matrix with row index set X and column index set Y. A GF(2)-*pivot* on a nonzero *pivot element* B_{xy} of a matrix B over GF(2) is carried out as follows.

$$(2.3.3.1) \quad \text{We replace for every } u \in (X - \{x\}) \text{ and every } w \in (Y - \{y\}), B_{uw} \text{ by } B'_{uw} = B_{uw} + (B_{uy} \cdot B_{xw}).$$

$(2.3.3)$

$$(2.3.3.2) \quad \text{We exchange the indices } x \text{ and } y. \text{ That is, } y \text{ becomes the index of the row originally indexed by } x, \text{ and } x \text{ becomes the index of the column originally indexed by } y.$$

For example, view B of (2.3.1) to be a matrix over GF(2). A GF(2)-pivot on $B_{ad} = 1$ may be displayed as follows.

$$(2.3.4) \qquad B = \begin{array}{c} \\ a \\ b \\ c \end{array}\begin{array}{|cccc|} \hline d & e & f & g \\ \textcircled{1} & 0 & 1 & 1 \\ 1 & 1 & 0 & 1 \\ 0 & 1 & 1 & 1 \\ \hline \end{array} \xrightarrow{\text{GF(2)-pivot}} B' = \begin{array}{c} \\ d \\ b \\ c \end{array}\begin{array}{|cccc|} \hline a & e & f & g \\ 1 & 0 & 1 & 1 \\ 1 & 1 & 1 & 0 \\ 0 & 1 & 1 & 1 \\ \hline \end{array}$$

Effect of GF(2)-pivot in B of (2.3.1)

Here and later we use a circle to highlight the pivot element.

Suppose we append an identity matrix I to B, getting A of (2.3.2), and do row operations in A to convert column d to a unit vector. We achieve this by adding row a to row b. Next, we exchange the columns a and d, including indices. Finally, we replace the row index a by d. Let A' be the resulting matrix. Below we display A and A'.

$$(2.3.5) \quad A = \begin{array}{c} \\ a \\ b \\ c \end{array}\begin{array}{|ccc|cccc|} \hline a & b & c & d & e & f & g \\ 1 & 0 & 0 & \textcircled{1} & 0 & 1 & 1 \\ 0 & 1 & 0 & 1 & 1 & 0 & 1 \\ 0 & 0 & 1 & 0 & 1 & 1 & 1 \\ \hline \end{array} \xrightarrow[\substack{\text{and} \\ \text{column exchange}}]{\text{row operations}} A' = \begin{array}{c} \\ d \\ b \\ c \end{array}\begin{array}{|ccc|cccc|} \hline d & b & c & a & e & f & g \\ 1 & 0 & 0 & 1 & 0 & 1 & 1 \\ 0 & 1 & 0 & 1 & 1 & 1 & 0 \\ 0 & 0 & 1 & 0 & 1 & 1 & 1 \\ \hline \end{array}$$

Effect of row operations and column exchange
in A of (2.3.2)

By definition $A = [I \mid B]$, and evidently $A' = [I \mid B']$. The latter conclusion holds in general, provided the row operations in A replace, in the pivot

column, the nonzeros other than the pivot element by 0s. The pivot rules
of (2.3.3) are thus nothing but an abbreviated method for obtaining the
submatrix B' of A' directly from B. Since A and A' are related by row
operations, every column index subset of A corresponding to a basis of A
also indexes a basis of A', and vice versa.

The above operations and conclusions can be extended to arbitrary
fields $\mathcal{F}$ as follows. Let B be a matrix over $\mathcal{F}$ with row index set X and
column index set Y. An $\mathcal{F}$-*pivot* on a nonzero *pivot element* B_{xy} of B is
defined as follows.

(2.3.6)

(2.3.6.1)   We replace for every $u \in (X - \{x\})$ and every
$w \in (Y - \{y\})$, B_{uw} by $B'_{uw} = B_{uw} + (B_{uy} \cdot B_{xw})/(-B_{xy})$.

(2.3.6.2)   We replace B_{xy} by $-B_{xy}$, and exchange the indices x and y.

Clearly, the GF(2)-pivot of (2.3.3) is a special case of the $\mathcal{F}$-pivot. We have
the following result for $\mathcal{F}$-pivots.

(2.3.7) Lemma.   *Let B' be derived from B by an $\mathcal{F}$-pivot as described
by (2.3.6). Append identities to both B and B' to get $A = [I \mid B]$ and
$A' = [I \mid B']$. Declare the row index sets of B and B' to become the column
index sets of the identity submatrices I of A and A', respectively. Then
every column index subset of A corresponding to a basis of A also indexes
a basis of A', and vice versa.*

The proof proceeds along the lines of the GF(2) case, except for a simple scaling argument. We omit the details. Pivots have several important
features. For the discussion below, let B, B_{xy}, and B' be the matrices just
defined.

First, when we $\mathcal{F}$-pivot in B' on B'_{yx}, we obtain B again.

Second, the pivot operation is symmetric with respect to rows versus
columns. Thus, the $\mathcal{F}$-pivot operation and the operation of taking the
transpose commute.

Third, we may use $\mathcal{F}$-pivots to compute determinants as follows. Suppose that B is square. If we delete row y and column x from B', then
the resulting matrix, say B'', satisfies $|\det_{\mathcal{F}} B''| = |(\det_{\mathcal{F}} B)/B_{xy}|$. Thus,
B is nonsingular if and only if this is so for B''. Obviously, this way of
computing determinants is nothing but the well-known method based on
row operations.

Fourth, pivots lead to a simple proof of the *submodularity* of the rank
function, to be covered next.

Submodularity of Matrix Rank Function

Let f be a function that takes the matrices over $\mathcal{F}$ to the nonnegative integers. Suppose for any matrix B over $\mathcal{F}$, and for any partition of B of the form

(2.3.8)
$$B = \begin{array}{|c|c|c|} \hline B^{11} & B^{12} & B^{13} \\ \hline B^{21} & B^{22} & B^{23} \\ \hline B^{31} & B^{32} & B^{33} \\ \hline \end{array}$$

Partitioned version of B

the values of f for the submatrices

(2.3.9)
$$D^1 = \begin{array}{|c|c|} \hline B^{11} & B^{12} \\ \hline B^{21} & B^{22} \\ \hline \end{array} \quad ; \quad D^2 = \begin{array}{|c|c|} \hline B^{22} & B^{23} \\ \hline B^{32} & B^{33} \\ \hline \end{array}$$

$$D^3 = \begin{array}{|c|c|c|} \hline B^{21} & B^{22} & B^{23} \\ \hline \end{array} \quad ; \quad D^4 = \begin{array}{|c|} \hline B^{12} \\ \hline B^{22} \\ \hline B^{32} \\ \hline \end{array}$$

Submatrices D^1, D^2, D^3, D^4 of B

satisfy the inequality

(2.3.10)
$$f(D^1) + f(D^2) \geq f(D^3) + f(D^4).$$

Then f is *submodular*. We have the following result.

(2.3.11) Lemma. *The $\mathcal{F}$-rank function is submodular.*

Proof. If B^{22} is nonzero, pivot in B^{22}. From the resulting matrix, say C, delete the pivot row and pivot column. This step converts the submatrices D^1, D^2, D^3, and D^4 of B to, say, C^1, C^2, C^3, C^4. Evidently for $k = 1, 2, 3, 4$, $\mathcal{F}$-rank $C^k = \mathcal{F}$-rank $D^k - 1$. Thus, the desired conclusion follows by induction once we handle the case where $B^{22} = 0$. In that situation, we have

(2.3.12)
$$\mathcal{F}\text{-rank } D^1 \geq \mathcal{F}\text{-rank } B^{21} + \mathcal{F}\text{-rank } B^{12}$$
$$\mathcal{F}\text{-rank } D^2 \geq \mathcal{F}\text{-rank } B^{32} + \mathcal{F}\text{-rank } B^{23}$$
$$\mathcal{F}\text{-rank } D^3 \leq \mathcal{F}\text{-rank } B^{21} + \mathcal{F}\text{-rank } B^{23}$$
$$\mathcal{F}\text{-rank } D^4 \leq \mathcal{F}\text{-rank } B^{12} + \mathcal{F}\text{-rank } B^{32}$$

Thus,

$$\mathcal{F}\text{-rank } D^1 + \mathcal{F}\text{-rank } D^2 \geq \mathcal{F}\text{-rank } B^{21} + \mathcal{F}\text{-rank } B^{12}$$

(2.3.13)
$$+ \mathcal{F}\text{-rank } B^{32} + \mathcal{F}\text{-rank } B^{23}$$

$$\geq \mathcal{F}\text{-rank } D^3 + \mathcal{F}\text{-rank } D^4$$

as desired.   □

The reader may want to prove the following result of linear algebra using the submodularity of the $\mathcal{F}$-rank function.

(2.3.14) Lemma. *Let A be a matrix over a field $\mathcal{F}$, with $\mathcal{F}$-rank $A = k$. If both a row submatrix and a column submatrix of A have $\mathcal{F}$-rank equal to k, then they intersect in a submatrix of A with the same $\mathcal{F}$-rank. In particular, any k $\mathcal{F}$-independent rows of A and any k $\mathcal{F}$-independent columns of A intersect in a $k \times k$ $\mathcal{F}$-nonsingular submatrix of A.*

Bipartite Graph of Matrix

Let A be any matrix over any field. Then $\mathrm{BG}(A)$ is the following bipartite graph. Each row and each column of A corresponds to a node. Each nonzero entry A_{xy} leads to an edge connecting row node x with column node y. In contrast to the earlier graph definitions, we do allow isolated nodes in connection with $\mathrm{BG}(A)$. We can afford to do so since we never attempt to apply reductions or extensions to $\mathrm{BG}(A)$.

Connected Matrix

We say that A is *connected* if $\mathrm{BG}(A)$ is connected. Suppose A is trivial, i.e., A is $k \times 0$ or $0 \times k$ for some $k \geq 1$. Then $\mathrm{BG}(A)$ and hence A are connected if and only if $k = 1$. Suppose A is empty, i.e., of size 0×0. Then $\mathrm{BG}(A)$ is the empty graph. By the earlier definition, the empty graph is connected. Thus, the empty matrix is connected. A *connected block* of a matrix is a maximal connected and nonempty submatrix.

Parallel Vectors

Two column or row vectors of A are *parallel* if they are nonzero and if one of them is a scalar multiple of the other one. Equivalently, the two vectors must be nonzero and must form a rank 1 matrix.

Eulerian Matrix

Define a $\{0, \pm 1\}$ matrix to be *column* (resp. *row) Eulerian* if the columns (resp. rows) sum to $0 \pmod 2$, or equivalently, if each row (resp. column) of the matrix has an even number of nonzeros. Declare a $\{0, \pm 1\}$ matrix to be *Eulerian* if it is both column and row Eulerian.

Display of Matrices

We employ a particular convention for the display of matrices. If in some region of a matrix we explicitly list some entries but not all of them, then the omitted entries are always to be taken as zeros. This convention unclutters the appearance of matrices with complicated structure.

2.4 Complexity of Algorithms

We cover elementary notions of the computational complexity of algorithms in a summarizing discussion. Define a *problem* to be any question about $m \times n$ $\{0, \pm 1\}$ matrices that is answered each time by "yes" or "no." Any such matrix represents a *problem instance*. Suppose some algorithm determines the correct answer for each problem instance. In the case of an affirmative answer, a second $m \times n$ $\{0, \pm 1\}$ matrix is possibly part of the output.

We measure the size of each problem instance by the size of a binary encoding of the input matrix and, if applicable, of the output matrix. Denote by s that measure. Up to constants, $m \cdot n$ or the total number of nonzeros in the input and output matrices constitutes an upper bound on s.

We may imagine the algorithm to be encoded as a computer program. The algorithm is *polynomial* if the run time of the computer program can, for some positive integers α, β, and γ, be uniformly bounded by a polynomial of the form $\alpha \cdot s^{\beta} + \gamma$. We also say that the algorithm is of *order β*, and we denote this by $O(s^{\beta})$.

Suppose there are positive integers δ, ϵ, and ζ such that the following holds. For each problem instance of size s and with an affirmative answer, a proof of "yes" exists whose binary encoding is bounded by $\delta \cdot s^{\epsilon} + \zeta$. Then the problem is said to be in $\mathcal{NP}$.

A problem P is *polynomially reducible* to a problem P' if there is a polynomial algorithm that transforms any instance of P into an instance of P'.

The class $\mathcal{NP}$ has a subclass of $\mathcal{NP}$-*complete* problems, which in some sense are the hardest problems of $\mathcal{NP}$. Specifically, a problem is $\mathcal{NP}$-*complete* if every problem in $\mathcal{NP}$ is polynomially reducible to it. Thus, existence of a polynomial solution algorithm for just one of the $\mathcal{NP}$-complete problems would imply existence of polynomial solution algorithms for every problem in $\mathcal{NP}$. It is an open question whether or not such polynomial algorithms exist.

Let P be a given problem. If some $\mathcal{NP}$-complete problem is polynomially reducible to P, then P is $\mathcal{NP}$-hard.

A polynomial algorithm is not necessarily useable in practice. The constants α and β of the upper bound $\alpha \cdot s^{\beta}$ on the run time may be huge, and the algorithm may require large run times even for small problem instances. The definition of "polynomial" completely ignores the magnitude of these constants.

The polynomial algorithms of this book always involve constants α and exponents β that are small enough to make the schemes useable in practice. Despite this fact, a note of caution is in order. We frequently accept some algorithmic inefficiency in the interest of simplicity and clarity of the exposition. Thus, most schemes of this book can be speeded up. The required modifications can be complex, but they also yield substantially faster algorithms.

In the next section, we provide references for the material of this chapter.

2.5 References

The introductory material of almost any book on graph theory — for example, Ore (1962), Harary (1969), Wilson (1972), or Bondy and Murty (1976) — covers most of the graph definitions of Section 2.2. The view of nodes as edge subsets is used in Truemper (1988); most computer programs for graph problems rely on the same viewpoint. The Tutte connectivity is due to Tutte (1966a). Most matrix definitions of Section 2.3 are included in any book on linear algebra, see for example Faddeev and Faddeeva (1963), Strang (1980), or Lancaster and Tismenetsky(1985). The definition of matrix submodularity is a translation of matroid submodularity (see Truemper (1985a)). Details about the computational complexity definitions may be found in Garey and Johnson (1979).

Chapter 3

From Graphs to Matroids

3.1 Overview

In this chapter, we construct matroids from graphs and matrices. In Section 3.2, we start with graphs. We encode them by certain binary matrices that lead to matroids we call graphic. For these matroids, we adapt a number of the graph concepts and operations of Chapter 2, for example the operation of taking minors. In Section 3.3, we generalize the construction of Section 3.2 by starting with arbitrary binary matrices, not just those arising from graphs. From the binary matrices, we deduce the binary matroids. In Section 3.4, we carry the generalization one step further. This time, we begin with abstract matrices, which represent a proper generalization of matrices over fields. From the abstract matrices, we define all matroids.

It is easy to determine matroids that cannot be produced from any graph, or from any binary matrix, or even from any matrix over any field. On the other hand, compact characterizations of the matroids that cannot be obtained from the matrices over some given field are usually difficult to find. We meet an exception in Section 3.5. There we characterize the matroids producible from the binary matrices by excluding a certain 4-element matroid, called U_4^2, as a minor. Sections 3.4 and 3.5 may be skipped by the reader who is only interested in binary matroids. Later, we occasionally refer to the material on general matroids to point out extensions. The final section, 3.6, lists references.

The chapter requires knowledge of the definitions of Chapter 2. To assist the reader, we will repeat certain definitions.

3.2 Graphs Produce Graphic Matroids

In this section, we deduce the graphic matroids from graphs by the following two-step process. In the first step, we encode graphs G by certain binary matrices called node/edge incidence matrices F. In the second step, we derive from the matrices F the graphic matroids M. For insight into the structure of the matroids M so created, we transform the node/edge incidence matrices F by elementary row operations into certain binary matrices B. The latter matrices contain in compact form all facts about M. Thus, we say that the matrices B represent M. We translate a number of graph definitions and concepts for G into statements about the matrices B and the graphic matroids M. In particular, we link trees, cycles, cotrees, and cocycles of G to features of B and M. We also describe the effect of taking minors in the graphs G on B and M, and characterize k-connectivity of G in terms of partitions of B and M. Finally, we establish the relationship between graphs that give rise to the same graphic matroid, and conclude with a handy procedure for deciding whether a certain 1-element binary extension of a graphic matroid is graphic.

Throughout this section, we assume G to be a connected graph with edge set E. The following graph will serve as an example.

(3.2.1)

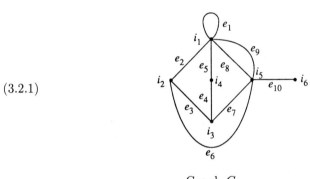

Graph G

Observe the node symbols i_1, i_2, ..., i_6. Recall that each one of these symbols stands for the subset of edges incident at the respective node.

Node/Edge Incidence Matrix

We may represent G by a binary matrix F called the *node/edge incidence matrix*. Each node of G corresponds to a row of that matrix, and each edge to a column. Suppose an edge e connects nodes i and j in G. Into column e of the matrix, we place one 1 into row i, a second 1 into row j, and 0s into the remaining rows. This rule accommodates all edges of G except for loops. There are several ways to treat loops. Here we decide to place only 0s into the columns of loops. For the example graph G of (3.2.1), the resulting matrix F is

(3.2.2)

$$
F = \begin{array}{c} \\ i_1 \\ i_2 \\ i_3 \\ i_4 \\ i_5 \\ i_6 \end{array}
\begin{array}{c} e_1\ e_2\ e_3\ e_4\ e_5\ e_6\ e_7\ e_8\ e_9\ e_{10} \\
\left[\begin{array}{cccccccccc}
0 & 1 & 0 & 0 & 1 & 0 & 0 & 1 & 1 & 0 \\
0 & 1 & 1 & 0 & 0 & 1 & 0 & 0 & 0 & 0 \\
0 & 0 & 1 & 1 & 0 & 0 & 1 & 0 & 0 & 0 \\
0 & 0 & 0 & 1 & 1 & 0 & 0 & 0 & 0 & 0 \\
0 & 0 & 0 & 0 & 0 & 1 & 1 & 1 & 1 & 1 \\
0 & 0 & 0 & 0 & 0 & 0 & 0 & 0 & 0 & 1
\end{array}\right]
\end{array}
$$

Node/edge incidence matrix F of G of (3.2.1)

Except for the endpoints of loops, we can reconstruct G from F. Thus, modulo that small defect, F represents G.

Recall from Chapter 2 the following definitions. A *cycle* of G is the edge set of a walk where all nodes are distinct and where one returns to the departure node. A *tree* of G is a maximal edge subset of E without cycles. A *cocycle* of G is a minimal set of edges that intersects every tree of G. A *cotree* of G is the set $E - X$ for some tree X of G. The *rank* of G is the cardinality of any one of its trees. Thus, it is the number of nodes of G minus 1.

In the discussion to follow, we always assume that the graph G has a cycle and a cocycle. Toward the end of this section, we address the special (actually elementary) situation where G has no cycle or no cocycle.

Over GF(2), the linear dependence of $n \geq 1$ vectors, say of $f^1, f^2, \ldots,$ f^n, is characterized as follows: That set is GF(2)-dependent if there exists a nonempty subset $J \subseteq \{1, 2, \ldots, n\}$ such that $\sum_{j \in J} f^j = 0$ (summation in GF(2), of course). Declare the vectors $f^1, \ldots, f^n$ to be *minimally* GF(2)-*dependent*, for short GF(2)-*mindependent*, if they are GF(2)-dependent, and if every proper subset of these vectors is GF(2)-independent.

For example, column e_1 of F of (3.2.2) is GF(2)-mindependent. So are the columns e_2, e_3, e_4, e_5. The first case corresponds to the loop e_1 of G, the second one to the cycle $\{e_2, e_3, e_4, e_5\}$. Indeed, by the just-given definition, a set of GF(2)-mindependent columns of the node/edge incidence matrix must correspond to a subgraph $\overline{G}$ of G with the following two properties. First, each node of $\overline{G}$ has even degree. Second, there is no subgraph of $\overline{G}$ but the empty one where each node has even degree. Evidently, the cycles of G are the only subgraphs of G with these two properties. This implies that the subgraphs of G without cycles correspond to the GF(2)-independent column subsets of F. In particular, the trees of G correspond to the bases of F, and the rank of G is the GF(2)-rank of F.

So far, we have interpreted cycles and trees of G in terms of F. How do cocycles and cotrees manifest themselves in F? We will answer that question in a moment. In the meantime, the reader may want to try obtaining an answer.

Graphic Matroid

Suppose we are just interested in the collection, say $\mathcal{I}$, of trees of G and their subsets. That interest may be surprising. But the set $\mathcal{I}$ contains a significant amount of information about G. Exactly how much, we will see later in this section. For example, with $\mathcal{I}$ we can decide whether or not an edge subset C of E is a cycle. The answer is "yes" if and only if C is not in $\mathcal{I}$ and every proper subset of C is in $\mathcal{I}$. On the other hand, we cannot decide with $\mathcal{I}$ which nodes are the endpoints of loops.

For the graph G of (3.2.1), $\mathcal{I}$ includes the sets $\{e_2, e_3, e_4, e_7, e_{10}\}$ and $\{e_3, e_5, e_6\}$. We know that each set in $\mathcal{I}$ is the index set of a column submatrix F' of F with GF(2)-independent columns. Conversely, every such index set is recorded in $\mathcal{I}$.

Still assume that we are just interested in $\mathcal{I}$. We are tempted to combine the edge set E of G and the set $\mathcal{I}$ to the ordered pair $M = (E, \mathcal{I})$. The set E is the *groundset* of M, and $\mathcal{I}$ is the collection of *independent sets* of M. Sometimes, we want to emphasize that M is deduced from G and denote it by $M(G)$. In subsequent sections of this chapter, we will see that $M(G)$ is a special case of what then will be called *binary matroids* or even just *matroids*. In the spirit of those definitions, we call $M(G)$ the *graphic matroid* of G.

Representation Matrix

We have established the collection $\mathcal{I}$ from the node/edge incidence matrix F of G. Elementary row operations performed on F do not affect GF(2)-independence of columns. Thus, we may determine $\mathcal{I}$ from any matrix derived from F by such operations. We discuss a special case of such row operations next.

First, we delete one row from F getting a matrix F'. Since each column of F has an even number of 1s, the sum of the rows of F is the zero vector. Hence, F and F' have same GF(2)-rank.

Second, we perform binary row operations to convert the column submatrix of F' corresponding to some tree of G to an identity matrix.

For the example matrix F of (3.2.2), we select $\{e_2, e_3, e_4, e_7, e_{10}\}$ as tree of G. When we apply the preceding two-step procedure to F, we get the matrix of (3.2.3) below. Note the row indices of the matrix. They are in agreement with the notation introduced in Section 2.3.

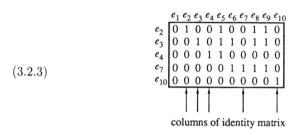

columns of identity matrix

(3.2.3)

Matrix obtained from F of (3.2.2)

The matrix of (3.2.3) is a bit difficult to read. We thus permute its columns to collect the identity submatrix at the left end. This change results in the matrix A below.

(3.2.4)

$$
A = \begin{array}{c} \\ e_2 \\ e_3 \\ e_4 \\ e_7 \\ e_{10} \end{array}
\begin{array}{ccccc|ccccc}
e_2 & e_3 & e_4 & e_7 & e_{10} & e_1 & e_5 & e_6 & e_8 & e_9 \\
1 & & & & & 0 & 1 & 0 & 1 & 1 \\
& 1 & & & & 0 & 1 & 1 & 1 & 1 \\
& & 1 & & & 0 & 1 & 0 & 0 & 0 \\
& & & 1 & & 0 & 0 & 1 & 1 & 1 \\
& & & & 1 & 0 & 0 & 0 & 0 & 0 \\
\end{array}
$$

Matrix A deduced from the matrix of (3.2.3)

The unspecified entries in the identity submatrix of A are to be taken as zeros, in agreement with the convention introduced in Section 2.3 about unspecified matrix entries.

In the case of a general graph G, let X be some tree of G, and Y be the corresponding cotree $E - X$. Then for some binary matrix B, the matrix A is of the form

(3.2.5)

$$
A = X \begin{array}{c|c} \begin{array}{cc} X & Y \end{array} \\ \hline I & B \end{array}
$$

Matrix A for general graph G with tree X

As explained in Section 2.3, the same information is conveyed by B with row index set X and column index set Y; that is, by

(3.2.6)

$$
\begin{array}{c|c} & Y \\ \hline X & B \end{array}
$$

Matrix B for general graph G with tree X

For our example case,

(3.2.7)

$$
B = X \begin{array}{c} e_2 \\ e_3 \\ e_4 \\ e_7 \\ e_{10} \end{array} \begin{array}{|ccccc|} \hline 0 & 1 & 0 & 1 & 1 \\ 0 & 1 & 1 & 1 & 1 \\ 0 & 1 & 0 & 0 & 0 \\ 0 & 0 & 1 & 1 & 1 \\ 0 & 0 & 0 & 0 & 0 \\ \hline \end{array}
$$

with column labels $e_1\ e_5\ e_6\ e_8\ e_9$ under Y.

Matrix B for graph G of (3.2.1)

The matrix B may be viewed as a binary encoding of the matroid $M = (E, \mathcal{I})$. Since M was defined to be a graphic matroid, we call B a *graphic matrix*. We also say that B *represents* M over $GF(2)$, or that B is a *representation matrix* of M. In the literature, the term *standard representation matrix* is sometimes used. The node/edge incidence matrix is then a *nonstandard representation matrix*. We omit "standard" since we almost always employ matrices like B to represent M.

Tree, Subgraph Rank

We show how trees, cycles, cotrees, cocycles, and the rank of subgraphs of G manifest themselves in B of (3.2.6). We repeatedly make use of some partition (X_1, X_2) of X, and of some partition (Y_1, Y_2) of Y. Typically, we just specify one set of X_1, X_2, and one set of Y_1, Y_2. For any such partitions, we assume B to be partitioned as

(3.2.8)

$$
B = \begin{array}{c|c|c} & Y_1 & Y_2 \\ \hline X_1 & B^1 & D^2 \\ \hline X_2 & D^1 & B^2 \end{array}
$$

Partitioned version of B

Let Z be a tree of G. Define $X_2 = Z \cap X$ and $Y_1 = Z \cap Y$. In A of (3.2.5), the submatrix $\overline{A}$ indexed by $Z = X_2 \cup Y_1$ is thus a $GF(2)$-basis of A of the form

(3.2.9)

$$
\overline{A} = \begin{array}{c|c|c} & X_2 & Y_1 \\ \hline X_1 & 0 & B^1 \\ \hline X_2 & \begin{smallmatrix} 1 & & \\ & \ddots & \\ & & 1 \end{smallmatrix} & D^1 \end{array}
$$

Submatrix of A indexed by $Z = X_2 \cup Y_1$

The submatrix B^1 of $\overline{A}$ is square, and, by cofactor expansion, GF(2)-nonsingular. Conversely, any square and GF(2)-nonsingular submatrix B^1 of B defined by (X_1, X_2) and (Y_1, Y_2) corresponds to a tree $Z = X_2 \cup Y_1$ of G. More generally, let the submatrix B^1 of B of (3.2.8) be of any size and with GF(2)-rank $B^1 = k$. Then the subgraph of G containing precisely the edges of $X_2 \cup Y_1$ has rank equal to $|X_2| + k$.

Cycle

We know that a cycle C of G corresponds to a column submatrix $\overline{A}$ of A with GF(2)-mindependent columns. Analogously to the tree case, let such a column submatrix be indexed by $X_2 = C \cap X$ and $Y_1 = C \cap Y$. Once more, (3.2.9) displays the general case. GF(2)-mindependence of the columns of $\overline{A}$ implies that they must add up to 0 in GF(2). Thus, each row of B^1 (resp. D^1) must contain an even (resp. odd) number of 1s. Note that this parity condition on the row sums of B^1 and D^1 uniquely determines X_1 and X_2 when Y_1 is specified. Furthermore, by GF(2)-mindependence, no nonempty proper column submatrix of $\overline{A}$, say indexed by $X_2' \subseteq X_2$ and $Y_1' \subseteq Y_1$, satisfies the analogous parity condition.

Fundamental Cycle

The special case of a cycle with $|Y_1| = 1$, say $Y_1 = \{y\}$, is of particular interest. If column y contains no 1s, then y is a loop. Otherwise, X_2 is the index set of the rows with a 1 in column y. The cycle is $X_2 \cup \{y\}$. It is the *fundamental cycle* that y forms with a subset of the tree X. By reversing these arguments, we obtain a quick way of constructing B. Let the tree X be given. We take each edge $y \in Y (= E - X)$ in turn and find the fundamental cycle C that y forms with a subset of X. Then column y of B is the characteristic vector of $C - \{y\}$. That is, the entry in column y and row $x \in X$ is 1 if $x \in (C - \{y\})$, and is 0 otherwise.

Parallel Elements

A cycle of cardinality 2, say $\{y, z\}$ with $y \in Y$, manifests itself in B by two parallel columns indexed by y and z, or by a unit vector column indexed by y and with the 1 in row z. We say that y and z are *in parallel* in M. We define any element to be parallel to itself. Then "is parallel to" is easily checked to be an equivalence relation. The equivalence classes are the *parallel classes* of M. They are precisely the parallel classes of G.

Cotree

Recall that a cotree of G is the set $E - Z$ for some tree Z of G. Consider the transpose of B of (3.2.8), i.e.,

(3.2.10)

$$B^t = \begin{array}{c|c|c|} & X_1 & X_2 \\ \hline Y_1 & (B^1)^t & (D^1)^t \\ \hline Y_2 & (D^2)^t & (B^2)^t \\ \hline \end{array}$$

Transpose of B of (3.2.8)

Suppose $Z = X_2 \cup Y_1$. The cotree $X_1 \cup Y_2$ corresponds to the GF(2)-nonsingular submatrix $(B^1)^t$ of B in the same way in which the tree Z is related to the GF(2)-nonsingular B^1 of B.

Append an identity to B^t, getting

(3.2.11)

$$A^* = \begin{array}{c|c|c|} & Y & X \\ \hline Y & I & B^t \\ \hline \end{array}$$

B^t with additional identity matrix

Evidently, every cotree indexes a GF(2)-basis of A^*, and vice versa.

Cocycle

We know that a cocycle is a minimal set that intersects every tree. Put differently, a cocycle is a minimal set that is not contained in any cotree. The latter definition shows that cocycles are related to cotrees in the same way that cycles are related to trees. Thus, for the interpretation of cocycles in terms of B, the previous discussion for cycles becomes applicable once we make suitable substitutions. The special cycle case with $|Y_1| = 1$ becomes the special cocycle case with $|X_1| = 1$, say with $X_1 = \{x\}$. If row x of B contains no 1s, then x is a coloop. Otherwise, let Y_1 be the index set of the columns with a 1 in row x. Then $Y_1 \cup \{x\}$ is a cocycle. It is the *fundamental cocycle* that x forms with a subset of the cotree Y. We may use fundamental cocycles to construct the rows of B. The process is analogous to the construction of the columns of B via fundamental cycles.

Coparallel or Series Elements

A cocycle of cardinality 2, say $\{x, z\}$ with $x \in X$, manifests itself in B by two parallel rows indexed by x and z, or by a unit vector row indexed by

x and with the 1 in column z. We also say that x and z are *coparallel*, or *in series*. Declare any element to be in series with itself. Then "is in series with" is an equivalence relation. The equivalence classes are the *coparallel* or *series classes* of M. They are precisely the series classes of G, according to the nonstandard definition of "series" for graphs in Section 2.2.

Dual of Graphic Matroid

For a given graph G, let $\mathcal{I}^*$ be the set of cotrees and their subsets. The fact that cycles are related to $\mathcal{I}$ in the same way in which cocycles are related to $\mathcal{I}^*$ suggests that we tie $\mathcal{I}$ and $\mathcal{I}^*$ together by duality. Specifically, we define the pair $M^* = (E, \mathcal{I}^*)$ to be the *dual matroid* of M. The prefix "co-" dualizes a term. For example, each set in $\mathcal{I}^*$ is *co-independent*. Consistent with the definition of *graphic matroid*, we call M^* the *cographic matroid* of G, and denote it by $M(G)^*$. By the above discussion, B^t represents M^*. For this reason, we call B^t *cographic*.

Let G be a connected plane graph, and G^* be its dual. In Section 2.2, the following is shown. The cotrees of G are the trees of G^*, and the cocycles of G are the cycles of G^*. Thus, in this special case, M^* is the graphic matroid of G^*. Consistent with graph terminology, we call M, as well as all of its representation matrices B, *planar*. This definition and the preceding observation imply that planarity of M implies graphicness of M and M^*, or equivalently, graphicness and cographicness of M.

Is it possible that the dual of a graphic matroid is not graphic? The answer is "yes." Toward the end of this section, we include two examples of a graphic M where M^* is not graphic. In Chapter 10, it is proved that M^* is graphic if and only if G is planar.

Pivot

According to the pivot rule (2.3.3) of Section 2.3, a GF(2)-pivot in B of (3.2.6), say with pivot element B_{xy}, is carried out as follows.

(3.2.12)

(3.2.12.1)   We replace for every $u \in (X - \{x\})$ and every $w \in (Y - \{y\})$, B_{uw} by $B_{uw} + (B_{uy} \cdot B_{xw})$ (operations in GF(2)).

(3.2.12.2)   We exchange the indices x and y.

Let B' be the matrix produced from B by the pivot. From Section 2.3, we know that this change corresponds to elementary row operations and one column exchange that transform the matrix A composed of B and an identity matrix, i.e.,

(3.2.13)

$$A = X \begin{array}{|c|c|} \hline X & Y \\ \hline I & B \\ \hline \end{array}$$

Matrix A derived from B

to

(3.2.14)

$$A' = X' \begin{array}{|c|c|} \hline X' & Y' \\ \hline I & B' \\ \hline \end{array}$$

Matrix A' derived from B'

where $X' = (X - \{x\}) \cup \{y\}$ and $Y' = (Y - \{y\}) \cup \{x\}$. Hence, by pivots we can deduce from B any matrix B'' that corresponds to a specified tree X'' of G. This fact is very useful. It permits us always to select a B'' that is particularly convenient for our purposes. One such case we see next, when we discuss the effects of edge deletions, additions, contractions, and expansions. We briefly review these graph operations.

A *deletion* of a noncoloop edge is the removal of that edge. A *deletion* of a coloop is accomplished by a contraction, which is defined next. A *contraction* of a nonloop edge is the contraction of the edge so that its endpoints become one vertex. A *contraction* of a loop is a deletion. An *addition* of an edge is the addition of an edge that does not become a coloop. An *expansion* by an edge involves splitting of a node into two nodes, which are joined by the new edge. The new edge cannot be a loop. A *reduction* is a deletion or a contraction. An *extension* is an addition or an expansion. For disjoint $U, W \subseteq E$, the *minor* $G/U \backslash W$ is the graph produced by contraction of the edges of U and deletion of the edges of W. The process is well defined because of an implicit ordering of the edges of G. No such ordering needs to be assumed when U contains no cycle and W no cocycle. Any graph producible from G by some sequence of reductions is obtainable as $G/U \backslash W$, where U and W obey the cycle/cocycle condition just stated.

Matroid Deletion, Addition, Contraction, Expansion, Minor

We translate the above graph operations into matroid language. We start with the taking of minors. So let $G/U \backslash W$ be one such minor. As just stated, we may assume U and W to obey the cycle/cocycle condition. We claim that G has a tree X and a cotree $Y = E - X$ such that $U \subseteq X$ and

$W \subseteq Y$. The proof is as follows. Since W contains no cocycle of G, the minor $G \backslash W$ has the same rank as G. Evidently, $G \backslash W$ contains U. Since U does not contain a cycle of G, U also does not contain a cycle of $G \backslash W$. Thus, in $G \backslash W$ we may extend U to a tree X of $G \backslash W$. Since $G \backslash W$ and G have same rank, X is also a tree of G. By the construction, the tree X of G contains U, and the cotree $Y = E - X$ of G contains W.

Let B be the matrix of (3.2.6) corresponding to X and Y, i.e., B is

(3.2.15)

$$
\begin{array}{c|c}
 & Y \\
\hline
X & B
\end{array}
$$

Matrix B for general graph G with tree X

Derive B' from B by deleting the rows indexed by U and the columns indexed by W. We claim that B' represents the graphic matroid of $G/U \backslash W$. We denote that matroid by $M/U \backslash W$, and we call it the *minor* of M obtained by *contraction* of U and *deletion* of W. The proof is as follows.

Each tree Z' of $G/U \backslash W$ is contained in $E - (U \cup W)$. Indeed, $Z' \cup U$ must be a tree $Z \subseteq (E - W)$ of G. Conversely, for every tree $Z \subseteq (E - W)$ of G with $U \subseteq Z$, the set $Z' = Z - U$ is a tree of $G/U \backslash W$. We know from the earlier discussion that $Z = X_2 \cup Y_1$, with $X_2 \subseteq X$ and $Y_1 \subseteq Y$, is a tree of G if and only if the square submatrix B^1 of B indexed by $X_1 = X - X_2$ and Y_1 is GF(2)-nonsingular. Thus, the trees Z' of $G/U \backslash W$ correspond precisely to the square GF(2)-nonsingular submatrices B^1 of B'. Hence, B' represents $M/U \backslash W$, the graphic matroid of $G/U \backslash W$.

Recall from Section 2.2 that the cycles and cocycles of G undergo the following changes as we go from G to $G/U \backslash W$. The cycles of $G/U \backslash W$ are the minimal nonempty sets of the collection $\{C - U \mid C \subseteq E - W; \ C = \text{cycle of } G\}$. The cocycles of $G/U \backslash W$ are the minimal nonempty sets of the collection $\{C^* - W \mid C^* \subseteq E - U; \ C^* = \text{cocycle of } G\}$. Correspondingly, the circuits and cocircuits of $M/U \backslash W$ may be derived from those of M.

We just proved that each deletion (resp. contraction) of an edge of G that is not a coloop (resp. loop) produces the removal of a column (resp. row) from a properly selected matrix B. Each addition (resp. expansion), the operation inverse to deletion (resp. contraction), corresponds to the adjoining of a column (resp. row) to B.

Suppose G is a plane graph, and G^* its dual. We know that B^t represents the graphic matroid of G^*. Now row vectors of B appear as column vectors in B^t, and column vectors of B appear as row vectors in B^t. A deletion in G induces a column removal in B, and thus a row removal in B^t. The latter removal corresponds to a contraction in G^*. By matroid arguments, we have proved that deletions in G correspond to contractions in G^*.

Up to this point, we have assumed that G has a cycle and a cocycle. Now suppose that this is not the case. Then G consists only of loops incident at one node, or of coloops, or is empty. In the first case, we define B to have no rows and as many columns as G has loops. In the second case, B is to have no columns and as many rows as G has coloops. In the third situation, B is to be the 0×0 matrix. By the definitions of Chapter 2, in the first two situations B is a trivial matrix, and in the third one the empty matrix. A trivial (resp. empty) B represents a *trivial* (resp. *empty*) *matroid*. The empty matroid has $E = \emptyset$, $\mathcal{I} = \mathcal{I}^* = \{\emptyset\}$, and has no circuits or cocircuits. We leave it to the reader to verify that the above reduction and extension results are valid for the special case of a trivial or empty B.

Separations and Connectivity

We turn to Tutte k-separations, for short k-separations, of G. We show how such separations manifest themselves in B of (3.2.15). Recall from Chapter 2 that a *k-separation* of G with $k \geq 1$ is a pair (E_1, E_2) of nonempty sets that partition E and that have the following properties. First, $|E_1|, |E_2| \geq k$. Second, the subgraph G_1 (resp. G_2) obtained from G by removal of the edges of E_2 (resp. E_1) must be connected. Third, identification of k nodes of G_1 with k nodes of G_2 must produce G. For $k \geq 2$, the graph G is *Tutte k-connected*, for short *k-connected*, if G has no l-separation for $1 \leq l < k$.

Given a k-separation (E_1, E_2) of G and given the matrix B of (3.2.15), define for $i = 1, 2$, $X_i = E_i \cap X$ and $Y_i = E_i \cap Y$. Thus, B can be partitioned as

(3.2.16)
$$B = \begin{array}{c|c|c|} & Y_1 & Y_2 \\ \hline X_1 & B^1 & D^2 \\ \hline X_2 & D^1 & B^2 \\ \hline \end{array}$$

Partitioned version of B

We claim that

(3.2.17)   GF(2)-rank D^1 + GF(2)-rank $D^2 = k - 1$.

For a proof, append to B an identity matrix, getting

(3.2.18)
$$A = \begin{array}{c|c|c|c|c|} & X_1 & X_2 & Y_1 & Y_2 \\ \hline X_1 & \begin{smallmatrix} 1 \\ & \ddots \\ & & 1 \end{smallmatrix} & 0 & B^1 & D^2 \\ \hline X_2 & 0 & \begin{smallmatrix} 1 \\ & \ddots \\ & & 1 \end{smallmatrix} & D^1 & B^2 \\ \hline \end{array}$$

Matrix B of (3.2.16) with additional identity matrix

Derive from A the matrix A^1 (resp. A^2) by deleting the columns indexed by $X_2 \cup Y_2$ (resp. $X_1 \cup Y_1$). Then

$$(3.2.19) \qquad \begin{aligned} \text{GF(2)-rank } A^1 &= |X_1| + \text{GF(2)-rank } D^1 \\ \text{GF(2)-rank } A^2 &= |X_2| + \text{GF(2)-rank } D^2 \end{aligned}$$

The matrix A may be obtained from the node/edge incidence matrix F of G by row operations. Thus, for $i = 1, 2$, the rank of G_i is equal to GF(2)-rank A^i. We combine this result with (3.2.19) and get

$$(3.2.20) \qquad \begin{aligned} (\text{rank of } G_1) &+ (\text{rank of } G_2) \\ &= \text{GF(2)-rank } A^1 + \text{GF(2)-rank } A^2 \\ &= |X_1| + |X_2| + \text{GF(2)-rank } D^1 + \text{GF(2)-rank } D^2 \\ &= (\text{rank of } G) + \text{GF(2)-rank } D^1 + \text{GF(2)-rank } D^2 \end{aligned}$$

Finally, the graphs G_1 and G_2 are connected, and identification of k nodes of these graphs produces G. For $i = 1, 2$, let the graph G_i have $n_i + k$ nodes. Then G has $n_1 + n_2 + k$ nodes. With these definitions,

$$(3.2.21) \qquad \begin{aligned} (\text{rank of } G_1) + (\text{rank of } G_2) &= (n_1 + k - 1) + (n_2 + k - 1) \\ &= (n_1 + n_2 + k - 1) + (k - 1) \\ &= (\text{rank of } G) + (k - 1) \end{aligned}$$

Then (3.2.17) follows directly from (3.2.20) and (3.2.21).

Definition of k-Separation and k-Connectivity

The preceding discussion motivates the following definitions. For any $k \geq 1$, the matrix B of (3.2.15) is k-separable if B can be partitioned as in (3.2.16) such that

$$(3.2.22) \qquad \begin{aligned} &|X_1 \cup Y_1|, |X_2 \cup Y_2| \geq k \\ &\text{GF(2)-rank } D^1 + \text{GF(2)-rank } D^2 \leq k - 1 \end{aligned}$$

The pair $(X_1 \cup Y_1, X_2 \cup Y_2)$ is a k-separation of B. For $k \geq 2$, the matrix B is k-connected if B has no l-separation for $1 \leq l < k$.

Connectivity in M is defined via that of B. That is, for $k \geq 1$, M is k-separable if B is k-separable. For $k \geq 2$, M is k-connected if M (equivalently, B) has no l-separation for $1 \leq l < k$. In particular, the empty matroid is k-connected for all $k \geq 2$.

The above definitions and observations validate the following lemma.

(3.2.23) Lemma.   *Let G be a connected graph, and M be the graphic matroid of G. For $k \geq 1$, any k-separation of G induces a k-separation of M.*

Let us naïvely attempt to prove the converse. That is, we assume a k-separation $(X_1 \cup Y_1, X_2 \cup Y_2)$ of M as given by (3.2.16) and (3.2.22). We let G_1 (resp. G_2) be the subgraph created from G by removal of the edges of $E_2 = X_2 \cup Y_2$ (resp. $E_1 = X_1 \cup Y_1$). Now we try to reverse the order of the arguments made earlier about (3.2.16)–(3.2.21).

The inequality GF(2)-rank D^1 + GF(2)-rank $D^2 \leq k - 1$ of (3.2.22) creates a first difficulty. We would like equality. This problem is easily avoided. We simply restrict ourselves to a k-separation of M with minimal k. Equivalently, we may demand M, and hence B, to be k-connected and k-separable.

By (3.2.20) and (3.2.21), we have

$$(3.2.24) \qquad (\text{rank of } G_1) + (\text{rank of } G_2) = (\text{rank of } G) + k - 1$$

We also know that $|E_1|, |E_2| \geq k$ by (3.2.22). We could prove (E_1, E_2) to be a k-separation of G if we could show G_1 and G_2 to be connected. Try as we might, this we cannot do. For good reason, since (E_1, E_2) is not always a k-separation of G, as we shall see.

Link between Graph and Matroid Separations

So let us scale down our goal. Let us simply strive to obtain a detailed description of the structure of G with k-separable M, k being minimal. To this end, we temporarily abandon our notion of nodes as edge subsets. Instead, we adopt the customary notion of nodes as points. Thus, a node of G may occur in several subgraphs of G. We retain the assumption that G is a connected graph with edge set E.

We need a few definitions to describe and prove the structure of G. Let $H_1, H_2, \ldots, H_p$, $p \geq 2$, be subgraphs of G whose edge sets partition E. Each H_i contains at least one edge. The *connecting vertices* of H_i are the vertices of H_i that also occur in some H_j, $j \neq i$. The remaining vertices of H_i are *internal*. The vertices that occur in both H_i and H_j, $i \neq j$, are the *common* vertices of H_i and H_j. The *sum* of $H_1, H_2, \ldots, H_r$, $r \leq p$, written as $H_1 + H_2 + \cdots + H_r$, is the not necessarily connected subgraph of G whose edge set is the union of the edge sets of $H_1, H_2, \cdots, H_r$.

Let $L(H_1, H_2, \ldots, H_p)$ be the following connected graph. Its nodes are labeled $H_1, H_2, \ldots, H_p$. As many parallel arcs connect node H_i with node H_j of $L(H_1, H_2, \ldots, H_p)$ as H_i and H_j have vertices in common. We declare the graphs $H_1, H_2, \ldots, H_p$ to be *connected in tree fashion* if $L(H_1, H_2, \ldots, H_p)$ is a tree. They are *connected in cycle fashion* if the latter graph is a cycle. To avoid confusion, we use "vertex" and "edge" in connection with G, and "node" and "arc" when $L(H_1, H_2, \ldots, H_p)$ is involved.

We are now ready to describe and prove the structure of G.

(3.2.25) Theorem. *Let M be the graphic matroid of a connected graph G. Assume (E_1, E_2) is a k-separation of M with minimal $k \geq 1$. Define G_1 (resp. G_2) from G by removing the edges of E_2 (resp. E_1) from G. Let $R_1, R_2, \ldots, R_g$ be the connected components of G_1, and $S_1, S_2, \ldots, S_h$ be those of G_2.*

(a) *If $k = 1$, then the R_i and S_j are connected in tree fashion.*
(b) *If $k = 2$, then the R_i and S_j are connected in cycle fashion.*
(c) *If $k \geq 3$, either (c.1) or (c.2) below holds.*

    (c.1) *Each of G_1 and G_2 is connected (thus $G_1 = R_1$ and $G_2 = S_1$) and contains a cycle or an internal vertex. The two graphs have exactly k vertices in common.*

    (c.2) *One of g and h is equal to 2, and the other one is equal to k. Without loss of generality assume $g = 2$ and $h = k$. Then S_1, $S_2, \ldots, S_h$ contain exactly one edge each. The union of the edge sets of the S_i (which is E_2) is a cocycle of G of cardinality k. The two connected components R_1 and R_2 of G_1 contain at least one cycle each.*

Proof. The edge sets of $R_1, R_2, \ldots, R_g$ and $S_1, S_2, \ldots, S_h$ partition E. Let m_i (resp. n_j) be the number of internal vertices of R_i (resp. S_j). Define p_{ij} to be the number of vertices R_i and S_j have in common. We accomplish the proof via a series of claims.

Claim 1. The graph $L(R_1, \ldots, R_g, S_1, \ldots, S_h)$ has $g + h$ nodes and

$$(3.2.26) \qquad \sum_{i,j} p_{ij} = g + h + k - 2$$

arcs, and thus is a tree plus $k - 1$ arcs.

Proof. The number of nodes of $L(R_1, \ldots, R_g, S_1, \ldots, S_h)$ is by definition $g + h$. Since the graphs R_i, S_j, and G are connected, we have (rank of R_i) = $m_i + \sum_j p_{ij} - 1$, (rank of S_j) = $n_j + \sum_i p_{ij} - 1$, and (rank of G) = $\sum_i m_i + \sum_j n_j + \sum_{i,j} p_{ij} - 1$. Furthermore, (rank of G_1) = $\sum_i (\text{rank of } R_i)$ = $\sum_i m_i + \sum_{i,j} p_{ij} - g$, and (rank of G_2) = $\sum_j (\text{rank of } S_j)$ = $\sum_j n_j + \sum_{i,j} p_{ij} - h$. By (3.2.24), (rank of G_1)+(rank of G_2) = (rank of G)+$k-1$, and thus $(\sum_i m_i + \sum_{i,j} p_{ij} - g) + (\sum_j n_j + \sum_{i,j} p_{ij} - h) = (\sum_i m_i + \sum_j n_j + \sum_{i,j} p_{ij} - 1) + k - 1$. Solving the latter equation for $\sum_{i,j} p_{ij}$, which is the number of arcs of $L(R_1, \ldots, R_g, S_1, \ldots, S_h)$, we obtain $\sum_{i,j} p_{ij} = g + h + k - 2$. Q. E. D. Claim 1

Claim 2. Parts (a) and (b) of the theorem hold.

Proof. Suppose $k = 1$. By Claim 1, $L(R_1, \ldots, R_g, S_1, \ldots, S_h)$ is a tree, and (a) follows. Let $k = 2$. Then $L(R_1, \ldots, R_g, S_1, \ldots, S_h)$ has exactly one cycle. If there are additional arcs, then $L(R_1, \ldots, R_g, S_1, \ldots, S_h)$ has a degree 1 node, and thus, G and M are 1-separable, a contradiction of the minimality of k. Thus, (b) follows.   Q. E. D. Claim 2

As a result of Claim 2, we may assume from now on that $k \geq 3$.

Claim 3. At least one of the graphs $R_1, \ldots, R_g, S_1, \ldots, S_h$ has at least as many edges as it has connecting vertices.

Proof. Assume otherwise. Thus, each subgraph R_i (resp. S_j) of G is a tree on $\sum_j p_{ij}$ (resp. $\sum_i p_{ij}$) vertices. Hence, G has $\sum_{i,j} p_{ij}$ vertices and $\sum_i (\sum_j p_{ij} - 1) + \sum_j (\sum_i p_{ij} - 1) = 2 \sum_{i,j} p_{ij} - g - h$ edges. Using $\sum_{i,j} p_{ij} = g + h + k - 2$ of (3.2.26), we see that G has $g + h + k - 2$ vertices and $g + h + 2(k - 2)$ edges. Since G is k-connected, the degree of each vertex of G is at least k. Thus,

$$(3.2.27) \qquad k \cdot (\text{number of vertices}) \leq 2 \cdot (\text{number of edges})$$

Accordingly, $k(g + h + k - 2) \leq 2(g + h + 2(k - 2))$, which implies $g + h \leq 4 - k \leq 1$, a contradiction.   Q. E. D. Claim 3

Claim 4. For all i and j,

$$(3.2.28) \qquad (\text{rank of } R_i) + (\text{rank of } S_j) \geq (\text{rank of } R_i + S_j)$$

The inequality is strict if and only if $p_{ij} \geq 2$.

Proof. Direct computation verifies the claim.   Q. E. D. Claim 4

By Claim 3, we may assume from now on that R_1 has at least as many edges as it has connecting vertices. Equivalently, R_1 has an internal vertex or a cycle.

Claim 5. If $g = 1$, i.e., if $G_1 = R_1$, then case (c.1) applies.

Proof. If each S_j has no internal vertex and is a tree, then G_2 has $|E_2| = \sum_j p_{1j} - h$ edges. Using (3.2.26) with $g = 1$, we have $|E_2| = k - 1$, which contradicts $|E_2| \geq k$. Thus, without loss of generality, S_1 has at least as many edges as it has vertices. Assume $h \geq 2$. Since G does not have a 1-separation, S_2 has at least two vertices in common with G_1. Shift the edge set of S_2 from E_2 to E_1. The resulting pair (E_1^0, E_2^0) corresponds to the subgraphs $R_1 + S_2$ and $S_1 + S_3 + \cdots + S_h$ of G and, with the aid of (3.2.28), is easily checked to be an l-separation of M with $l < k$. But this contradicts the minimality of k. Thus, $h = 1$. By (3.2.26), $G_1 = R_1$ and $G_2 = S_1$ have k vertices in common, so case (c.1) holds.   Q. E. D. Claim 5

From now on, we assume $g \geq 2$.

Claim 6. If in $L(R_1, \ldots, R_g, S_1, \ldots, S_h)$ the node S_j has degree 2, then the subgraph S_j of G contains exactly one edge.

Proof. By the assumption, the subgraph S_j of G has two vertices in common with $R_1, \ldots, R_g$. Suppose the edge set of S_j, say E', has at least two edges. Then $(E', E - E')$ is a 2-separation of M, which is not possible since $k \geq 3$.   Q. E. D. Claim 6

Claim 7. Case (c.2) applies.

Proof. Shift the edge sets of $R_2, \ldots, R_g$ from E_1 to E_2. The resulting partition (E_1^1, E_2^1) of E corresponds to graphs R_1 and $G_2^1 = R_2 + \cdots + R_g + S_1 + \cdots + S_h$. By (3.2.28), that partition is easily seen to be an l-separation of M with $l \leq k$. By the minimality of k, we have $l = k$. Then by Claim 5, R_1 and G_2^1 have exactly k vertices in common, and G_2^1 is connected. Suppose $S_1, S_2, \ldots, S_t$ are the S_j graphs that have vertices in common with R_1. Since G_2^1 is connected, so is the graph $L(R_2, \ldots, R_g, S_1, \ldots, S_h)$. Indeed, the latter graph is obtained from $L(R_1, \ldots, R_g, S_1, \ldots, S_h)$ by the removal of node R_1, and thus, by Claim 1, has $g + h - 1$ nodes and $g + h - 2$ arcs. Hence, $L(R_2, \ldots, R_g, S_1, \ldots, S_h)$ is a tree. Any tip node of that tree must correspond to some S_j, $1 \leq j \leq t$, since otherwise G and M are 1-separable. For the same reason, $t \geq 2$. Let S_1 be a tip node of the tree. Then the subgraph S_1 of G has exactly one vertex in common with $R_2 + \cdots + R_g + S_1 + \cdots + S_h$.

Similar arguments show that the graphs $R_1 + S_1$ and $R_2 + \cdots + R_g + S_2 + \cdots + S_h$ also correspond to a k-separation of G and M. Thus, we conclude the following. R_1 and $R_2 + \cdots + R_g + S_1 + \cdots + S_h$ have exactly k vertices in common. $R_1 + S_1$ and $R_2 + \cdots + R_g + S_2 + \cdots + S_h$ also have exactly k vertices in common. S_1 and $R_2 + \cdots + R_g + S_2 + \cdots + S_h$ have exactly one vertex in common. The last three statements imply that R_1 and S_1 have exactly one vertex in common. Hence, the node S_1 in $L(R_1, \ldots, R_g, S_1, \ldots, S_h)$ has degree 2. By Claim 6, the subgraph S_1 of G contains just an edge.

Inductively, we now show that each of the subgraphs $S_2, \ldots, S_t$ contains just one edge. Specifically, we assume for some $r < t$, that $S_1, \ldots, S_r$ have just one edge each. We then prove that one of the subgraphs S_j of G, $r + 1 \leq j \leq t$, say S_{r+1}, has just one edge. For the case $r + 1 = t$, we also show that $t = h = k$ and $g = 2$. We omit the detailed arguments since they are very similar to the above ones.

It remains to be shown that each of R_1 and R_2 contains a cycle. We know that each subgraph $S_1, \ldots, S_h$ has just one edge. So if R_1 or R_2 has no cycle, i.e., is a tree, then G has a degree 2 vertex, and G and M are 2-separable, a contradiction. Thus, (c.2) holds.   Q. E. D. Claim 7

The above claims clearly establish Theorem (3.2.25).   □

Theorem (3.2.25) has several important results as corollaries. They are the main reason why, in this book, we prefer Tutte connectivity over vertex or cycle connectivity for graphs.

(3.2.29) Corollary.   *Let M be the graphic matroid of a connected graph G. For any $k \geq 2$, M is k-connected if and only if this is so for G.*

Proof.   By Lemma (3.2.23), any l-separation of G is an l-separation of M. Via Theorem (3.2.25), we argue that any l-separation of M with minimal l implies that G has an l-separation, as follows. The cases (a), (b), and (c.1) of the theorem are straightforward. For case (c.2), the graphs $R_1 + S_1 + \cdots + S_h$ and R_2 produce the desired l-separation of G.   □

(3.2.30) Corollary.   *Let G be a connected plane graph. Then for any $k \geq 2$, G is k-connected if and only if this is so for the dual graph G^*.*

Proof.   By Corollary (3.2.29) and matroid duality, the graphs G and G^* and the matroids M and M^* are all k-connected if this is so for any one of these graphs and matroids.   □

Recall from Chapter 2 that a matrix A is connected if the associated bipartite graph $BG(A)$ is connected. It is easily seen that A is not connected if and only if A is a trivial matrix, i.e., of size $m \times 0$ or $0 \times m$, for some $m \geq 1$, or A has a row or column without 1s, or A has a block decomposition. Let us apply this result to a binary matrix B representing a graphic matroid M. It is easily checked via (3.2.16) that B is 2-connected if and only if B is connected. Correspondingly, we define the matroid M to be *connected* if it is 2-connected. Note that "G is connected" is a statement quite different from "B is connected" or "M is connected." The latter two statements are by Corollary (3.2.29) equivalent to "G is 2-connected." Admittedly, the use of "connected" has become a bit confusing by the above definitions. But that use is so well accepted in matroid theory that we employ it here, too. The next corollary summarizes the above relationship for future reference.

(3.2.31) Corollary.   *Let G be a connected graph and B be a representation matrix of the graphic matroid M of G. Then G is 2-connected if and only if B (and hence M) is connected.*

Finally, we have the following characterization of a 3-connected graph in terms of any representation matrix of its graphic matroid.

(3.2.32) Corollary.   *The following statements are equivalent for a connected graph G with edge set E and for any representation matrix B of the graphic matroid M of G.*

(i) *G is 3-connected.*
(ii) *If $|E| \geq 2$: G has no loops or coloops.*
   If $|E| \geq 4$: G has no parallel or series edges. Furthermore, deletion of at most two stars does not produce a disconnected graph.

(iii) M *is 3-connected.*

(iv) *B is connected, has no parallel or unit vector rows or columns, and has no partition as in* (3.2.16) *with* GF(2)-*rank* $D^1 = 1$, $D^2 = 0$, *and* $|X_1 \cup Y_1|, |X_2 \cup Y_2| \geq 3$.

(v) *Same as* (iv), *but* $|X_1 \cup Y_1|, |X_2 \cup Y_2| \geq 5$.

Proof. Corollary (3.2.29) proves (i) $\Longleftrightarrow$ (iii). The implications (i) $\Longleftrightarrow$ (ii) and (iii) $\Longleftrightarrow$ (iv) are easily seen, and (iv) $\Longrightarrow$ (v) is trivial. The possibly surprising (v) $\Longrightarrow$ (iv) is established by a straightforward checking process as follows. In B of (3.2.16), assume GF(2)-rank $D^1 = 1$ and $D^2 = 0$. If the length of B^1 is 3 or 4, then B can be seen to have a zero column or row, or parallel or unit vector rows or columns. Any such case is already excluded by the first part of (iv). Thus, it suffices to require $|X_1 \cup Y_1| \geq 5$, and by duality, $|X_2 \cup Y_2| \geq 5$.   □

In the remainder of this section, we address the following questions: How are the graphs related that produce a given graphic matroid? How can one obtain a graph that generates a given graphic matroid? When does a binary matrix correspond to a graphic matroid?

Graph 2-Isomorphism

We begin with the first question. So let M be a graphic matroid. Declare any two connected graphs G and H that produce M to be *2-isomorphic*. Necessarily, G and H have the same edge set, say E. Each one of the following sets of edge subsets of G or H completely determines M, and thus must be the same for G and H: the set of trees, the set of cycles, the set of cotrees, and the set of cocycles. For the same reason, G and H have the same rank function, the same k-separations for any $k \geq 1$, and the same connectivity. Despite the numerous relationships between G and H, these graphs may be quite different. For example, in (3.2.34) below, the first graph and the last graph of the sequence are quite distinct, yet will be shown to be 2-isomorphic.

We start with the case where M is 1-separable. We claim that the 2-connected components of G, say $G_1, G_2, \ldots, G_t$, are connected in tree fashion. By Theorem (3.2.25), G consists of $p \geq 2$ connected subgraphs that are connected in tree fashion. Select a case with p maximum. If one of the subgraphs is not 2-connected, then that subgraph itself consists of $q \geq 2$ subgraphs that are connected in tree fashion. Evidently, this contradicts the maximality of p. By 2-isomorphism, the edge set of each 2-connected component G_i of G must be that of a 2-connected component, say H_i, of H. Thus, $H_1, H_2, \ldots, H_t$ are the 2-connected components of H, which are also connected in tree fashion. We emphasize that the tree structure produced by $G_1, \ldots, G_t$ may be entirely different from that of $H_1, \ldots, H_t$.

If for all i, we had $G_i = H_i$, then we would have completely explained the difference between G and H. But this may not be so. Thus, we must understand the relationships between 2-connected but different G_i and H_i. To simplify the notation, we assume M, G, and H to be themselves 2-connected. The next lemma shows that G is equal to H if we strengthen that assumption to 3-connectedness.

(3.2.33) Lemma. *Let G and H be 3-connected and 2-isomorphic graphs. Then $G = H$.*

Proof. By trivial checking, we may assume that G has at least six edges. Let Z be any star of G. By Corollary (3.2.32) (ii), Z is a cocycle of G and $G \backslash Z$ is 2-connected. Thus, Z is a cocycle of H and $H \backslash Z$ is 2-connected. This is only possible if Z is a star of H. We conclude that each star of G is one of H, and vice versa. Then $G = H$. $\qquad \square$

By Lemma (3.2.33) and the earlier discussion, just one case remains, where M, G, and H are 2-connected and 2-separable. Take a 2-separation of G. It induces two subgraphs G' and G''. Assume that identification of nodes k and l of G' with nodes m and n, respectively, of G'' produces G. Instead, let us identify k of G' with n of G'', and l of G' with m of G''. Here is an example of this operation.

(3.2.34)

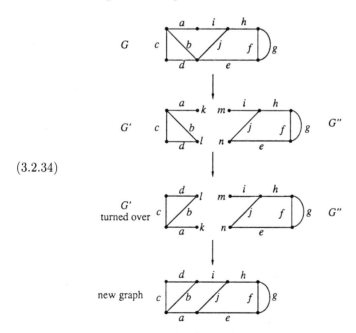

Example of switching operation

Roughly speaking, we have switched the nodes of attachment of G' with G''. For this reason, we say that the new graph is obtained from G by a *switching*. It is easy to see that G and the new graph have the same set of cycles. Thus, they are 2-isomorphic.

Let G' and G'' be the just-defined subgraphs of a 2-separation of G. The graph G' may not be 2-connected. By Theorem (3.2.25), the 2-connected components of G' are connected in tree fashion. That tree must be a path, since otherwise G is 1-separable. Apply the same arguments to G''. Combine the two observations. Thus, G has 2-connected subgraphs, say $G_1, G_2, \ldots, G_t$, for some $t \geq 2$, that are connected in cycle fashion. In the notation of Theorem (3.2.25), $G_1 + G_2 + \cdots + G_t = G$. By 2-isomorphism, H also has 2-connected subgraphs, say $H_1, H_2, \ldots, H_t$, where for all i, G_i and H_i have the same edge set. Clearly, $H_1 + H_2 + \cdots + H_t = H$. We now establish how the H_i are linked in H.

(3.2.35) Lemma. $H_1, H_2, \ldots, H_t$ *are connected in cycle fashion.*

Proof. Each 2-connected G_i with at least two edges constitutes one side of a 2-separation of G. By 2-isomorphism and Theorem (3.2.25), this also holds for the 2-connected H_i and H. Thus, H_i has exactly two nodes in common with the remaining subgraphs H_j of H. The same conclusion holds trivially if G_i and H_i have just one edge. Let C be any cycle of G that includes at least one edge of G_i and at least one edge of some G_j, $j \neq i$. Then C must include at least one edge each from $G_1, G_2, \ldots, G_t$. The analogous fact holds for the H_i. These observations imply that $H_1, H_2, \ldots, H_t$ are connected in cycle fashion. ☐

We now link 2-isomorphism and switching.

(3.2.36) Theorem.   *Let G and H be 2-connected and 2-isomorphic graphs. Then H may be obtained from G by switchings.*

Proof. If G is 3-connected, Lemma (3.2.33) applies. Otherwise, as explained above, for some $t \geq 2$, let $G_1, G_2, \ldots, G_t$ be 2-connected subgraphs of G linked in cycle fashion and satisfying $G_1 + G_2 + \cdots + G_t = G$. By Lemma (3.2.35), the corresponding 2-connected subgraphs $H_1, H_2, \ldots, H_t$ of H are connected in cycle fashion as well.

Consider G_1 by itself. Join the two nodes that G_1 has in common with the remaining G_i, $i \geq 2$, by an edge u. Let G'_1 be the resulting 2-connected graph. Analogously, add an edge u to H_1, getting H'_1. By the structure of G and H, the graphs G'_1 and H'_1 are 2-isomorphic. By induction, G'_1 can by switchings be transformed to H'_1. The same switchings can be performed in G when we view the edge u of G'_1 as representing $G_2 + \cdots + G_t$. Thus, by certain switchings, every G_i of G becomes H_i. Finally, the subgraphs H_i of the new G can by switchings be so positioned that H results. ☐

Chapter 10 includes a polynomial algorithm that for a given graphic matroid M finds a graph G producing M. The algorithm can even determine whether an arbitrary binary matroid is graphic. The algorithm essentially consists of two subroutines. One of them is called the graphic-ness testing subroutine. We describe and validate it next, using Theorem (3.2.36).

Graphicness Testing Subroutine

The input to the subroutine consists of a matrix B' given by

(3.2.37)

$$
B' = X \begin{array}{c|c|c} & Y & z \\ \hline & B & b \end{array}
$$

Input matrix for graphicness testing subroutine

The submatrix B indexed by X and Y is known to be graphic. Also given is a graph G that produces B. That graph is known to be 3-connected or to be a subdivision of a 3-connected graph. In the subroutine, we want to decide whether B' is graphic. We first analyze the relationships among B', B, b, and G.

We know that the row index set X of B is a tree of G. Let Z be the set of rows $x \in X$ for which $b_x = 1$. In the tree X of G, paint each edge of Z red. We leave the remaining edges of X unpainted.

Suppose B' is graphic. Let H' be a graph for B', with the edges of Z painted red as in G. In that graph, the set X is also a tree. Moreover, according to the column vector b and the painting rule, the red edges must form a fundamental cycle with the edge z. Thus, the red edges form a path in H' as well as in $H'\backslash z = H$. The graph H generates B, as does G. Hence, the 2-connected G and H are 2-isomorphic. By Theorem (3.2.36), H is obtainable from G by switchings. Thus, graphicness of B' implies that G can by switchings be transformed to a graph where the red edges form a path.

Conversely, assume that G can by switchings be changed into a graph H where the red edges form a path. Add an edge z to H whose endpoints are those of the path. Let H' be the resulting graph. Then in H' the fundamental cycle that z forms with X is $\{z\} \cup Z$, and thus, H' generates B'. Therefore, B' is graphic.

Recall that any graph G to be processed by the graphicness testing subroutine either is 3-connected or is a subdivision of a 3-connected graph. In the first case, no switchings are possible. Thus, B' is graphic if and only if the red edges form a path in G. Assume the second case, i.e., G is a subdivision of a 3-connected graph. Then in any 2-separation of G, the edge set of one side is readily seen to be a subset of some series class of G. Evidently, any switching amounts to a resequencing of some edges of the series class. Conversely, any resequencing of the edges of a series class can clearly be accomplished by switchings. It is a trivial matter to check whether resequencing of series class edges can result in a red path. Thus, we can readily decide whether B' is graphic. We leave it to the reader to write down the rules formally.

Suppose B' is found to be graphic. The subroutine then outputs the graph H' and stops. If B' has been determined to be not graphic, then the subroutine says so and stops.

We demonstrate the subroutine with four examples. In the first case, we have

(3.2.38)

$$B' = \begin{array}{c} X \left\{ \begin{array}{c} \\ \\ \\ \\ \end{array} \right. \end{array} \begin{array}{c} \overset{\displaystyle \longleftarrow Y \longrightarrow}{\begin{array}{cccccc} e & f & g & h & i & z \end{array}} \\ \begin{array}{c} a \\ b \\ c \\ d \end{array} \begin{array}{|ccccc|c|} \hline 1 & 0 & 0 & 1 & 1 & 0 \\ 1 & 1 & 0 & 0 & 0 & 1 \\ 0 & 1 & 1 & 0 & 1 & 0 \\ 0 & 0 & 1 & 1 & 0 & 1 \\ \hline \end{array} \end{array}$$

Example 1 for graphicness test

The graph G

(3.2.39)

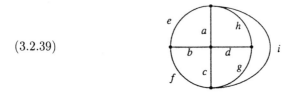

Graph G for Example 1

produces the submatrix B of B' indexed by X and Y. Evidently, G is isomorphic to the wheel W_4 plus one additional edge and is 3-connected. According to our graphicness test for B', the edges of X corresponding to the 1s in column z of B' must be painted red. Thus, we paint the edges b and d. These red edges form a path, so B' is graphic. In G, we join the two endpoints of that path by an additional edge z to obtain the following graph for B'.

(3.2.40)

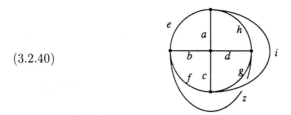

Graph H' for Example 1

That graph is isomorphic to K_5, the complete graph on five vertices. Thus, B' of (3.2.38) represents up to index changes $M(K_5)$, the graphic matroid of K_5.

The second example involves the matrix

(3.2.41)

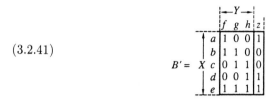

Example 2 for graphicness test

A graph G for the submatrix B indexed by X and Y is given by

(3.2.42)

Graph G for Example 2

Evidently, the graph is a subdivision of the 3-connected wheel W_3. This time, we must paint the edges a, d, and e red. The red edges do not form a path, but we can obtain a path by resequencing a and f as well as d and h. Following such sequencing, we join the endpoints of the resulting red path by a new edge z to obtain the graph

(3.2.43)

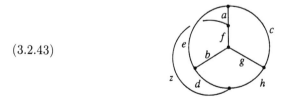

Graph H' for Example 2

Thus, B' is graphic. We leave it to the reader to verify that the graph for B' is isomorphic to $K_{3,3}$, the complete bipartite graph with three vertices on either side. We conclude that B' of (3.2.41) represents up to index changes $M(K_{3,3})$, the graphic matroid of $K_{3,3}$.

For the third example, we re-index and repartition the transpose of the matrix of $(3.2.38)$ to get

$(3.2.44)$

$$B' = \begin{array}{c} \\ X \end{array} \begin{array}{c|ccc|c} & g & h & i & z \\ \hline a & 1 & 0 & 1 & 0 \\ b & 1 & 1 & 0 & 0 \\ c & 0 & 1 & 1 & 0 \\ d & 1 & 0 & 0 & 1 \\ e & 0 & 1 & 0 & 1 \\ f & 0 & 0 & 1 & 1 \end{array}$$

$$\overset{\longmapsto Y \longrightarrow}{}$$

Example 3 for graphicness test

A graph G producing B is

$(3.2.45)$

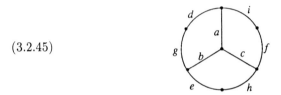

Graph G for Example 3

The graph is clearly a subdivision of the 3-connected wheel W_3. This time we must paint the edges d, e, and f red. Obviously, no resequencing of d and g, of e and h, and of f and i can result in a red path. Thus, B' is not graphic. Since it is up to indexing the transpose of the matrix of $(3.2.38)$, it represents up to a change of indices $M(K_5)^*$. Thus, that matroid is not graphic.

For the fourth case, we re-index and repartition the transpose of the matrix of $(3.2.41)$ as

$(3.2.46)$

$$B' = \begin{array}{c} \\ X \end{array} \begin{array}{c|cccc|c} & e & f & g & h & z \\ \hline a & 1 & 0 & 0 & 1 & 1 \\ b & 1 & 1 & 0 & 0 & 1 \\ c & 0 & 1 & 1 & 0 & 1 \\ d & 0 & 0 & 1 & 1 & 1 \end{array}$$

$$\overset{\longmapsto Y \longrightarrow}{}$$

Example 4 for graphicness test

The graph G of $(3.2.47)$ below produces the submatrix B of B' indexed by X and Y. Evidently, G is isomorphic to the 3-connected wheel W_4.

(3.2.47)

Graph G for Example 4

According to column z of B', we paint in G all edges of X red. The red edges do not form a path in G. Thus, B' is not graphic. Since B' is up to a re-indexing the transpose of the matrix of (3.2.41), it represents up to a change of indices $M(K_{3,3})^*$. Thus, that matroid is not graphic.

The last two examples establish the following result.

(3.2.48) Lemma.   *The matroids $M(K_5)^*$ and $M(K_{3,3})^*$ are not graphic.*

There are easier ways to prove Lemma (3.2.48). The matroid $M(K_5)^*$ has ten elements and rank 6. Thus, a graph for that matroid has ten edges and seven nodes. Since $M(K_5)^*$ is 3-connected, the degree of each node of the graph is at least 3. But then the graph must have at least eleven edges, a contradiction. The matroid $M(K_{3,3})^*$ has nine elements and rank 4. Since contraction of any edge in $K_{3,3}$ reduces that graph to the 3-connected wheel graph W_4, deletion of any element from $M(K_{3,3})^*$ must result in a 3-connected minor. We conclude that a graph for $M(K_{3,3})^*$ must have nine edges and five nodes, and deletion of any edge from that graph must produce a 3-connected minor. There is only one candidate graph. It is K_5 minus one edge. But that graph has two vertices of degree 3, and the deletion of some edge produces a 2-separable minor, a contradiction.

We conclude this section by recording the existence of the polynomial graphicness testing subroutine.

(3.2.49) Lemma.   *There is a polynomial algorithm, called the graphicness testing subroutine, for the following problem. Input is a binary matrix $B' = [B \mid b]$, where B is graphic. Also given is a graph G for B. It is known that G is 3-connected or is a subdivision of a 3-connected graph. The algorithm decides whether B' is graphic. In the affirmative case, a graph H for B' is also produced.*

We mentioned earlier that the graphicness testing subroutine will in Chapter 10 be combined with a second subroutine to a polynomial test for graphicness of binary matroids. That second subroutine carries out the following task. It analyzes the connectivity of the binary matroid for which graphicness is to be decided. In doing so, the subroutine converts the given problem into a sequence of subproblems, each of which can be solved by the above graphicness testing subroutine.

In the next section, we extend the class of graphic matroids to the class of binary matroids.

3.3 Binary Matroids Generalize Graphic Matroids

So far, we have used graphs to create the graphic matroids. In the process, we have developed quite a few matroid concepts. In this section, we generalize the graphic matroids to the binary ones. Indeed, the definitions and concepts for graphic matroids introduced in the preceding section are already so rich that most of them need at most a trivial adjustment to make them suitable for binary matroids. Thus, this section is long on definitions and short on motivation of concepts and explanations.

The reader familiar with matroid theory will surely notice that many results of this section hold for general matroids, not just binary ones. In the next section, 3.4, we include a list of these results. We cover binary matroids here in such detail for two reasons. First, we want to exhibit how features and properties of matroids are motivated by elementary linear algebra arguments. Binary matrices and matroids are the perfect vehicle to display that relationship. Second, the techniques and arguments of this section are used in much more complicated settings in subsequent chapters. The discussion of this chapter thus sets the stage and prepares the reader for that material.

We proceed as follows. We define the binary matroids via binary matrices, along with fundamental concepts such as base, circuit, cobase, cocircuit, rank function, and representation matrix. We introduce matroid minors via the reduction operations of deletion and contraction and explain the effect these operations have on bases, circuits, cobases, and cocircuits.

Next, we describe (Tutte) k-separations and k-connectivity. We link these concepts to the related ones for graphs. At that time, we are ready for a census of the 3-connected binary matroids with at most eight elements.

In this book, the presentation relies heavily on what we call the matrix viewpoint of binary matroids. In the remainder of the section, we show that other viewpoints are just as important. In particular, we introduce the submodularity of the rank function and prove with that concept a basic 3-connectivity result. We now begin the detailed presentation.

Binary Matroid

Let F be a binary matrix with a column index set E. Let $\mathcal{I}$ be the collection

of subsets $Z \subseteq E$ such that the column submatrix of F indexed by Z has
GF(2)-independent columns. We consider $Z = \emptyset$ to be in $\mathcal{I}$. The sets
Z of $\mathcal{I}$ are *independent*. Declare $M = (E, \mathcal{I})$ to be the *binary matroid*
generated by F. The set E is the *groundset* of the matroid. A *base* X
of M is a maximum cardinality subset of $\mathcal{I}$. Equivalently, X indexes the
columns of a GF(2)-basis of F. A *circuit* C of M is a minimal subset of E
that is not contained in any base of M. Equivalently, C indexes a GF(2)-
mindependent column submatrix of F. A *cobase* Y of M is the set $E - X$
for some base X. A *cocircuit* C^* of M is a minimal subset of E that is not
contained in any cobase of M. The *rank* of a subset $Z \subseteq E$, denoted by
$r(Z)$, is the cardinality of a maximal independent subset contained in Z.
The function $r(\cdot)$ is the *rank function* of M. Collect in a set $\mathcal{I}^*$ all cobases
Y of M and all their subsets. The sets Z^* of $\mathcal{I}^*$ are *co-independent*. The
pair $M^* = (E, \mathcal{I}^*)$ is the *dual matroid* of M.

   Suppose M is the graphic matroid of a connected graph G with edge
set E. Then a base (resp. a circuit, a cobase, a cocircuit, the rank function)
of M is a tree (resp. a cycle, a cotree, a cocycle, the rank function) of G.

Representation Matrix

Suppose a matrix A is deduced from F by elementary row operations.
Clearly, GF(2)-independence of columns is not affected by such a change.
Thus, we may determine $\mathcal{I}$ from A instead of F. A special case is as follows.
First, we delete GF(2)-dependent rows from F, getting, say, F'. Second,
we perform binary row operations to convert the column submatrix of F'
indexed by some base X to an identity. With $Y = E - X$, we thus have
for some binary matrix B,

(3.3.1)
$$A = X \begin{array}{|c|c|} \hline X & Y \\ \hline I & B \\ \hline \end{array}$$

Matrix A for matroid M with base X

We allow the special cases $X = \emptyset$ or $Y = \emptyset$. B is then a trivial or empty
matrix. The information contained in A is also conveyed by the submatrix
B of A, which has the form

(3.3.2)
$$X \begin{array}{|c|} \hline Y \\ \hline B \\ \hline \end{array}$$

Matrix B for matroid M with base X

The binary B is a *representation matrix* of M. We also say that B represents M over GF(2). In the literature, the term *standard representation*

matrix is sometimes used for A or B. The matrix F is then a *nonstandard representation matrix*. But we almost always work with B, so the abbreviated terminology suffices.

Bases, circuits, and the rank function of M manifest themselves in B as follows. For any partition of X into X_1 and X_2 and of Y into Y_1 and Y_2, we assume B to be partitioned as

(3.3.3)

$$B = \begin{array}{c|c|c} & Y_1 & Y_2 \\ \hline X_1 & B^1 & D^2 \\ \hline X_2 & D^1 & B^2 \end{array}$$

Partitioned version of B

Base, Rank Function

A set $Z \subseteq E$ is a base of M if and only if $X_2 = Z \cap X$ and $Y_1 = Z \cap Y$ induce a partition in B where B^1 is square and GF(2)-nonsingular. More generally, let Z be an arbitrary subset of E. Then B^1 defined via $X_2 = Z \cap X$ and $Y_1 = Z \cap Y$ has GF(2)-rank k if and only if Z has rank $r(Z) = |X_2| + k$ in M.

Circuit

Let $C \subseteq E$. Define $X_2 = C \cap X$ and $Y_1 = C \cap Y$. Then C is a circuit of M if and only if in B of (3.3.3), the number of 1s in the rows of B^1 (resp. D^1) is even (resp. odd), and for any proper subset $C' \subset C$, the corresponding $(B^1)'$ and $(D^1)'$, defined by $X_2' = C' \cap X$ and $Y_1' = C' \cap Y$, do not satisfy that parity condition. Note that $Y_1 = C \cap Y$ and the parity condition uniquely determine X_2, and thus B^1 and D^1.

Recall that a $\{0,1\}$ matrix is column Eulerian if each row contains an even number of 1s, or equivalently, if the columns sum (in GF(2)) to 0. By the above discussion, any circuit of M indexes a column submatrix of $A = [I \mid B]$ that is column Eulerian. Conversely, any column submatrix of A that is column Eulerian and that does not contain a proper column submatrix with that property corresponds to a circuit of M.

We describe three special cases of circuits using the above notation.

Loop

Suppose $|C| = 1$, say $C = \{y\}$. The element y is a *loop* of M. Necessarily, $Y_1 = \{y\}$, and column y of B must be a zero vector. Conversely, any zero column vector of B corresponds to a loop.

Parallel Elements, Triangle

Suppose $|C| = 2$. The two elements of C, say y and z, are said to be *parallel*. Both elements of C cannot be in X, so we may assume $y \in Y$. Two cases are possible. We have either $X_2 = \{z\}$ and $Y_1 = \{y\}$, or $X_2 = \emptyset$ and $Y_1 = \{y, z\}$. In the first case, column y of B is a unit vector with 1 in row z. In the second case, columns y and z of B are parallel. Conversely, a column unit vector or two parallel columns of B correspond to two parallel elements of M. If $|C| = 3$, then C is a *triangle*.

Fundamental Circuit

Suppose $|C| \geq 2$ and $|Y_1| = 1$, say $Y_1 = \{y\}$. Then X_2 is the index set of the rows of B with 1s in column y, and $C = X_2 \cup \{y\}$. The circuit C is called the *fundamental circuit* the element y forms with the base X of M. The fundamental circuits C_y that the elements $y \in Y$ form with X allow a fast construction of B. Indeed, each column $y \in Y$ of B is the characteristic vector of $C_y - \{y\}$. That is, column y has 1s in the rows indexed by $C_y - \{y\}$, and 0s elsewhere.

Cobase, Cocircuit

Cobases and cocircuits of M are exhibited by B as follows. Let $Z \subseteq E$. As before, define $X_2 = Z \cap X$, $X_1 = X - X_2$, $Y_1 = Z \cap Y$, and $Y_2 = Y - Y_1$. By the earlier definition of base and cobase, the set Z is a base of M and $Z^* = E - Z = X_1 \cup Y_2$ is a cobase of M if and only if in the transpose of B,

(3.3.4)
$$B^t = \begin{array}{c|c|c|} & X_1 & X_2 \\ \hline Y_1 & (B^1)^t & (D^1)^t \\ \hline Y_2 & (D^2)^t & (B^2)^t \\ \hline \end{array}$$

Transpose of B of (3.3.3)

the submatrix $(B^1)^t$ is square and GF(2)-nonsingular. Append to B^t an identity matrix, getting

(3.3.5)
$$A^* = \begin{array}{c|c|c|} & Y & X \\ \hline Y & I & B^t \\ \hline \end{array}$$

B^t with additional identity matrix

Evidently, Z^* is a cobase of M if and only if Z^* indexes the columns of a GF(2)-basis of A^*.

Let $C^* \subseteq E$, $X_1 = C^* \cap X$, $X_2 = X - X_1$, $Y_2 = C^* \cap Y$, and $Y_1 = Y - Y_2$. Then C^* is a cocircuit of M if and only if the earlier described circuit condition holds for C^* and B^t instead of C and B. We describe three special cases in terms of B.

Coloop

Suppose $|C^*| = 1$, say $C^* = \{x\}$. The element x is a *coloop* of M. Necessarily, $X_1 = \{x\}$, and row x of B must be a zero vector. Conversely, any zero row vector of B corresponds to a coloop.

Coparallel or Series Elements, Triad

Suppose $|C^*| = 2$. The two elements of C^*, say x and z, are said to be *coparallel* or *in series*. Both elements cannot be in Y, so we may assume $x \in X$. Two cases are possible. We have either $X_1 = \{x\}$ and $Y_2 = \{z\}$, or $X_1 = \{x, z\}$ and $Y_2 = \emptyset$. In the first case, row x of B is a unit vector with 1 in column z. In the second case, rows x and z of B are parallel. Conversely, a row unit vector or two parallel rows of B correspond to two series elements of M. If $|C^*| = 3$, then C^* is a *triad*.

Fundamental Cocircuit

Suppose $|C^*| \geq 2$, and $|X_1| = 1$, say $X_1 = \{x\}$. Then Y_2 is the index set of the columns of B with 1s in row x, and $C^* = Y_2 \cup \{x\}$. The cocircuit C^* is called the *fundamental cocircuit* the element x forms with the cobase Y of M. Analogously to the circuit case, the fundamental cocircuits permit a fast construction of B.

Binary Spaces

We relate circuits and cocircuits of M to binary spaces on E, i.e., to the linear subspaces of GF(2)^E. Specifically, consider the nullspace S of A of (3.3.1), which is given by $\{s \mid A \cdot s = 0\}$. Evidently, the vectors $s \in S$ with minimal support are exactly the characteristic vectors of the circuits of M. Any basis of these vectors generates S.

By definition of M^* from A^* of (3.3.5), the minimal support vectors of the nullspace $S^* = \{s^* \mid A^* \cdot s^* = 0\}$ of A^* are exactly the characteristic vectors of the cocircuits of M^*.

So, in a way, the circuits of M generate S. Correspondingly, the cocircuits produce S^*, which is well known (and easily proved) to be the orthogonal complement of S.

Intersection of Circuits and Cocircuits

Frequently, a judicious choice of the base X and the cobase Y of M produces a B that simplifies the proof of some result. Equivalently, we may want to proceed by GF(2)-pivots from a given representation matrix of M to some other one to exhibit a particular aspect of M. The pivots were covered in Section 2.3, and also in (3.2.12). Thus, we omit details about that operation.

Here is an example result that is easily proved with a clever choice of X and Y. That result ties circuits to cocircuits by a parity condition.

(3.3.6) Lemma. *Let C (resp. C^*) be a circuit (resp. cocircuit) of a binary matroid M. Then $|C \cap C^*|$ is even.*

Proof. Choose a cobase Y that contains all elements of C^* save one, say x. Let B be the related matrix. Thus, row x of B is the characteristic vector of $C^* - \{x\}$. If $x \in C$ (resp. $x \notin C$), then by the previously described parity condition for circuits, row x of B has an odd (resp. even) number of 1s in the columns indexed by $C \cap Y$. Either case proves $|C \cap C^*|$ to be even. ☐

Of course, Lemma (3.3.6) also follows trivially from the just-cited orthogonality of the binary spaces S and S^* produced by the circuits and cocircuits, respectively, of M.

Symmetric Difference of Circuits

We should mention the following result about circuits.

(3.3.7) Lemma. *Let C_1 and C_2 be two circuits of a binary matroid M. Then $(C_1 \cup C_2) - (C_1 \cap C_2)$, the symmetric difference of C_1 and C_2, is a disjoint union of circuits of M.*

Proof. Given B, define $A = [I \mid B]$. As observed earlier, the column submatrix of A indexed by C_1 (resp. C_2) is column Eulerian. The same fact holds for the column submatrix indexed by $Z = (C_1 \cup C_2) - (C_1 \cap C_2)$. Thus, the columns indexed by Z are GF(2)-dependent. Hence, Z contains a circuit C. Then $Z - C$ is also column Eulerian, and the desired conclusion follows by induction. ☐

Note that the proof of Lemma (3.3.7) remains valid when each of C_1 and C_2 is a disjoint union of circuits. Thus, we have the following seemingly more general result.

(3.3.8) Lemma. *Let each of C_1 and C_2 be a disjoint union of circuits of a binary matroid M. Then the symmetric difference of C_1 and C_2 is a disjoint union of circuits of M.*

Deletion, Contraction, Reduction

We turn to *deletions* and *contractions* of binary matroids. Any such operation is a *reduction*.

The operations are defined as follows. Let B be the matrix of (3.3.2) representing M. Thus, B is

(3.3.9)

$$
\begin{array}{c|c}
 & Y \\
\hline
X & B \\
\end{array}
$$

Matrix B for matroid M with base X

A *deletion* of an element $w \in E$ from M leads to a matroid $\overline{M}$ represented as follows. If w is not a coloop of M, select any representation matrix B of M where $w \in Y$. Delete column w from B. The resulting matrix $\overline{B}$ represents $\overline{M}$. It is easy to verify that the same $\overline{M}$ results, regardless of which specific B is selected to determine $\overline{B}$. The proof consists of showing that all possible $\overline{B}$ are obtainable from each other by GF(2)-pivots. If w is a coloop of M, we declare the deletion to be a contraction, which is covered next. A *contraction* of an element $u \in E$ in M leads to a matroid $\overline{M}$ represented as follows. If u is not a loop of M, select any representation matrix B of M where $u \in X$. Delete row u from B. The resulting matrix $\overline{B}$ represents $\overline{M}$. Here, too, the outcome does not depend on the selection of the specific B. If u is a loop, declare the contraction to be a deletion.

Even after the discussion of deletions and contractions for graphs and graphic matroids in Sections 2.2 and 3.2, the reader may still be a bit puzzled that we have declared the deletion of a coloop of M to be a contraction, and the contraction of a loop of M to be a deletion. Below, we motivate these rules using the matrix $A = [I \mid B]$. Suppose we intend to delete an element w from M. Our goal is to transform A to a matrix $\overline{A} = [I \mid \overline{B}]$ so that the index sets of independent column submatrices of A are precisely the index sets of independent column submatrices of A that do not include w. Note that this goal is a generalization of the goal for edge deletions in graphs, as covered in Section 2.2. There we relied on tree subsets instead of independent column submatrices. It turns out that we must consider three cases of A to determine the desired $\overline{A}$. In the first case, w indexes a column of B. Then deletion of column w from A, and thus from B, gives the desired $\overline{A}$ and $\overline{B}$. This case is covered by the deletion rule given above. In the second case, w indexes a column of the identity I, and row w of B is nonzero. Then by row operations in A, or equivalently by a GF(2)-pivot in B, we achieve the first case, again in conformance with the deletion rule given above. The third case is like the second one, except that row w of B is zero. Note that w is then a coloop of M. Suppose we delete column w

from A. Then row w of the resulting matrix A' is zero and has no influence on the independence of column subsets of A'. Thus, we might as well drop that row from A'. But these two steps, deletion of column w from A followed by deletion of row w, correspond precisely to the deletion of row w from B, and thus to the contraction of the coloop w. The situation for the contraction of a loop may be explained in the same manner using $A^* = [I \mid B^t]$ instead of A.

Uniqueness of Reductions

Let U and W be two disjoint subsets of E. Suppose we contract the elements of U and delete the elements of W. In a moment, we outline a proof showing that the outcome is not affected by the order in which the reductions are carried out. Assuming that result, we are justified in denoting the resulting unique matroid by $M/U \backslash W$. Any such matroid is a *minor* of M. For convenience, we consider M itself to be a minor of M. Analogously to the case of $G/U \backslash W$ in Section 2.2, we may simplify the notation for $M/U \backslash W$. Thus, we may write M/U, $M \backslash W$, M/u when $U = \{u\}$, etc.

That the order of the reductions is irrelevant may be shown by induction as follows: One reverses the sequence of two successive reduction steps and proves that this change has no effect on the outcome. We leave it to the reader to carry out the elementary case analysis. Note that the preceding uniqueness result for reduction sequences is at variance with the situation for graphs. According to Section 2.2, the ordering of reduction sequences does matter when graphs are involved. Indeed, in that section we introduced a technical device to enforce a certain ordering. We have just seen that the sequence of the reductions is irrelevant in binary matroids, and thus in graphic matroids. This implies that the graphs producible by differing sequences must all correspond to the same matroid minor. Thus, all such graphs are 2-isomorphic. In this book, we almost always look at graphs from the matroid standpoint, and can afford to ignore differences between 2-isomorphic graphs. Indeed, any one graph of a collection of 2-isomorphic candidates may be used to carry out proofs or to develop ideas about matroids. These facts motivated our choice of the technical device of Section 2.2.

Circuit/Cocircuit Condition

Analogously to the case for graphs, for any minor $\overline{M}$ of M, there are disjoint $U, W \subseteq E$ such that U contains no circuit, W contains no cocircuit, and $\overline{M} = M/U \backslash W$. We say that such U and W satisfy the *circuit/cocircuit condition*. The proof of the existence of U and W is almost trivial. We know that a representation matrix for $\overline{M}$ can be deduced from one for M by a sequence of operations, each of which is a GF(2)-pivot or the deletion of a row or column. We could perform all GF(2) pivots initially, then carry out the deletion of rows and columns. Let U be the index set of the rows so deleted, and W be that of the columns. Clearly, U and W satisfy the circuit/cocircuit condition, and $\overline{M} = M/U \backslash W$ as desired.

Bases, Cobases, Circuits, and Cocircuits of Minors

Sometimes, it is convenient to assume the circuit/cocircuit condition. An instance comes up next. We want to express bases, cobases, circuits, and cocircuits of a minor $\overline{M} = M/U \backslash W$ in terms of the related sets of M. We claim that the formulation below suffices if U and W satisfy the circuit/cocircuit condition.

(3.3.10)   The set of bases of $\overline{M} = M/U \backslash W$ is
$\{X - U \mid U \subseteq X \subseteq E - W; \quad X = \text{base of } M\}$;
the set of cobases of $\overline{M}$ is
$\{X^* - W \mid W \subseteq X^* \subseteq E - U; \quad X^* = \text{cobase of } M\}$

(3.3.11)   The set of circuits of $\overline{M} = M/U \backslash W$ consists of
the minimal members of
$\{C - U \mid C \subseteq E - W; \quad C = \text{ circuit of } M\}$;
the set of cocircuits of $\overline{M}$ consists of
the minimal members of
$\{C^* - W \mid C^* \subseteq E - U; \quad C^* = \text{ cocircuit of } M\}$

Validation of (3.3.10) and (3.3.11) is not difficult. By the circuit/cocircuit condition on U and W, contraction of U and deletion of W may be translated to submatrix-taking in some matrix B of (3.3.9) representing M. Thus, one only needs to examine the effect of such submatrix-taking on bases, cobases, circuits, and cocircuits. We omit the arguments since they closely follow the presentation of Section 3.2 about minors of graphic matroids.

Display of Minor, Visible Minor

Let M with representation matrix B have $\overline{M}$ as a minor. Suppose $\overline{B}$ represents $\overline{M}$. If $\overline{B}$ is a submatrix of B, then we say that B *displays* $\overline{M}$ *via* $\overline{B}$, or more briefly, that B *displays* $\overline{M}$. Note that the submatrix of B claimed to be $\overline{B}$ must match not only the numerical entries, but also the row and column index sets of $\overline{B}$. We also say that the minor $\overline{M}$ is *visible* by the display of $\overline{B}$ in B. By the definition of minor via pivots and row/column deletions, we have the following simple but useful lemma.

(3.3.12) Lemma. *Let M be a binary matroid with a minor $\overline{M}$, and $\overline{B}$ be a representation matrix of $\overline{M}$. Then M has a representation matrix B that displays $\overline{M}$ via $\overline{B}$ and thus makes the minor $\overline{M}$ visible.*

Two special cases of Lemma (3.3.12) involve so-called contraction and deletion minors, to be discussed next.

Contraction Minor

Define a minor $\overline{M}$ of M on a set E to be a *contraction minor* of M if for some $U \subseteq E$, $\overline{M} = M/U$. If M is the graphic matroid of a graph G, then any minor of G corresponding to a contraction minor of M is called a *contraction minor* of G. The next lemma shows how a contraction minor $\overline{M}$ manifests itself in representation matrices of M displaying $\overline{M}$.

(3.3.13) Lemma. *The following statements are equivalent for any binary matroid M and any minor $\overline{M}$ of M. Let $\overline{B}$ be a representation matrix of $\overline{M}$.*

(i) *$\overline{M}$ is a contraction minor of M.*

(ii) *M has a representation matrix B displaying $\overline{M}$ via $\overline{B}$, where B is of the form*

(3.3.14)

$$
B = \begin{array}{c} \\ X \left\{ \begin{array}{c} X \\ U \end{array} \right. \end{array}
\begin{array}{c} \overset{\longleftarrow Y \longrightarrow}{\overline{Y} \quad W} \\ \boxed{\begin{array}{c|c} \overline{B} & 0 \\ \hline 0/_1 \end{array}} \end{array}
$$

Matrix B displaying contraction minor $\overline{M}$

(iii) *Every representation matrix B of M displaying $\overline{M}$ via $\overline{B}$ is of the form given by (3.3.14).*

Proof. Assuming (i), we deduce (iii) as follows. Let B display $\overline{M}$ via $\overline{B}$. Thus, B is of the form given by (3.3.14), except that possibly the submatrix of B indexed by $\overline{X}$ and W may not be zero. Assume that submatrix to be nonzero. Now $\overline{M} = M/(U \cup W)$ by assumption. Then by the rule for contractions, a matrix for $\overline{M}$ is produced from B by deletion of the rows of U, followed by one or more pivots and deletion of one or more rows, and finally by deletion of some zero columns. But the resulting matrix has fewer rows than $\overline{B}$ and cannot represent $\overline{M}$, a contradiction. Trivially, (iii) implies (ii). Finally, the contraction rule proves that (ii) implies (i). $\quad\square$

Suppose we have a representation matrix B of a binary matroid M. Assume that B displays a minor $\overline{M}$ of M. Then parts (ii) and (iii) of Lemma (3.3.13) provide a simple way of ascertaining whether or not $\overline{M}$ is a contraction minor of M. In the notation of (3.3.14), the answer is "yes" if and only if the submatrix of B indexed by $\overline{X}$ and W is 0.

Deletion Minor

A minor $\overline{M} = M\backslash W$ is a *deletion minor* of M. Correspondingly to the contraction case, we define *deletion minors* of a graph G via those of the graphic matroid of G. By duality, Lemma (3.3.13) implies the following result.

(3.3.15) Lemma. *The following statements are equivalent for any binary matroid M and any minor $\overline{M}$ of M. Let $\overline{B}$ be a representation matrix of $\overline{M}$.*

 (i) *$\overline{M}$ is a deletion minor of M.*
 (ii) *M has a representation matrix B displaying $\overline{M}$ via $\overline{B}$, where B is of the form*

(3.3.16)

Matrix B displaying deletion minor $\overline{M}$

 (iii) *Every representation matrix B of M displaying $\overline{M}$ via $\overline{B}$ is of the form given by (3.3.16).*

Addition, Expansion, Extension

Addition and *expansion* are the inverse operations of deletion and contraction. Thus, an addition (resp. expansion) corresponds to the adjoining of a column (resp. row) to a given B. An *extension* is an addition or expansion. We denote additions by "+" and expansions by "&." For example, a matrix for $\overline{M}\&\overline{U}+\overline{W}$ is obtained from one for $\overline{M}$ by adjoining rows indexed by $\overline{U}$ and columns indexed by $\overline{W}$. Suppose the matroid so created is a minor of some other matroid. Then without further specification, the entries in the added rows and columns are well defined. For example, suppose $\overline{M} = M/U\backslash W$ where U and W observe the circuit/cocircuit condition. Let $\overline{U} \subseteq U$ and $\overline{W} \subseteq W$. Then $\overline{M}\&\overline{U}+\overline{W}$ is taken to be $M/(U-\overline{U})\backslash(W-\overline{W})$. We use a simplified notation for extensions analogously to that for reductions. Thus, we may write $M\&U$, $M+W$, $M\&u$ when $U = \{u\}$, etc.

Deletion, Contraction, Addition, Expansion in Dual Matroid

By definition, deletions (resp. contractions) of M correspond to the removal of columns (resp. rows) from an appropriately chosen B. Recall that

B^t represents M^*. Thus, deletions (resp. contractions) of M correspond to contractions (resp. deletions) in M^*. Put differently, for any disjoint subsets U and W of E, $(M/U \backslash W)^* = M^*/W \backslash U$. Furthermore, additions (resp. expansions) in M correspond to expansions (resp. additions) in M^*.

Matroid Isomorphism

Two matroids are *isomorphic* if they become equal upon a suitable relabeling of the elements. Analogously to the case of graph minors, a given matroid $\overline{M}$ may be a minor of a matroid M, or may only be isomorphic to a minor of M. In the first situation, we say, as expected, that $\overline{M}$ is a minor of M, or that M has $\overline{M}$ as a minor. But in the second case, we frequently abbreviate "M has a minor isomorphic to $\overline{M}$" to "M has an $\overline{M}$ minor."

k-Separation and k-Connectivity

We turn to separations and the connectivity of M. Let B be partitioned as in (3.3.3), i.e.,

(3.3.17)

$$B = \begin{array}{c|c|c} & Y_1 & Y_2 \\ \hline X_1 & B^1 & D^2 \\ \hline X_2 & D^1 & B^2 \end{array}$$

Partitioned version of B

If for some $k \geq 1$,

(3.3.18)
$$|X_1 \cup Y_1|, |X_2 \cup Y_2| \geq k$$
$$\text{GF}(2)\text{-rank } D^1 + \text{GF}(2)\text{-rank } D^2 \leq k - 1$$

then $(X_1 \cup Y_1, X_2 \cup Y_2)$ is a *Tutte k-separation*, for short *k-separation*, of B and M. The k-separation is *exact* if the rank condition of (3.3.18) holds with equality. B and M are *Tutte k-separable*, for short *k-separable*, if they have a k-separation. For $k \geq 2$, B and M are *Tutte k-connected*, for short *k-connected*, if they have no l-separation for $1 \leq l < k$. When M is 2-connected, we also say that M is *connected*. Let M be the graphic matroid of a graph G. If M is connected (i.e., 2-connected), then by Corollary (3.2.29), G is 2-connected. Thus, connectedness of M is *not* the same as connectedness of G. We mentioned this problem previously in Section 3.2.

The two connectivity corollaries (3.2.31) and (3.2.32) for graphs and graphic matroids have easy extensions to the general binary case, as follows.

(3.3.19) Lemma.   *Let M be a binary matroid with a representation matrix B. Then M is connected if and only if this is so for B.*

Proof. Via (3.3.17) and (3.3.18), it is easily checked that B is connected if and only if it is 2-connected. Thus, M is 2-connected, and hence connected, if and only if B is connected.   □

(3.3.20) Lemma.   *The following statements are equivalent for a binary matroid M with set E and a representation matrix B of M.*

(i) *M is 3-connected.*

(ii) *B is connected, has no parallel or unit vector rows and columns, and has no partition as in (3.3.17) with $\mathrm{GF}(2)$-rank $D^1 = 1$, $D^2 = 0$, and $|X_1 \cup Y_1|, |X_2 \cup Y_2| \geq 3$.*

(iii) *Same as (ii), but $|X_1 \cup Y_1|, |X_2 \cup Y_2| \geq 5$.*

Proof. (i) $\Longleftrightarrow$ (ii) follows directly from the definition of 3-connectivity, and (ii) $\Longrightarrow$ (iii) is trivial. Finally, (iii) $\Longrightarrow$ (ii) is established as follows. In B of (3.3.17), assume $\mathrm{GF}(2)$-rank $D^1 = 1$ and $D^2 = 0$. If the length of B^1 is 3 or 4, then B can be seen to have a zero column or row, or parallel or unit vector rows or columns. Any such case is already excluded by the first part of (ii). Thus, it suffices to require $|X_1 \cup Y_1| \geq 5$, and by duality, $|X_2 \cup Y_2| \geq 5$.   □

Census of Small Binary Matroids

It is instructive that we include a census of small 3-connected binary matroids, say on $n \leq 8$ elements. In that census, we refer to graphic and cographic matroids and their graphic and cographic representation matrices. We have discussed such matroids in detail in Section 3.2. We also refer to regular and nonregular matroids, which are defined via a property of real matrices termed total unimodularity. Let us take up the latter concepts one by one. A real matrix A is *totally unimodular* if every square submatrix D of A has $\det_{\mathbb{R}} D = 0$ or ± 1. In particular, all entries of a totally unimodular matrix must be 0 or ± 1. A binary matroid M is *regular* if in some binary representation matrix B of M the 1s can be so signed so that a $\{0, \pm 1\}$ real totally unimodular matrix results.

We shall motivate these definitions in Chapter 9. For the time being, we just state without proof two of the many important properties of regular matroids. First, a binary matroid M is regular if and only if every representation matrix B can be signed to become a real totally unimodular matrix. Second, every graphic or cographic matroid is regular. Because of the first property, it makes sense that we define a binary matrix to be *regular* if its 1s can be signed so that a real totally unimodular matrix results. That property also implies that the dual matroid and every minor of a regular matroid are regular. Finally, we know from Section 3.2 that graphicness and cographicness are maintained under minor-taking.

Here is the promised census of the 3-connected binary matroids with $n \leq 8$ elements. It may be verified by straightforward enumeration of cases.

$n = 0$: M is the empty matroid.

$n = 1$: M consists of a loop or a coloop; the representation matrix B is the 0×1 or 1×0 trivial matrix, respectively.

$n = 2$: M consists of two parallel elements. We may also consider the two elements to be in series. At any rate, $B = [\, 1 \,]$.

$n = 3$: M is a triangle, i.e., a circuit with three elements, or a triad, i.e., a cocircuit with three elements. In the first case, $B = [\, 1/1 \,]$. In the second case, $B = [\, 1\ 1 \,]$.

$n = 4, 5$:   There is no 3-connected binary matroid with four or five elements.

$n = 6$: M is the graphic matroid $M(W_3)$, where W_3 is the wheel with three spokes. Up to pivots, B is the matrix

(3.3.21)
$$\begin{bmatrix} 1 & 0 & 1 \\ 1 & 1 & 0 \\ 0 & 1 & 1 \end{bmatrix}$$

Matrix representing graphic matroid $M(W_3)$

Note that up to this point, all matroids have been graphic and cographic.

$n = 7$: M is the *Fano matroid* F_7 given by

(3.3.22)
$$B^7 = \begin{bmatrix} 1 & 0 & 1 & 1 \\ 1 & 1 & 0 & 1 \\ 0 & 1 & 1 & 1 \end{bmatrix}$$

Matrix representing Fano matroid F_7

or its dual F_7^* given by $(B^7)^t$. We claim that F_7, and hence F_7^*, are not regular. Indeed, if B^7 is signed so that all 2×2 submatrices have $\{0, \pm 1\}$ real determinants, then up to column and row scaling, B^7 is already that matrix. But the first three columns of B^7 define a 3×3 matrix with real determinant equal to 2. The name is based on the fact that the matroid is the Fano plane, which is the projective geometry $PG(2,2)$. The seven elements of the matroid are the points of the geometry.

$n = 8$:   If M is regular, then M is the graphic matroid of $M(W_4)$, where W_4 is the wheel with four spokes. Up to pivots, B is the matrix

(3.3.23)

$$\begin{array}{|cccc|} \hline 1 & 0 & 0 & 1 \\ 1 & 1 & 0 & 0 \\ 0 & 1 & 1 & 0 \\ 0 & 0 & 1 & 1 \\ \hline \end{array}$$

Matrix representing graphic matroid $M(W_4)$

If M is nonregular, then M is one of two matroids given by

(3.3.24)

$$B^{8.1} = \begin{array}{|cccc|} \hline 1 & 1 & 1 & 1 \\ 1 & 1 & 0 & 0 \\ 1 & 0 & 1 & 0 \\ 1 & 0 & 0 & 1 \\ \hline \end{array}$$

Matrix $B^{8.1}$ representing nonregular matroid
with eight elements, case 1

and

(3.3.25)

$$B^{8.2} = \begin{array}{|cccc|} \hline 0 & 1 & 1 & 1 \\ 1 & 0 & 1 & 1 \\ 1 & 1 & 0 & 1 \\ 1 & 1 & 1 & 0 \\ \hline \end{array}$$

Matrix $B^{8.2}$ representing nonregular matroid
with eight elements, case 2

The matroid represented by $B^{8.2}$ is the affine geometry AG(2,3), which has eight points corresponding to the eight binary vectors with three entries each. A subset of the points is (affinely) GF(2)-mindependent if the corresponding subset of vectors has even cardinality and is linearly GF(2)-dependent, and if it is minimal with respect to these two conditions. For that matroid, contraction (resp. deletion) of any element produces the nonregular matroid F_7 (resp. F_7^*) as a minor. Thus, that matroid is highly nonregular. In contrast, deletion of any row or column from $B^{8.1}$ produces a matrix that represents a regular matroid.

So far, we have stressed what one might call the *matrix viewpoint* of binary matroids. As we shall see in Section 3.4, that notion can be extended to general matroids using matrices termed *abstract*. The main advantage of that notion is the fact that a binary matrix (or in general, an abstract matrix) displays numerous bases, circuits, cobases, and cocircuits of the matroid simultaneously. For this reason, we view matrices (binary, or over other fields, or abstract) to be an important tool of matroid theory.

There are, however, other and equally important viewpoints. One of them considers the rank function $r(\cdot)$ of M as a main tool. Recall that this function was defined via B of (3.3.17) as follows. For any subset $Z \subseteq E$, let $X_2 = Z \cap X$ and $Y_1 = Z \cap Y$. By (3.3.17), $X_1 = X - X_2$ and Y_1 index the submatrix B^1 of B. Then declare the rank of Z to be $r(Z) = |X_2| + \mathrm{GF}(2)$-rank B^1. The utility of the rank function largely rests upon a property called *submodularity*. Section 2.3 includes a definition of submodularity for functions defined on matrices. There it is shown that the matrix rank function is submodular. Shortly, we define submodularity for functions that map the subsets of a set E to the nonnegative integers. We then prove the matroid rank function to be submodular.

The matrix viewpoint, the rank function viewpoint, as well as several others not mentioned so far (e.g., the geometric viewpoint, the lattice viewpoint — see the books cited in Chapter 1), have advantages and disadvantages. Intuitively speaking, matrices very conveniently display bases, circuits, cobases, and cocircuits that differ just by a few elements. On the other hand, the relationships between radically different bases, circuits, etc., are not well exhibited. The rank function approach, as well as several others, treats all such cases evenly. Indeed, for the solution of several problems involving radically different bases, circuits, etc., the rank function seems particularly suitable.

The results described in this book rely largely on the matrix viewpoint. But the reader should not be misled by this fact. It just turns out that the results of this book are nicely treatable by matrices. But there are a number of problems of matroid theory where other approaches, in particular ones relying on the rank function, are superior to the matrix technique employed here. Below, we describe two simple instances that exhibit the power of the rank function approach. But first we express the defining k-separation conditions of (3.3.18) in terms of $r(\cdot)$.

k-Separation Condition for Rank Function

(3.3.26) Lemma. *Let M be a binary matroid on a set E and with rank function $r(\cdot)$. Suppose E_1 and E_2 partition E. Then (a) and (b) below hold.*

(a) *(E_1, E_2) is a k-separation of M if and only if*

$$(3.3.27) \qquad \begin{aligned} &|E_1|, |E_2| \geq k \\ &r(E_1) + r(E_2) \leq r(E) + k - 1 \end{aligned}$$

(b) (E_1, E_2) is an exact k-separation of M if and only if

(3.3.28)
$$|E_1|, |E_2| \geq k$$
$$r(E_1) + r(E_2) = r(E) + k - 1$$

Proof. (a) To establish the "only if" part, take any representation matrix B of M, say indexed by X and Y as before. Define for $i = 1, 2$, $X_i = E_i \cap X$ and $Y_i = E_i \cap Y$. Let these sets partition B as in (3.3.17). Since $|E_1|, |E_2| \geq k$, we have $|X_1 \cup Y_1|, |X_2 \cup Y_2| \geq k$. By the definition of $r(\cdot)$, we have for $i = 1, 2$, $r(E_i) = |X_i| + \mathrm{GF}(2)$-rank D^i, and $r(E) = |X_1| + |X_2|$. Since $r(E^1) + r(E^2) \leq r(E) + k - 1$, we have $|X_1| + \mathrm{GF}(2)$-rank $D^1 + |X_2| + \mathrm{GF}(2)$-rank $D^2 \leq |X_1| + |X_2| + k - 1$, or $\mathrm{GF}(2)$-rank $D^1 + \mathrm{GF}(2)$-rank $D^2 \leq k - 1$. Thus, by (3.3.18), $(X_1 \cup Y_1, X_2 \cup Y_2) = (E_1, E_2)$ is a k-separation of M. For the proof of the "if" part, one reverses the above arguments.

(b) This follows from the proof of (a) by suitable replacement of some inequalities by equations. ☐

We now define the submodularity property and prove that the rank function $r(\cdot)$ has that property.

Submodularity of Rank Function

A function $f(\cdot)$ from the set of subsets of a finite set E to the nonnegative integers is *submodular* if for any subsets $S, T \subseteq E$,

(3.3.29)
$$f(S) + f(T) \geq f(S \cup T) + f(S \cap T)$$

(3.3.30) Lemma. $r(\cdot)$ *is a submodular function.*

Proof. Take any binary representation matrix of M. Let A^1, A^2, A^3, and A^4 be the column submatrices of $A = [I \mid B]$ corresponding to the column index sets S, T, $S \cup T$, and $S \cap T$, respectively. Evidently, A^4 is a submatrix of A^1, A^2, and A^3, and both A^1 and A^2 are submatrices of A^3. Let B^4 be a basis of A^4. For $i = 1, 2, 3$, extend B^4 to a basis of A^i, say by adjoining B^i. Evidently, the number of columns of $[B^4 \mid B^1]$, $[B^4 \mid B^2]$, $[B^4 \mid B^3]$, and B^4 is the matroid rank of S, T, $S \cup T$, and $S \cap T$, respectively. Thus, the submodularity inequality (3.3.29) holds if and only if B^1 and B^2 together have at least as many columns as B^3. The latter condition holds since by the construction, $[B^4 \mid B^3]$ is a basis of A^3, while $[B^4 \mid B^1 \mid B^2]$ spans all columns of A^3. ☐

We use submodularity of $r(\cdot)$ in the proof of the next result. Define $M \textcircled{c} z$ to be the matroid obtained from M/z by deletion of the elements of each parallel class except for one representative of each class. Similarly, derive $M \textcircled{d} z$ from $M \backslash z$ by contracting the elements of each series class except for one representative of each class.

(3.3.31) Lemma.   *Let M be a 3-connected binary matroid on a set E. Take z to be any element of E. Then $M\mathbin{\text{\ooalign{$\bigcirc$\cr$c$\cr}}}z$ or $M\mathbin{\text{\ooalign{$\bigcirc$\cr$d$\cr}}}z$ is 3-connected.*

Proof. The arguments below rely repeatedly on three observations that follow directly from (3.3.10) and (3.3.11). First, if a set is not a circuit (resp. cocircuit) of M, then it cannot be a circuit (resp. cocircuit) in any minor derived from M by deletions only (resp. contractions only). In particular, the 3-connected matroid M has no parallel or series elements, and thus, for any element z, the minor $M\backslash z$ (resp. M/z) has no parallel (resp. series) elements. Second, if an element z is not a coloop (resp. loop) of M, then $M\backslash z$ (resp. M/z) has the same rank as M. Third, the rank of a set $Z \subseteq E$ drops at most by 1 when an element $z \notin Z$ is contracted, and stays the same when $z \notin Z$ is deleted.

Assume the lemma fails, i.e., for some 3-connected M and $z \in E$, both $M\mathbin{\text{\ooalign{$\bigcirc$\cr$c$\cr}}}z$ and $M\mathbin{\text{\ooalign{$\bigcirc$\cr$d$\cr}}}z$ are 2-separable. Let (P', Q') be a 2-separation of $M\mathbin{\text{\ooalign{$\bigcirc$\cr$c$\cr}}}z$. We know that no two elements of $M\mathbin{\text{\ooalign{$\bigcirc$\cr$c$\cr}}}z$ are in parallel. Thus, $|P'| \geq 3$ or P' contains two series elements. The obvious assignment of the parallel elements by which M/z and $M\mathbin{\text{\ooalign{$\bigcirc$\cr$c$\cr}}}z$ differ converts (P', Q') to a 2-separation (P, Q) of M/z where $P \supseteq P'$ and $Q \supseteq Q'$. Since M is 3-connected, M/z has no series elements. If $|P'| = 2$ and $P' = P$, then P must be a set of two series elements in M/z, which is impossible. Thus, $|P'| = 2$ implies $|P| \geq 3$. We conclude $|P| \geq 3$ in general, and $|Q| \geq 3$ by symmetry. Using duality, $M\mathbin{\text{\ooalign{$\bigcirc$\cr$d$\cr}}}z$ must have a 2-separation (R, S) with $|R|, |S| \geq 3$.

Assume $|P \cap R| \leq 1$. Then $|P \cap S|, |Q \cap R| \geq 2$. The same conclusion holds if $|Q \cap S| \leq 1$. Thus, we may assume $|P \cap R|, |Q \cap S| \geq 2$ or $|P \cap S|, |Q \cap R| \geq 2$. By symmetry, we may suppose that the former situation holds.

Denote by $r(\cdot)$, $r_{M/z}(\cdot)$, and $r_{M\backslash z}(\cdot)$ the rank functions of M, M/z, and $M\backslash z$, respectively. Since (P, Q) and (R, S) are 2-separations of M/z and $M\backslash z$, respectively, we have

$$(3.3.32) \qquad \begin{aligned} r_{M/z}(P) + r_{M/z}(Q) &\leq r_{M/z}(E - \{z\}) + 1 \\ r_{M\backslash z}(R) + r_{M\backslash z}(S) &\leq r_{M\backslash z}(E - \{z\}) + 1 \end{aligned}$$

Now $r_{M/z}(P) \geq r(P \cup \{z\}) - 1$, $r_{M/z}(Q) \geq r(Q \cup \{z\}) - 1$, and $r_{M/z}(E - \{z\}) = r(E) - 1$. Also, $r_{M\backslash z}(R) = r(R)$, $r_{M\backslash z}(S) = r(S)$, and $r_{M\backslash z}(E - \{z\}) = r(E)$. We use these relationships to deduce from (3.3.32) the inequalities

$$(3.3.33) \qquad \begin{aligned} r(P \cup \{z\}) + r(Q \cup \{z\}) &\leq r(E) + 2 \\ r(R) \phantom{\cup \{z\}} + r(S) \phantom{\cup \{z\}} &\leq r(E) + 1 \end{aligned}$$

By submodularity, we also have

$$(3.3.34) \qquad \begin{aligned} r(P \cap R) + r(P \cup R \cup \{z\}) &\leq r(P \cup \{z\}) + r(R) \\ r(Q \cap S) + r(Q \cup S \cup \{z\}) &\leq r(Q \cup \{z\}) + r(S) \end{aligned}$$

Add the two inequalities of (3.3.33). Similarly, add the two inequalities of (3.3.34). The resulting two inequalities imply

$$
(3.3.35) \qquad
\begin{aligned}
&[r(P \cap R) + r(Q \cup S \cup \{z\})] \\
&+ [r(Q \cap S) + r(P \cup R \cup \{z\})] \le 2r(E) + 3
\end{aligned}
$$

But then at least one of the pairs $(P \cap R, Q \cup S \cup \{z\})$ and $(Q \cap S, P \cup R \cup \{z\})$ must be a 2-separation of M, which is not possible. $\qquad\square$

Along the same lines, but with much simpler arguments, one can prove the following lemma. We leave the proof to the reader.

(3.3.36) Lemma.   *Let M be a connected binary matroid on a set E. Take z to be any element of M. Then M/z or $M\backslash z$ is connected.*

We conclude this section by proving that submodularity of the $r(\cdot)$ function and submodularity of the GF(2)-rank function are equivalent. To show this, we assume S and T to be subsets of E for a binary matroid M with a binary representation matrix B. Let the customary index sets X and Y of B be partitioned into X_0, X_1, X_2, X_3, and Y_0, Y_1, Y_2, Y_3, respectively, so that

$$
(3.3.37) \qquad
\begin{aligned}
X_0 &= X - (S \cup T) \\
X_1 &= X \cap (S - T) \\
X_2 &= X \cap (S \cap T) \\
X_3 &= X \cap (T - S) \\
Y_0 &= Y - (S \cup T) \\
Y_1 &= Y \cap (S - T) \\
Y_2 &= Y \cap (S \cap T) \\
Y_3 &= Y \cap (T - S)
\end{aligned}
$$

Let the partitions of X and Y induce the following partition of B.

(3.3.38)

Partitioned version of B

Define submatrices D^1, D^2, D^3, D^4 of B as in (2.3.9), i.e.,

(3.3.39)

$$D^1 = \begin{array}{|c|c|} \hline B^{11} & B^{12} \\ \hline B^{21} & B^{22} \\ \hline \end{array} \; ; \quad D^2 = \begin{array}{|c|c|} \hline B^{22} & B^{23} \\ \hline B^{32} & B^{33} \\ \hline \end{array}$$

$$D^3 = \begin{array}{|c|c|c|} \hline B^{21} & B^{22} & B^{23} \\ \hline \end{array} \; ; \quad D^4 = \begin{array}{|c|} \hline B^{12} \\ \hline B^{22} \\ \hline B^{32} \\ \hline \end{array}$$

Submatrices D^1, D^2, D^3, D^4 of B

By the discussion following (3.3.3), the equations below relate the GF(2)-rank of the D^i to the matroid rank of S, T, $S \cup T$, and $S \cap T$.

(3.3.40)
$$r(S) = |X_1| + |X_2| + \text{GF(2)-rank } D^1$$
$$r(T) = |X_2| + |X_3| + \text{GF(2)-rank } D^2$$
$$r(S \cup T) = |X_1| + |X_2| + |X_3| + \text{GF(2)-rank } D^3$$
$$r(S \cap T) = |X_2| + \text{GF(2)-rank } D^4$$

Then clearly the submodularity inequality for $r(\cdot)$, which is $r(S) + r(T) \geq r(S \cup T) + r(S \cap T)$, holds if and only if this is so for the submodularity inequality for the GF(2)-rank function, which is GF(2)-rank D^1 + GF(2)-rank $D^2 \geq$ GF(2)-rank D^3 + GF(2)-rank D^4.

In the next section, we move from binary matrices to abstract ones. Correspondingly, we obtain all matroids instead of just the binary ones.

3.4 Abstract Matrices Produce All Matroids

Binary matrices produce the binary matroids. In a natural extension, we could consider matrices over fields other than GF(2) and the matroids produced by them. We skip that step. Instead, we move in this section directly to abstract concepts of independence, bases, circuits, and rank, and in the process create the entire class of matroids. We also introduce abstract matrices as a generalization of matrices over fields, and we describe some of their features. Routine arguments prove that the abstract matrices generate all matroids, and that the matroids produce all abstract matrices. Indeed, one may view abstract matrices as one way of encoding matroids. Abstract matrices exhibit many features of linear algebra. They also display several properties of matroids rather conveniently, and we have found them to be very useful. In particular, they often help one to detect and prove new structural results that are hidden from view when, instead, one thinks of a matroid as a construction via certain sets, functions, geometries, or operators.

Abstract matrices behave to quite an extent like binary matrices. This fact explains why so many seemingly special results for binary matroids hold for the general case. To support that claim, toward the end of this section we list results of Section 3.3 for binary matroids that, sometimes after an elementary modification, hold for general matroids. We are now ready for the detailed discussion.

Definition of General Matroid

Let E be a set of vectors over some field $\mathcal{F}$. A central result of linear algebra says that for any given subset of vectors of E, the bases of that subset have the same cardinality. We abstract from this fact the axioms for general matroids as follows. A *matroid* M on a *ground set* E is a pair $(E, \mathcal{I})$, where $\mathcal{I}$ is a certain subset of the power set of E. The set $\mathcal{I}$ is the set of *independent* subsets of M. A subset of E that is not in $\mathcal{I}$ is called *dependent*. The set $\mathcal{I}$ must observe the following axioms.

(3.4.1)

	(i)	The null set is in $\mathcal{I}$.
	(ii)	Every subset of any set in $\mathcal{I}$ is also in $\mathcal{I}$.
	(iii)	For any subset $\overline{E} \subseteq E$, the maximal subsets of $\overline{E}$ that are in $\mathcal{I}$ have the same cardinality.

The cardinality of any maximal independent subset of any $\overline{E} \subseteq E$ is called the *rank* of $\overline{E}$. A *base* of M is a maximal independent subset of E. A *circuit* is a minimal dependent subset of E. A *cobase* is the set $E - X$ for some base X. Let $\mathcal{I}^*$ be the set of cobases and their subsets. The pair $M^* = (E, \mathcal{I}^*)$ is a matroid, as is easily checked. It is called the *dual matroid* of M. A *cocircuit* of M is a circuit of M^*.

One can axiomatize matroids in terms of bases, circuits, and other subsets of E, or by certain functions, geometries, and operators. It is usually a simple, though at times tedious, exercise to prove equivalence of these systems. Here we just include the axioms that rely on bases, circuits, and the rank function.

Axioms Using Bases, Circuits, Rank Function

For bases, the axioms are as follows. Let $\mathcal{B}$ be a set of subsets of E. Suppose

$\mathcal{B}$ observes the following axioms.

(3.4.2)

    (i) $\mathcal{B}$ is nonempty.

    (ii)   For any sets $B_1, B_2 \in \mathcal{B}$ and any $x \in (B_1 - B_2)$, there is a $y \in (B_2 - B_1)$ such that $(B_1 - \{x\}) \cup \{y\}$ is in $\mathcal{B}$.

Then $\mathcal{B}$ is the set of bases of a matroid on E.

Via circuits, we may define matroids as follows. Let $\mathcal{C}$ be the empty set, or be a set of nonempty subsets of E observing the following axioms.

(3.4.3)

    (i)   For any $C_1, C_2 \in \mathcal{C}$, C_1 is not a proper subset of C_2.

    (ii)   For any two $C_1, C_2 \in \mathcal{C}$ and any $z \in (C_1 \cap C_2)$, there is a set $C_3 \in \mathcal{C}$ where $C_3 \subseteq (C_1 \cup C_2) - \{z\}$.

Then $\mathcal{C}$ is the set of circuits of a matroid on E.

With the rank function, we specify a matroid as follows. Let $r(\cdot)$ be a function from the power set of E to the nonnegative integers. Assume $r(\cdot)$ satisfies the following axioms for any subsets S and T of E.

(3.4.4)

    (i) $r(S) \leq |S|$.

    (ii) $S \subseteq T$ implies $r(S) \leq r(T)$.

    (iii) $r(S) + r(T) \geq r(S \cup T) + r(S \cap T)$.

Then $r(\cdot)$ is the rank function of a matroid on E.

We omit the proofs of equivalence of the systems. It is instructive, though, to express each one of $\mathcal{I}$, $\mathcal{B}$, $\mathcal{C}$, and $r(\cdot)$ in terms of the other ones. We do this next.

Suppose $\mathcal{I}$ is given. Then $\mathcal{B}$ is the set of $Z \in \mathcal{I}$ with maximum cardinality. $\mathcal{C}$ is the set of the minimal $C \subseteq E$ that are not in $\mathcal{I}$. For any $\overline{E} \subseteq E$, $r(\overline{E})$ is the cardinality of a maximal set $Z \subseteq \overline{E}$ that is in $\mathcal{I}$.

Suppose $\mathcal{B}$ is given. Then $\mathcal{I}$ is the set of all $X \in \mathcal{B}$ plus their subsets. $\mathcal{C}$ is the set of the minimal $C \subseteq E$ that are not contained in any $X \in \mathcal{B}$. For any $\overline{E} \subseteq E$, $r(\overline{E})$ is the cardinality of any maximal set $X \cap \overline{E}$ where $X \in \mathcal{B}$.

Suppose $r(\cdot)$ is given. Then $\mathcal{I}$ is the set of $\overline{E} \subseteq E$ where $r(\overline{E}) = |\overline{E}|$. $\mathcal{B}$ is the set of $Z \subseteq E$ for which $|Z| = r(E)$. $\mathcal{C}$ is the set of the minimal $C \subseteq E$ for which $r(C) = |C| - 1$.

Abstract Matrix

We take a detour to introduce *abstract matrices*. We want to acquire a good understanding of such matrices, since they not only represent matroids, but also display a lot of structural information about matroids that other ways do not.

An *abstract matrix* B is a $\{0,1\}$ matrix with row and column indices plus a function called *abstract determinant* and denoted by det. The function det associates with each square submatrix D of the $\{0,1\}$ matrix the value 0 or 1, i.e., det D is 0 or 1. Note that numerically identical square submatrices with differing row or column index sets may have different determinants. The reader should not be misled by the symbols 0 and 1. Indeed, for the moment, we do not view abstract matrices as part of some algebraic structure. It turns out, though, that 0 and 1 allow a rather appealing use of linear algebra terms. For example, we call D *nonsingular* if det $D = 1$, and *singular* otherwise.

The function det must obey several conditions. First, if D is the 1×1 matrix $[\,0\,]$ (resp. $[\,1\,]$), then det $D = 0$ (resp. det $D = 1$).

Second, for any nonempty submatrix B^1 of B, the maximal nonsingular submatrices must have the same size. This condition may be rephrased as follows. Start with some nonsingular submatrix of B^1. Iteratively add a row and a column such that each time another nonsingular submatrix results. Stop when no further enlargement is possible. The above maximality condition demands that the order of the final nonsingular submatrix is the same regardless of the choice of the initial nonsingular submatrix and of the rows and columns added to it. The order of any such final nonsingular submatrix is called the *rank* of B^1. For the case where B^1 is trivial or empty, we declare rank B^1 to be 0. Upon deletion of a column or row, we demand that the rank drop at most by the rank of that row or column.

Third, the rank function of B must behave much like the rank function of matrices over fields. In particular, for any partition of any submatrix of B of the form

(3.4.5)
$$B = \begin{array}{|c|c|c|} \hline B^{11} & B^{12} & B^{13} \\ \hline B^{21} & B^{22} & B^{23} \\ \hline B^{31} & B^{32} & B^{33} \\ \hline \end{array}$$

Partitioned submatrix of B

the submatrices

$$D^1 = \begin{array}{|c|c|} \hline B^{11} & B^{12} \\ \hline B^{21} & B^{22} \\ \hline \end{array} \quad ; \quad D^2 = \begin{array}{|c|c|} \hline B^{22} & B^{23} \\ \hline B^{32} & B^{33} \\ \hline \end{array}$$

(3.4.6)

$$D^3 = \begin{array}{|c|c|c|} \hline B^{21} & B^{22} & B^{23} \\ \hline \end{array} \quad ; \quad D^4 = \begin{array}{|c|} \hline B^{12} \\ \hline B^{22} \\ \hline B^{32} \\ \hline \end{array}$$

Submatrices D^1, D^2, D^3, D^4

must observe

(3.4.7) $\text{rank } D^1 + \text{rank } D^2 \geq \text{rank } D^3 + \text{rank } D^4.$

We call this property the *submodularity* of the rank function.
    We summarize the above requirements as follows.

Axioms for Abstract Matrix

(3.4.8)
(i)   If $D = [B_{xy}]$, then $\det D = B_{xy}$.
(ii)   For all submatrices B^1 of B: The maximal non-singular submatrices of B^1 have the same size, called the rank of B^1. When a row or column is deleted from B^1, the rank drops at most by the rank of that row or column.
(iii) The rank function is submodular.

The *transpose* of an abstract matrix B is B^t with determinants defined as follows. For any square submatrix B^1 of B, $\det B^1$ is the determinant value for the submatrix $(B^1)^t$ of B. By symmetry of the conditions of (3.4.8), B^t with its determinants is an abstract matrix, as expected.

    We may create abstract matrices in several ways. In the simplest case, we start with a matrix A over some field $\mathcal{F}$. Then we declare B to be the support matrix of A. Thus, B is a $\{0, 1\}$ matrix. We turn B into an abstract matrix as follows. We declare any square submatrix D of B to be nonsingular if the corresponding submatrix of A is $\mathcal{F}$-nonsingular, and to be singular otherwise. Well-known linear algebra results plus Lemma (2.3.11) imply that the axioms of (3.4.8) are satisfied.

Representation of Abstract Matrix

Suppose an abstract matrix B can be derived by the above construction from a matrix A over some field $\mathcal{F}$. We say that B is *represented* by A over $\mathcal{F}$. As an example, let A be the matrix

(3.4.9)
$$\begin{array}{c|cccc} & e & f & g & h \\ \hline a & 0 & 1 & 1 & 1 \\ b & 1 & 0 & 1 & 1 \\ c & 1 & 1 & 0 & 1 \\ d & 1 & 1 & 1 & 0 \end{array}$$

Matrix A producing an abstract matrix B

over GF(2). Then B is numerically identical to A, and the determinants
for the submatrices of B are given by the GF(2)-determinants of A. For
example, the submatrix D of B given by

(3.4.10)
$$D = \begin{array}{c} \\ a \\ b \end{array}\begin{array}{cc} g & h \\ \hline \boxed{\begin{array}{cc} 1 & 1 \\ 1 & 1 \end{array}} \end{array}$$

Submatrix D of abstract matrix B

has $\det D = 0$, since the related submatrix of A has GF(2)-determinant 0.

The example may be modified to produce an abstract matrix that is
not representable over any field. Let B and its determinants be as just
defined. Then change the determinant of D of (3.4.10) from 0 to 1. One
may check by enumeration that the new B observes the axioms of (3.4.8).
For a proof that the new B is not representable, suppose A over some field
$\mathcal{F}$ represents B. Then one readily shows that the rows and columns of A
can be scaled so that the matrix of (3.4.9) results. For D of (3.4.10) as
submatrix of the scaled A, we have $\det_{\mathcal{F}} D = 0$. But for D as submatrix
of B, we have $\det D = 1$, a contradiction.

For certain abstract matrices, the axioms of (3.4.8) completely determine the rank. For example, by axiom (ii) of (3.4.8), zero matrices have
rank 0. A more interesting instance is given in the next result.

(3.4.11) Lemma. *Let B be an $m \times m$ triangular abstract matrix. Then*
$\det B = 1$ *if and only if the diagonal of B contains only 1s.*

Proof. The case $m = 1$ is immediate by (i) of (3.4.8). Thus, consider the
case $m \geq 2$.

Assume that the diagonal of B has only 1s. For an inductive proof,
we partition B as

(3.4.12)
$$B = \begin{array}{c} \\ x \\ \overline{X} \end{array}\begin{array}{|c|c|} \multicolumn{1}{c}{y} & \multicolumn{1}{c}{\overline{Y}} \\ \hline 1 & 0 \\ \hline {}^{0}\!/_{1} & \overline{B} \\ \hline \end{array}$$

Triangular B with 1s on the diagonal

where $\overline{B}$ is triangular and has only 1s on the diagonal. Apply the submodularity condition (3.4.7) as follows. Declare B^{22} (resp. B^{23}, B^{32},
B^{33}) to be the submatrix of B indexed by x and $\overline{Y}$ (resp. x and y, $\overline{X}$
and $\overline{Y}$, $\overline{X}$ and y); all other B^{ij} are trivial or empty. Then by (3.4.6),
rank D^1 = rank $B^{22} = 0$, rank D^2 = rank B, rank $D^3 = 1$, and by induction, rank $D^4 = m - 1$. By submodularity, rank $B \geq 1 + (m - 1) = m$, so
$\det B = 1$.

Assume now that the diagonal of B contains a 0. Thus, we may partition B as

(3.4.13)

$$B = \begin{array}{c} \\ X_1 \\ x \\ X_2 \end{array} \begin{array}{|c|c|c|} \hline Y_1 & y & Y_2 \\ \hline B^1 & & 0 \\ \hline & 0 & \\ \hline 0/1 & & B^2 \\ \hline \end{array}$$

Triangular B with a zero on the diagonal

where both B^1 and B^2 are square. By axiom (ii) of (3.4.8), the rows of B indexed by $X_1 \cup \{x\}$ have rank of at most $|Y_1|$. The remaining rows of B have rank of at most $|X_2|$. Thus, again by axiom (ii) of (3.4.8), the rank of B is at most $|Y_1| + |X_2| = |X_1| + |X_2| \leq m - 1$. Hence, $\det B = 0$.   □

(3.4.14) Corollary. *The abstract matrices*

(3.4.15) $\boxed{0} \; ; \; \begin{bmatrix} 0 & 0 \\ 0 & 0 \end{bmatrix} \; ; \; \begin{bmatrix} 1 & 0 \\ 0 & 0 \end{bmatrix} \; ; \; \begin{bmatrix} 1 & 1 \\ 0 & 0 \end{bmatrix} \; ; \; \begin{bmatrix} 1 & 0 \\ 1 & 0 \end{bmatrix}$

Small singular abstract matrices

are singular. The matrices

(3.4.16) $\boxed{1} \; ; \; \begin{bmatrix} 1 & 0 \\ 0 & 1 \end{bmatrix} \; ; \; \begin{bmatrix} 1 & 1 \\ 1 & 0 \end{bmatrix}$

Small nonsingular abstract matrices

are nonsingular. The matrix

(3.4.17) $\begin{bmatrix} 1 & 1 \\ 1 & 1 \end{bmatrix}$

Abstract matrix with determinant 0 or 1

may be singular or nonsingular.

Proof. Lemma (3.4.11) handles the cases of (3.4.15) and (3.4.16). A matrix over GF(3) with support given by (3.4.17) may be GF(3)-singular or GF(3)-nonsingular. This fact validates the claim about (3.4.17).   □

Note that the GF(2)-determinants of the matrices of Corollary (3.4.14) agree with the abstract determinants, except possibly for the matrix of (3.4.17).

Abstract Matrices Encode Matroids

We claim that abstract matrices are nothing but one way of encoding matroids. For a proof, we first assume that an abstract matrix B with row index set X and column index set Y is at hand. We want to show that B encodes some matroid. Define a rank function $r(\cdot)$ on $E = X \cup Y$ as follows. Given $Z \subseteq E$, the sets $X_2 = Z \cap X$ and $Y_1 = Z \cap Y$ induce a partition of B of the form

(3.4.18)

$$B = \begin{array}{c|c|c|}
 & Y_1 & Y_2 \\
\hline
X_1 & B^1 & D^2 \\
\hline
X_2 & D^1 & B^2 \\
\hline
\end{array}$$

Partitioned version of B

Then define

(3.4.19) $$r(Z) = |X_2| + \text{rank } B^1.$$

(3.4.20) Lemma. $r(\cdot)$ is the rank function of a matroid.

Proof. We verify the rank axioms (3.4.4). Let $S, T \subseteq E$. Clearly $r(S) \leq |S|$. A simple case analysis and axiom (ii) of (3.4.8) confirm that $S \subseteq T$ implies $r(S) \leq r(T)$. Finally, $r(S) + r(T) \geq r(S \cup T) + r(S \cap T)$ is argued as at the end of Section 3.3, except that we use the abstract rank function instead of the GF(2)-rank function.   □

Matroids Produce Abstract Matrices

Given a matroid M on E with rank function $r(\cdot)$, we want to show that M may be encoded by some abstract matrix. Define X to be a base of M and $Y = E - X$. For any $y \in Y$, the set $X \cup \{y\}$ must contain at least one circuit. Indeed, there must be precisely one circuit in $X \cup \{y\}$, say C_y, since otherwise there is a base X' contained in $X \cup \{y\}$ with cardinality $|X'| \leq |X| - 1$. The circuit C_y is called the *fundamental circuit* that y creates with X. Declare the characteristic vectors of the fundamental circuits C_y, $y \in Y$, to be the columns of a matrix B. Thus, we have

(3.4.21)

$$\begin{array}{c|c|}
 & Y \\
\hline
X & B \\
\hline
\end{array}$$

Matrix B for matroid M with base X

where for $x \in X$ and $y \in Y$, we have $B_{xy} = 1$ if and only if $x \in C_y$. We endow B with determinants as follows. Let B^1 be a square submatrix of B, say indexed by $X_1 \subseteq X$ and $Y_1 \subseteq Y$ as in (3.4.18). Then declare $\det B^1 = 1$ if $X_2 \cup Y_1$ is a base of M, and to be 0 otherwise.

(3.4.22) Lemma. *B and its determinants constitute an abstract matrix.*

Proof. First, $\det[B_{xy}] = B_{xy}$ for all $x \in X$ and $y \in Y$, due to the equivalence of the following statements separated by semicolons: $B_{xy} = 1$; x is in the fundamental cycle C_y; $(X \cup \{y\}) - \{x\}$ does not contain a circuit; $(X \cup \{y\}) - \{x\}$ is a base of M; $\det[B_{xy}] = 1$.

Next, let B^1 be an arbitrary submatrix of B indexed by some $X_1 \subseteq X$ and $Y_1 \subseteq Y$. Define $X_2 = X - X_1$ and $Y_2 = Y - Y_1$. We prove that the maximal square nonsingular submatrices D of B^1 have same order. Any such matrix D corresponds to a base of M that contains X_2, that is contained in $X_1 \cup X_2 \cup Y_1$, and that, subject to these two conditions, intersects Y_1 as much as possible. By the independence axioms (3.4.1), the cardinality of any such maximal intersection is the same regardless of which independent subset of Y_1 one starts with. We conclude that the maximal nonsingular submatrices of B^1 have same order. That order is the rank of B^1, denoted by rank B^1.

The rank function $r(\cdot)$ of M is thus related to rank B^1 by $r(X_2 \cup Y_1) = |X_2| + \text{rank } B^1$. We apply the independence axioms of (3.4.1) to M and verify that removal of a column or row from B^1 reduces the rank at most by the rank of the removed column or row. Finally, the arguments at the end of Section 3.3 linking submodularity of $r(\cdot)$ with that of the GF(2)-rank function are easily adapted to prove that the rank function for B is submodular.

Thus, the axioms of (3.4.8) are satisfied, and B is an abstract matrix as claimed. □

The constructions of the proof of Lemmas (3.4.20) and (3.4.22) are inverses of each other in the following sense. Let M be a matroid with a base X. Define B to be the abstract matrix constructed from M in the proof of Lemma (3.4.22). The matroid deduced from B in the proof of Lemma (3.4.20) is M again. For this reason, we say that B *represents* M.

At this point, we have established an axiomatic link between abstract matrices and matroids. We could explore the ways in which matroid concepts manifest themselves in abstract matrices. But that discussion would largely duplicate the material of Section 3.3 about binary matroids. So we just sketch the definitions and relationships, and omit all proofs. Throughout, B is an abstract matrix with row index set X and column index set Y. The related matroid M on $E = X \cup Y$ has X as a base.

Column $y \in Y$ of B is the characteristic vector of $C_y - \{y\}$, where C_y is the *fundamental circuit* that y forms with X. Row $x \in X$ of B is the characteristic vector of $C_x^* - \{x\}$, where C_x^* is the *fundamental cocircuit* that x forms with Y. Two *parallel* (resp. *coparallel* or *series*) elements of M manifest themselves in two nonzero columns (resp. rows) of B of rank 1, or in a column (resp. row) unit vector. A *loop* (resp. *coloop*) of M is indicated by a column (resp. row) of 0s. The transpose of B represents the *dual matroid* M^* of M. Let U and W be disjoint subsets of E. Assume U contains no circuit and W contains no cocircuit. Then there is a B where $X \supseteq U$ and $Y \supseteq W$. Delete from B the rows of U and the columns of W, getting an abstract matrix $\overline{B}$. Correspondingly, *contraction* of U and *deletion* of W *reduce* M to the *minor* $\overline{M} = M/U\backslash W$, which is represented by $\overline{B}$. Furthermore, *expansion* by U and addition of W *extend* $\overline{M}$ to M.

Partition B as in (3.4.18). If for some $k \geq 1$, $|X_1 \cup Y_1|$, $|X_2 \cup Y_2| \geq k$ and rank D^1 + rank $D^2 \leq k - 1$, then $(X_1 \cup Y_1, X_2 \cup Y_2)$ is a k-*separation* of B and M. Let $k \geq 2$. If B and M have no l-separation, $1 \leq l < k$, then they are k-*connected*. M is *connected* if it is 2-connected. Consistent with Section 2.3, B is *connected* if the graph $\mathrm{BG}(B)$ is connected.

Abstract Pivot

We introduce pivots in abstract matrices. For comparison purposes, we rewrite the GF(2)-pivot of (3.2.12) as follows. Given is the pivot element $B_{xy} = 1$.

(3.4.23.1)   We replace each B_{uw}, $u \in (X - \{x\})$, $w \in (Y - \{y\})$, by $\det_2 D^{uw}$, where D^{uw} is the submatrix of B given by

(3.4.23)

$$
D^{uw} = \begin{array}{c} \\ x \\ u \end{array}\!\!\begin{array}{cc} y & w \\ \left[\begin{array}{cc} B_{xy} & B_{xw} \\ B_{uy} & B_{uw} \end{array}\right] \end{array}
$$

(3.4.23.2)   We exchange the indices x and y.

Let B be the abstract matrix for M as before. A pivot on B_{xy} is to produce the abstract matrix B' corresponding to the base $(X - \{x\}) \cup \{y\}$ of M. We claim that the following procedure generates the desired B'.

(3.4.24.1)   We replace each B_{uw}, $u \in (X - \{x\})$, $w \in (Y - \{y\})$, by $\det D^{uw}$, where D^{uw} is the submatrix of B as given above.

(3.4.24)

(3.4.24.2)   We exchange the indices x and y. Let B' be the resulting matrix.

(3.4.24.3)   We endow the square submatrices D' of B' with determinants. Let U' be the row index set of one such D', and W' be the column index set. Then $\det D' = \det D$ for the submatrix D of B indexed by U and W as specified below.

If $y \in U'$, $x \in W'$: $U = U' - \{y\}$, $W = W' - \{x\}$.
If $y \in U'$, $x \notin W'$: $U = (U' - \{y\}) \cup \{x\}$, $W' = W$.
If $y \notin U'$, $x \in W'$: $U = U'$, $W = (W' - \{x\}) \cup \{y\}$.
If $y \notin U'$, $x \notin W'$: $U = U' \cup \{x\}$, $W = W' \cup \{y\}$.

Validity of the procedure is argued as follows. By step (3.4.24.2), $X' = (X - \{x\}) \cup \{y\}$ and $Y' = (Y - \{y\}) \cup \{x\}$ index the rows and columns, respectively, of B'. By step (3.4.24.1), for $u \in (X - \{x\})$ and $w \in (Y - \{y\})$, $B'_{uw} = \det B^{uw}$. Now $\det D^{uw} = 1$ if and only if $(X - \{x, u\}) \cup \{y, w\}$ is a base of M. The latter set is $(X' - \{u\}) \cup \{w\}$, so B'_{uw} is correctly computed. The entries B'_{yw} $(= B_{xw})$, $w \in (Y - \{y\})$, are correct since $(X - \{x\}) \cup \{w\} = (X' - \{y\}) \cup \{w\}$. Similarly, the entries B'_{ux} $(= B_{uy})$, $u \in (X - \{x\})$, are correct since $(X - \{u\}) \cup \{y\} = (X' - \{u\}) \cup \{x\}$. Finally, B'_{yx} $(= B_{yx} = 1)$ since $(X' - \{y\}) \cup \{x\} = X$, the assumed base of M. Analogous arguments involving the sets U and W, instead of the elements u and w, validate (3.4.24.3). We conclude that B' is the abstract matrix corresponding to the base $X' = (X - \{x\}) \cup \{y\}$ of M as claimed.

Except for the computationally tedious step (3.4.24.3), the pivot is almost identical to the GF(2)-pivot of (3.4.23). Indeed, for 2×2 matrices, Corollary (3.4.14) establishes an almost complete agreement between GF(2)-determinants and abstract determinants. Thus, the step (3.4.24.1) of an abstract pivot looks very much like the step (3.4.23.1) of a GF(2)-pivot. Informally, one is tempted to say that general matroids behave locally to quite an extent like binary matroids.

There remains, of course, the cumbersome step (3.4.24.3). But we can always imagine that this step is implicitly carried out, without our actually having to write down the list of determinant values. When we take that attitude, the pivot operation becomes simple and useful.

Some Matroid Results

The similarity of local behavior of binary and general matroids is easily demonstrated. We do this here by listing a number of results for general matroids that by trivial modifications may be obtained from the results for binary matroids proved in Section 3.3. In each case, the proof as given in Section 3.3 suffices, or in that proof one simply substitutes an abstract matrix whenever a binary one is employed. With each result, we cite in parentheses the related result of Section 3.3.

(3.4.25) Lemma (see Lemma (3.3.6)). *Let C (resp. C^*) be a circuit (resp. cocircuit) of a matroid M. Then $|C \cap C^*| \neq 1$.*

(3.4.26) Lemma (see Lemma (3.3.12)). *Let M be a matroid with a minor $\overline{M}$, and $\overline{B}$ be an abstract representation matrix of $\overline{M}$. Then M has an abstract representation matrix B that displays $\overline{M}$ via $\overline{B}$ and thus makes the minor $\overline{M}$ visible.*

(3.4.27) Lemma (see Lemma (3.3.13)). *The following statements are equivalent for any matroid M and any minor $\overline{M}$ of M. Let $\overline{B}$ be an abstract representation matrix of $\overline{M}$.*

(i) $\overline{M}$ is a contraction minor of M.

(ii) M has an abstract representation matrix B displaying $\overline{M}$ via $\overline{B}$, where B is of the form

(3.4.28)

$$
B = \begin{array}{c} \\ X \Big\{ \\ \end{array}
\begin{array}{c|c|c}
 & \multicolumn{2}{c}{\overset{\longleftarrow Y \longrightarrow}{}} \\
 & \overline{Y} & W \\
\hline
X & \overline{B} & 0 \\
\hline
U & \multicolumn{2}{c}{0/1} \\
\end{array}
$$

Matrix B displaying contraction minor $\overline{M}$

(iii) Every abstract representation matrix B of M displaying $\overline{M}$ via $\overline{B}$ is of the form given by (3.4.28).

(3.4.29) Lemma (see Lemma (3.3.15)).   The following statements are equivalent for any matroid M and any minor $\overline{M}$ of M. Let $\overline{B}$ be an abstract representation matrix of $\overline{M}$.

(i) $\overline{M}$ is a deletion minor of M.

(ii) M has an abstract representation matrix B displaying $\overline{M}$ via $\overline{B}$, where B is of the form

(3.4.30)

$$
B = X \left\{
\begin{array}{c|c|c}
 & \multicolumn{2}{c}{\overset{\longleftarrow Y \longrightarrow}{}} \\
 & \overline{Y} & W \\
\hline
\overline{X} & \overline{B} & \\
\cline{1-2}
U & 0 & 0/1 \\
\end{array}
\right.
$$

Matrix B displaying deletion minor $\overline{M}$

(iii) Every abstract representation matrix B of M displaying $\overline{M}$ via $\overline{B}$ is of the form given by (3.4.30).

(3.4.31) Lemma (see Lemma (3.3.19)).   Let M be a matroid with an abstract representation matrix B. Then M is connected if and only if this is so for B.

(3.4.32) Lemma (see Lemma (3.3.20)).   The following statements are equivalent for a matroid M with set E and an abstract representation matrix B.

(i) M is 3-connected.

(ii) B is connected, has no parallel or unit vector rows or columns, and has no partition as in (3.4.18) with rank $D^1 = 1$, $D^2 = 0$, and $|X_1 \cup Y_1|, |X_2 \cup Y_2| \geq 3$.

(3.4.33) Lemma (see Lemma (3.3.26)). *Let M be a matroid on a set E and with rank function $r(\cdot)$. Suppose E_1 and E_2 partition E. Then (a) and (b) below hold.*

(a) (E_1, E_2) *is a k-separation of M if and only if*

$$(3.4.34) \qquad \begin{array}{c} |E_1|, |E_2| \geq k \\ r(E_1) + r(E_2) \leq r(E) + k - 1 \end{array}$$

(b) (E_1, E_2) *is an exact k-separation of M if and only if*

$$(3.4.35) \qquad \begin{array}{c} |E_1|, |E_2| \geq k \\ r(E_1) + r(E_2) = r(E) + k - 1 \end{array}$$

(3.4.36) Lemma (see Lemma (3.3.31)). *Let M be a 3-connected matroid on a set E. Take z to be any element of E. Then $M \textcircled{c} z$ or $M \textcircled{d} z$ is 3-connected.*

(3.4.37) Lemma (see Lemma (3.3.36)). *Let M be a connected matroid on a set E. Take z to be any element of E. Then M/z or $M \backslash z$ is connected.*

Aspects of Representability

In the remainder of this section and in the next one, we examine several aspects of the representability of abstract matrices. Recall that an abstract matrix B is represented by a matrix A of the same size and over some field $\mathcal{F}$ if the following holds. For every square submatrix of B, the determinant of that submatrix is 1 if and only if the $\mathcal{F}$-determinant of the corresponding submatrix of A is nonzero. We want to establish a direct connection between the matroid M, defined by B, and any matrix A representing B over some field $\mathcal{F}$. To this end, we partition such A in agreement with the partition of B of (3.4.18). Thus,

$$(3.4.38) \qquad A = \begin{array}{c|c|c} & Y_1 & Y_2 \\ \hline X_1 & A^1 & C^2 \\ \hline X_2 & C^1 & A^2 \end{array}$$

Partitioned version of matrix A over field $\mathcal{F}$

Evidently, the submatrix A^1 of A corresponds to B^1 of B. Assume these submatrices to be square. We know that $\det_{\mathcal{F}} A^1 \neq 0$ if and only if $\det B^1 = 1$. We also know that $\det B^1 = 1$ if and only if $X_2 \cup Y_1$ is a base of M. Thus,

$\det A^1 \neq 0$ if and only if $X_2 \cup Y_1$ is a base of M. When this relationship holds, we also say that A over $\mathcal{F}$ *represents* M.

We have seen that some abstract matrices are not representable over any field. By definition, the same holds for some matroids. In particular, the abstract matrix deduced from (3.4.9) by modifying the determinant of D of (3.4.10) produces a matroid that is not representable over any field.

Deciding whether or not a matroid is representable over some field is a nontrivial problem. Usually, one assumes that the matroid is given by a *black box* or *oracle* that in unit time tells whether or not a subset of E is independent in the matroid. No additional information about the matroid is available. Under these assumptions, even representability over GF(2) cannot be tested in polynomial time. Indeed, the same conclusion can be drawn for a great many representability questions. There is one extraordinary exception. One can test in polynomial time whether or not a matroid is representable over *every* field. That representability problem is intimately linked to the problem of deciding whether a matrix is totally unimodular. Details are covered in Chapter 11.

Suppose an abstract matrix B and the related matroid M are representable over some field $\mathcal{F}$. Let A be a matrix over $\mathcal{F}$ proving this fact. Then every matrix B' derivable from B by one or more pivots is also representable over $\mathcal{F}$. For a proof, one carries out a pivot in B, say on $B_{xy} = 1$, and a second $\mathcal{F}$-pivot in A on A_{xy}. Let B' and A' result. It is easily seen that A' establishes B' to be representable over $\mathcal{F}$. Combine this result with the trivial observation that every proper submatrix of B and the transpose of B are representable over $\mathcal{F}$. We conclude that every minor of M and the dual M^* of M are representable over $\mathcal{F}$.

It is easy to check that all matroids with at most three elements are representable over every field. Couple this observation with the fact that the taking of minors maintains representability. Evidently, a matroid M not representable over a given field $\mathcal{F}$ must have a minor that also is not representable over $\mathcal{F}$, but all of whose proper minors are representable over $\mathcal{F}$. We call such a minor a *minimal violator* of representability over $\mathcal{F}$. Not much is known about the minimal violators for the various fields $\mathcal{F}$. Complete lists of the minimal violators exist only for the fields GF(2) and GF(3). There is also a complete description for the case when representability over every field is demanded. Beyond these cases, only incomplete results are known.

For the case of GF(2), there is just one minimal violator. It is a matroid on four elements called U_4^2. In the next section, we define that matroid and prove the claim we just made. In Chapter 9, we determine the minimal violators for GF(3) and for the case of representability over every field. In both cases, there are just two minimal violators plus their duals. As we shall see, abstract matrices and abstract pivots are useful for the proof of the results for GF(2) and GF(3).

3.5 Characterization of Binary Matroids

In this section, we characterize the binary matroids in several ways. In particular, we prove that a certain matroid on four elements called U_4^2 is the only minimal violator of representability over GF(2). Accordingly, a matroid is binary if and only if it has no U_4^2 minors. To begin, we define, for $1 \le m \le n$, U_n^m to be the matroid on n elements where every subset of cardinality m is a base. It is easily checked that U_n^m is indeed a matroid. It is the *uniform matroid* of rank m on n elements.

Let M be a matroid on a set E. For an arbitrary base X of M, compute the abstract matrix B. Suppose M is representable over $\mathcal{F}$. By definition, there is a matrix A over $\mathcal{F}$ with B as support matrix, such that for all corresponding submatrices D of B and D' of A, $\det D = 1$ if and only if $\det_{\mathcal{F}} D' \ne 0$. If $\mathcal{F}$ is GF(2), evidently A is numerically identical to B.

Consider U_n^m, for $2 \le m \le n - 2$. Since every set of cardinality m is a base of U_n^m, every abstract matrix B for that matroid is of size $m \times n$, contains only 1s, and has only nonsingular square submatrices. In the related binary matrix A, all square submatrices of order at least 2 are GF(2)-singular. Thus, U_n^m is nonbinary. The smallest nonbinary case has $2 = m = n - 2$, i.e., the matroid is U_4^2. Representability over any field $\mathcal{F}$ is maintained under minor-taking. Thus, a binary matroid cannot possibly have U_4^2 minors. We now prove that absence of U_4^2 minors implies representability over GF(2).

Let M be a nonbinary matroid all of whose proper minors are binary. Select any abstract representation B for M, and let A be the associated binary matrix. Then there are minimal submatrices D of B and D' of A in the same position such that exactly one of $\det D$ and $\det_2 D'$ is 0. Since every proper minor of M is binary, we must have $D = B$ and $D' = A$. Since the entries of B agree with those of A, the order of B must be at least 2.

If B is a zero matrix, then both $\det B$ and $\det_2 A$ are zero, a contradiction. Thus, B contains a 1, say $B_{xy} = 1$. If the order of B is greater than 2, we perform a pivot in B and the corresponding GF(2)-pivot in A. In both cases, we delete the pivot row and pivot column. Let B' and A' result. By the minimality assumptions on B' and A' and the rules (3.4.24.1) and (3.4.23.1), the two matrices must agree numerically. By (3.4.24.3) and the analogous rule of linear algebra for A, exactly one of $\det B'$ and $\det_2 A'$ is zero. Thus, we have proved that a proper minor of M is not binary, a contradiction. Hence, B is a 2×2 matrix. By Corollary (3.4.14), there is only one choice for B and A. That is, B is the abstract matrix of (3.4.17), and A is also that matrix when viewed as binary. We include the two matrices below in (3.5.1). Since exactly one of the determinants of B and A is nonzero and $\det_2 A = 0$, we must have $\det B = 1$.

(3.5.1)
$$B = \begin{bmatrix} 1 & 1 \\ 1 & 1 \end{bmatrix} ; \quad A = \begin{bmatrix} 1 & 1 \\ 1 & 1 \end{bmatrix}$$

Abstract matrix B of minimal nonbinary matroid
and related binary matrix A

Evidently, M has four elements, and every 2-element subset is a base. Thus, M is isomorphic to U_4^2. We record this conclusion and state and prove a related corollary for future reference.

(3.5.2) Theorem.   *A matroid M is binary if and only if M does not have U_4^2 minors.*

(3.5.3) Corollary.   *Let an abstract matrix B represent a nonbinary matroid M. Let N be the binary matroid represented by the binary matrix A that is numerically identical to B. Then M has a base that is not a base of N.*

Proof. Carry out the earlier proof, except that the proper minors of M are not assumed to be binary. Accordingly, D and D' may be proper submatrices of B and A, respectively. By the pivot argument, we know that $\det D = 1$ and $\det_2 D' = 0$. This implies that the base of M corresponding to D is not a base of N.   ☐

The proof of Theorem (3.5.2) implies the following statement. A matroid M is nonbinary if and only if M has an abstract representation matrix B with a 2×2 submatrix D that is nonsingular and that contains only 1s. This fact leads to an elementary proof of the following theorem.

(3.5.4) Theorem.   *The following statements are equivalent for a matroid M on a set E.*

(i) *M is binary.*
(ii) *For any circuit C and cocircuit C^*, $|C \cap C^*|$ is even.*
(iii) *The symmetric difference of two circuits is a disjoint union of circuits.*
(iv) *The symmetric difference of two disjoint unions of circuits is a disjoint union of circuits.*
(v) *The symmetric difference of two distinct circuits contains a circuit.*
(vi) *Given any distinct circuits C_1 and C_2, and any two elements $e, f \in (C_1 \cap C_2)$, there is a circuit $C_3 \subseteq [(C_1 \cup C_2) - \{e, f\}]$.*
(vii) *For any base X and $Y = E - X$, any circuit C is the symmetric difference of the fundamental circuits C_y corresponding to X and with $y \in (C \cap Y)$.*

Proof. Statement (i) plus Lemmas (3.3.6), (3.3.7), (3.3.8) imply (ii)–(vi). Statement (vii) follows from (i) by the previous characterization of circuits in terms of column submatrices of the representation matrix B. We prove the converse implications by contradiction. Thus, we assume M to be a nonbinary matroid. For M, we select an abstract matrix B with the 2×2 submatrix D as described above and show by inspection and routine arguments that none of (ii)–(vii) holds. $\Box$

3.6 References

The basic material on graphic, binary, and general matroids relies on Whitney (1935), and Tutte (1965), (1966b), (1971). Theorem (3.2.25) is proved in Truemper (1987b). Corollary (3.2.29) was first established in Tutte (1966b); a short proof appears in Cunningham (1981), together with other connectivity concepts (see also Inukai and Weinberg (1981), Oxley (1981a), and Wagner (1985b)). Corollary (3.2.31) and its generalization to general matroids are included in Cunningham (1973); the results also appear in Duchamp (1974) and Krogdahl (1977).

The 2-isomorphism results (Lemma (3.2.33) and Theorem (3.2.36)) are due to Whitney (1933a). Shorter proofs are given in Truemper (1980a), Wagner (1985a), and Kelmans (1987), together with an upper bound of $n - 2$ switchings for 2-connected graphs with n vertices. The generalization to directed graphs is covered in Thomassen (1989). For a variation of the 2-isomorphism problem, define a matroid from the node/edge incidence matrix of a graph as in Section 3.2, except that the matrix is viewed to be over $\mathbb{R}$ instead of GF(2). Just as in the GF(2) case, several graphs may produce the same matroid. A partial analysis of the graphs generating the same matroid is given in Wagner (1988). Another variation, called vertex 2-isomorphism, is treated in Swaminathan and Wagner (1990).

The graphicness testing subroutine is due to Löfgren (1959). Well implemented, it has led to the presently most efficient algorithms for that problem (see Fujishige (1980), and Bixby and Wagner (1988)). Other relevant references are Gould (1958), Auslander and Trent (1959), (1961), Tutte (1960), (1964), Iri (1968), Tomizawa (1976a), Bixby and Cunningham (1980), Wagner (1983), and Bixby (1984a). The first polynomial test for graphicness of binary matroids was given by Tutte (1960). An efficient graphicness test for matroids not known to be binary is described in Seymour (1981c); see also Bixby (1982a), and Truemper (1982a).

The notion of submodularity of the rank function as one of the central tools of matroid theory is due to Edmonds (see, e.g. Edmonds (1970)). In this book, we use the submodularity concept rather infrequently. An excellent survey of the many facets and applications is given in Fujishige (1984). For optimization and decomposition results, see, e.g., Fujishige (1983), (1989), (1991), and Frank (1990b). Seymour (1988) analyses an unusual way of specifying a matroid. Define the *connectivity function* of a matroid on a set E and with rank function $r(\cdot)$ to be the function $c(X) = r(X) + r(E - X) - r(E)$, $X \subseteq E$. Clearly, the connectivity function determines the matroid at most up to duality. In Seymour (1988), it is shown that the connectivity function generally does not determine the matroid up to duality, but that this is so when the matroid is binary.

Lemma (3.3.31) and its generalization (Lemma (3.4.36)) have also been independently proved by others (Seymour (1981b), Bixby (1982b)). The special case of graphs was first treated in Seymour (1980b).

The concept of abstract matrices is based on that of partial representation matrices of Truemper (1984). The example of a nonrepresentable abstract matrix is taken from that reference. Early examples of nonrepresentable matroids are in MacLane (1936), Lazarson (1958), Ingleton (1959), (1971), and Vamos (1968). Additional references about the representability problem are included in Section 9.5.

Theorem (3.5.2) is one of the key contributions of Tutte (1958). That theorem set the stage for and motivated a number of subsequent results on representability. The proof and Corollary (3.5.3) are taken from Truemper (1982b). The conditions of Theorem (3.5.4) are from Whitney (1935), Rado (1957), Lehman (1964), Tutte (1965), Minty (1966), and Fournier (1981). Additional characterizations of the binary matroids may be found in Bixby (1974), Duchamp (1974), and White (1987).

For additional material on binary matroids or matroids representable over fields with characteristic 2, see Gerards (1988), Hassin (1988), (1990a), (1990b), (1991), Oxley (1990b), Lemos (1991), and Ziegler (1991).

Chapter 4

Series-Parallel and Delta-Wye Constructions

4.1 Overview

This chapter is the first of three on matroid tools. Here, we construct graphs and binary matroids with elementary procedures. For graphs, the constructions involve addition of a parallel edge, or subdivision of an edge into two series edges, or substitution of a triangle by a 3-star, or substitution of a 3-star by a triangle. We call the first two operations *series-parallel exten-sion steps*, for short SP *extension steps*. Either one of the triangle/3-star substitution steps is a *delta-wye step*, for short ΔY *step*. These operations have a natural translation to operations on binary matroids.

The power of SP extension steps is quite limited. Suppose in the graph case one starts with a cycle with just two edges and applies SP extension steps. Then rather simple graphs are produced. They are usually called *series-parallel graphs*, for short SP *graphs*. In the binary matroid case, let us start with a circuit containing just two parallel elements. Then we produce nothing else but the graphic matroids of the SP graphs. These results and some related material are described in Section 4.2.

The situation changes dramatically when we mix SP extension steps with ΔY steps. In the graph case, suppose we start again with a cycle with two edges. Then we produce all 2-connected planar graphs and more. How much more is a difficult open question. Similarly, suppose that in the binary matroid case we start with a circuit with two edges. Then we produce the graphic matroids of the just-described graphs, as well as nongraphic binary matroids. Here, too, the class of matroids so obtained is not well understood. In Chapter 11, it is proved that every matroid of that class is regular.

One may, of course, start a sequence of SP extension steps and ΔY steps with any collection of graphs, not just from the cycle with two edges. Little is known about the classes of graphs so obtained. The same goes for binary matroids, with one exception. The class of almost regular matroids, which we define later, is generated when one starts with two binary matroids on seven and eleven elements, respectively.

The cited material on SP extension steps and ΔY steps in graphs and binary matroids is covered in Sections 4.3 and 4.4, respectively. The final Section 4.5 contains applications, extensions of some of the matroid results to general matroids, and relevant references.

The material of this chapter builds upon Chapters 2 and 3.

4.2 Series-Parallel Construction

Start with the cycle with just two edges. In that small graph, replace one edge by two parallel edges or by two series edges. To the resulting graph apply either one of these two operations to get a third graph, and so on. Here is an example with three such extension steps.

(4.2.1)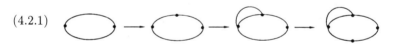

Series-parallel extension steps

Each iteration is a *series-parallel extension step*, for short SP *extension step*. The class of graphs producible this way are the *series-parallel graphs*, for short SP *graphs*. The inverse of an SP extension step is an SP *reduction step*.

In this section, we analyze the structure of SP graphs. In particular, we characterize them in terms of forbidden minors. We begin with some elementary lemmas.

(4.2.2) Lemma. *Every SP graph is 2-connected and planar. Any minor of an SP graph is also an SP graph, provided the minor has at least two edges and is 2-connected.*

Proof. The cycle with two edges is 2-connected and planar. An SP extension step in a graph with at least two edges cannot introduce a 1-separation or destroy planarity. By induction, the SP graphs are 2-connected and planar.

For the proof of the second part, paint in a given SP graph the edges of a given 2-connected minor red. Reduce the SP graph by SP reduction steps until the cycle with two edges is obtained. We examine a single reduction step and apply induction. We must consider two cases for that step: deletion of a parallel edge, and contraction of one of two edges with a common degree 2 endpoint. Consider the deletion case. If both edges are red, then the reduction is also an SP reduction in the minor. If exactly one edge is not red, then that edge is deleted. The minor must still be present, since contraction of that edge would turn the red edge into a loop, contrary to the assumption that the minor is 2-connected. If both edges are not red, then the minor is still present after deletion of one of these edges. For if both edges must be contracted to produce the minor, then the second contraction involves a loop, and thus is a deletion. The contraction case is handled analogously.   □

Recall that K_4 is the complete graph on four vertices.

(4.2.3) Lemma. *Every 3-connected graph G with at least six edges has a K_4 minor.*

Proof. Take any cycle C of G of minimal length. Since G is 3-connected and has at least six edges, it must have a node that does not lie on C. By Menger's Theorem, there are three internally node-disjoint paths from that additional node to three distinct nodes of C. Suitable deletions and contractions eliminate all other edges and reduce the cycle and three paths to a K_4 minor.   □

(4.2.4) Lemma. *K_4 is not an SP graph.*

Proof. K_4 does not have series or parallel edges.   □

(4.2.5) Lemma. *No SP graph has a K_4 minor.*

Proof. Presence of a K_4 minor would contradict Lemmas (4.2.2) and (4.2.4).   □

We are ready to characterize the SP graphs in terms of excluded minors.

(4.2.6) Theorem. *A 2-connected graph is an SP graph if and only if it has no K_4 minor.*

Proof. The "only if" part is handled by Lemma (4.2.5). We thus prove the converse. Let G be a 2-connected graph without K_4 minors. Simple checking validates the small cases with up to five edges. So assume G has at least six edges. By Lemma (4.2.3), G must be 2-separable. Choose a 2-separation so that for the two corresponding graphs G_1 and G_2, we have G_1 with minimal number of edges.

Suppose G_1 has exactly two edges. These edges must be parallel or incident at a degree 2 node of G. Thus, we can reduce and apply induction. Suppose G_1 has at least three edges. Let k and l be the nodes of G_1 that must be identified with two nodes of G_2 to produce G. Suppose G_1 has an edge z connecting k and l. That edge can be shifted from G_1 to G_2. The corresponding new 2-separation contradicts the minimality assumption on the edge set of G_1. Similarly, the nodes k and l cannot have degree 1 in G_1.

Add an edge e to G_1 connecting nodes k and l. The new graph G_1' is isomorphic to a proper minor of G. By induction, G_1' is an SP graph. By the above discussion, in G_1' the edge e is not parallel to another edge, and it does not have an endpoint of degree 2. Thus, any SP reduction step in G_1' can be carried out in G as well. We perform one such step in G, and invoke induction for the reduced graph.   □

One may reformulate the construction of SP graphs as follows. Start with some cycle. Iteratively enlarge the graph on hand as follows. Select a path in the graph where all internal nodes have degree 2. Let k and l be the endpoints of the path. Then add to the graph a path from k to l. Evidently, this construction creates precisely all SP graphs. It also allows a short proof of the following result.

(4.2.7) Lemma. *An SP graph without parallel edges either is a cycle with at least three edges, or has two internally node-disjoint paths with the following properties. Each path has at least two edges. Each intermediate node of the two paths has degree 2 in the graph, while the endpoints have degree of at least 3.*

Proof. Consider the alternate construction. The initial cycle must have at least three edges, since otherwise the final graph has parallel edges. When the first path is adjoined to the initial cycle, the lemma is satisfied, or the final graph has parallel edges. The same conclusion holds by induction after each additional path augmentation.   □

Lemma (4.2.7) has the following corollary.

(4.2.8) Corollary. *An SP graph with at least four edges and without parallel edges has at least two nonadjacent nodes with degree 2.*

Proof. If the SP graph is a cycle, then the conclusion is immediate. So assume that the SP graph is not a cycle. Then each one of the two paths postulated in Lemma (4.2.7) has at least one intermediate degree 2 node. Thus, the graph has two nonadjacent degree 2 nodes.   □

We introduce two interesting subclasses of the class of SP graphs by excluding certain graphs as minors. One of the excluded graphs we already know. It is $K_{2,3}$, the complete bipartite graph with two vertices on one side and three on the other one. The second excluded graph is the *double triangle*, obtained from the triangle by replacing each edge by two parallel edges. We denote that graph by C_3^2. We want to characterize first the SP graphs without $K_{2,3}$ minors, and then those without C_3^2 minors. To this end, define a graph to be *outerplanar* if it can be drawn in the plane so that all vertices lie on the infinite face.

(4.2.9) Theorem.   *The following statements are equivalent for a 2-connected graph G with at least two edges.*

(i) *G has no K_4 or $K_{2,3}$ minors.*
(ii) *G is an SP graph without $K_{2,3}$ minors.*
(iii) *G is outerplanar.*

Proof. By Theorem (4.2.6), G is an SP graph if and only if it has no K_4 minors. Thus, (i) $\Longleftrightarrow$ (ii). To show (ii) $\Longrightarrow$ (iii), let G be an SP graph without $K_{2,3}$ minors. Define C to be a cycle of G of maximum length. Suppose G has a node v that does not lie on C. Since G is 2-connected, there exist two paths from node v to distinct nodes i and j on C so that these paths have only the node v in common. If i and j are connected by an edge of C, then C can be extended to a longer cycle using the two paths, a contradiction. If i and j are not joined by an edge of C, then C and the two paths are easily reduced to a $K_{2,3}$ minor of G, another contradiction. Thus, all nodes of G occur on C. For the proof of outerplanarity, we may assume that G has no parallel edges. Draw C in the plane, say using a circle. Then draw the remaining edges, each time placing the edge inside the circle as a straight line segment. If any two such edges cross, then these edges plus C can be reduced to a K_4 minor of G, a contradiction. Thus, no edges cross, and we have produced an outerplanar drawing of G. For (iii) $\Longrightarrow$ (i), we note that K_4 and $K_{2,3}$ are not outerplanar and that outerplanarity is maintained under minor-taking. $\square$

For the second subclass of the class of SP graphs, we define a *suspended tree* to be any graph generated by the following process. Start with a tree. Create an additional vertex called the non-tree vertex. From that vertex, add one arc to each tip node of the tree, plus at most one arc to any other node of the tree. An example is given below. The initial tree is drawn with bold lines.

(4.2.10)

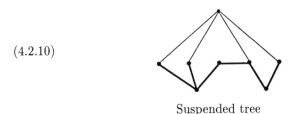

Suspended tree

We permit the degenerate case where the initial tree is just one node.

We have the following theorem, which turns out to be nothing but the dual version of Theorem (4.2.9).

(4.2.11) Theorem.   *The following statements are equivalent for a 2-connected graph G with at least two edges.*

(i) *G has no K_4 or C_3^2 minors.*
(ii) *G is an SP graph without C_3^2 minors.*
(iii) *Except for parallel edges, G is 2-isomorphic to a suspended tree.*

Proof. We use graphic matroids and duality to link Theorem (4.2.9) and the one at hand. Specifically, for each of the statements (i)–(iii) of Theorem (4.2.9), we carry out the following process. From that statement, we deduce the matroid version, then dualize that matroid statement, and finally show that the latter statement, when expressed in terms of graphs, yields the statement of Theorem (4.2.11) with the same number. Then one reverses the sequence of arguments, going from each statement of Theorem (4.2.11) to the corresponding one of Theorem (4.2.9). The proof is complete once the following two observations are made. First, the just-described dualization process takes any K_4 (resp. $K_{2,3}$) minor of one graph to a K_4 (resp. C_3^2) minor of the other one. Second, the dualization process takes the property of being an SP graph (resp. of being outerplanar) to the property of being an SP graph (resp. of being 2-isomorphic to a suspended tree up to parallel edges), and vice versa.   □

We relate the above material to binary matroids. We begin with the following construction. Start with the 1×1 binary matrix $B = [\,1\,]$. Iteratively enlarge that matrix by adjoining a binary column or row vector parallel to an existing column or row, or by adjoining a column or row unit vector. We call each such iteration a *matroid* SP *extension step*. Define the SP *matroids* to be the binary matroids represented by the matrices that can be so produced. The inverse of a matroid SP extension step is a *matroid* SP *reduction step*. The similarity of terminology with the graph case is no accident, as we shall see next.

(4.2.12) Lemma.   *Every SP matroid is the graphic matroid of an SP graph, and vice versa.*

Proof. The key relationships are provided in Section 3.2. The matrix $B = [\,1\,]$ represents the graphic matroid of a cycle with two edges. Suppose after some iterations of the construction process, a matrix B is on hand. Let B represent the graphic matroid of an SP graph G. Then the adjoining of a column z parallel to a given column y of B can be translated to adding an edge z parallel to edge y in G. Similarly, any other extension of B can be translated in G to an addition of parallel edges or to a subdivision of edges into series edges. Thus, every matrix produced by matroid SP extension steps represents the graphic matroid of some SP graph. With similar ease, one proves that the graphic matroid of any SP graph is represented by some B generated by SP extension steps.   □

With the aid of Lemma (4.2.12), we translate Theorem (4.2.6) to the following result. Recall that by Corollary (3.2.31), 2-connectedness in a graph G is equivalent to connectedness in the graphic matroid $M(G)$.

(4.2.13) Theorem. *A connected binary matroid is an SP matroid if and only if it has no $M(K_4)$ minor.*

Theorems (4.2.9) and (4.2.11) also have an interesting translation. Since these results are dual to each other, it suffices that we translate Theorem (4.2.11). To simplify matters, we rule out parallel elements. Observe that the matrices

$$(4.2.14) \qquad \begin{bmatrix} 1 & 0 & 1 \\ 1 & 1 & 0 \\ 0 & 1 & 1 \end{bmatrix}; \qquad \begin{bmatrix} 1 & 1 & 1 & 0 \\ 1 & 1 & 0 & 1 \end{bmatrix}$$

Representation matrices of $M(K_4)$ and $M(C_3^2)$

represent the graphic matroids $M(K_4)$ and $M(C_3^2)$, respectively. In the first case, the corresponding base of $M(K_4)$ is a star of K_4. In the second case, the corresponding base of $M(C_3^2)$ contains any two nonparallel edges of C_3^2.

(4.2.15) Theorem. *The following statements are equivalent for a connected binary matroid M with at least three elements, no two of which are parallel.*

(i) *M has no $M(K_4)$ or $M(C_3^2)$ minors.*
(ii) *M is an SP matroid without $M(C_3^2)$ minors.*
(iii) *No representation matrix of M has any one of the matrices of (4.2.14) as submatrix.*
(iv) *M is the graphic matroid of a suspended tree.*
(v) *M is a minor of the matroid M' represented by a binary matrix B' that is the node/edge incidence matrix of a tree.*

Proof. Equivalence of (i), (ii), (iii), and (iv) follows directly from Theorem (4.2.11). To show that (iv) implies (v), let M be the graphic matroid of a suspended tree G. Add edges, if necessary, so that every tree node is connected by exactly one arc with the extra node. The graphic matroid M' of that enlarged graph G' has M as a minor. The edges of G' incident at the extra node form a tree X. Thus, X is a base of M. It is easy to see that the representation matrix B' of M' corresponding to X is nothing but the node/edge incidence matrix of the tree. Hence, (v) holds. The arguments are easily reversed to prove that (v) implies (iv).   □

For subsequent reference, we include the following lemma.

(4.2.16) Lemma. *Let M be a connected binary matroid. Then any binary matroid obtained by an SP extension step from M is connected.*

Proof. By Lemma (3.3.19), connectedness of a binary matroid is equivalent to connectedness of any one of its representation matrices. Clearly, any SP extension step maintains connectedness of representation matrices. Thus, the resulting matroid is connected. ☐

4.3 Delta-Wye Construction for Graphs

The simplicity of SP graphs gives way to far more complicated graphs when we permit two operations in addition to SP extensions. One of them is the replacement of a triangle by a 3-star, and the second one is the inverse of that step. Either operation we call a ΔY *exchange.* Define a sequence of SP extensions and ΔY exchanges to be a ΔY *extension sequence.* The inverse sequence is a ΔY *reduction sequence.* A 2-connected graph is ΔY *reducible* if there is a ΔY reduction sequence that converts the graph to a cycle with just two edges. In this section, we show that ΔY extension sequences applied to such a cycle create all 2-connected planar graphs and more. Any graph so producible is a ΔY *graph.*

As an example for ΔY reduction sequences, we reduce K_5, the complete graph on five vertices, to the cycle with two parallel edges.

(4.3.1)

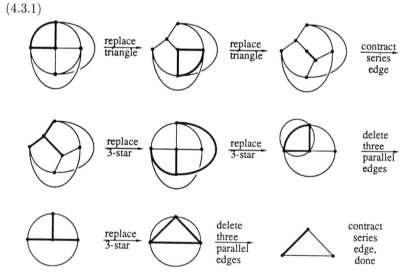

ΔY reduction sequence for K_5

In each graph of the reduction sequence, the triangle or 3-star involved in a ΔY exchange is indicated by bold lines. Similarly, we emphasize the series or parallel edges of SP reductions. We know from Lemma (3.2.48) that $M(K_5)$ is not cographic. Thus, K_5 is nonplanar, and the example demonstrates that ΔY graphs may be nonplanar.

A ΔY exchange may not preserve 2-connectedness. For example, when a vertex of a triangle in a 2-connected graph has degree 2, then replacement of that triangle by a 3-star produces a 1-separable graph. The next lemma gives the conditions under which 2-connectedness is maintained.

(4.3.2) Lemma. *Let G be a 2-connected graph. Then a triangle to 3-star exchange (resp. 3-star to triangle exchange) in G produces a 2-connected graph G' if and only if the triangle (resp. 3-star) does not contain two edges in series (resp. in parallel).*

Proof. Consider the triangle to 3-star exchange. The "only if" part has been argued above. For proof of the "if" part, suppose a triangle does not contain two series edges. Equivalently, the triangle does not contain a cocycle. Thus, G has a tree that does not include any edges of the triangle. With the aid of this tree, it is easy to see that the graph G' derived from the 2-connected graph G has no 1-separation.

If a 3-star has two edges in parallel, then a 3-star to triangle exchange is not possible. Thus, the "only if" part is trivially satisfied. The "if" part is easily checked, analogously to the case of a triangle to 3-star exchange.   □

The asymmetry of arguments in the proof of Lemma (4.3.2) is due to the fact that a triangle is always a cycle of a graph, while a 3-star is not always a cocycle. Note that the conditions of Lemma (4.3.2) are automatically satisfied in ΔY reduction sequences where ΔY exchanges are done only when an SP reduction is not possible.

Our goal is to show that the class of ΔY graphs includes all 2-connected planar graphs. That goal is restated in the next theorem.

(4.3.3) Theorem. *Every 2-connected planar graph is ΔY reducible.*

The proof relies on three lemmas. They show, in fact, that any 2-connected plane graph, i.e., an embedding of a planar graph in the plane, is ΔY reducible under the following restriction. We permit a triangle to 3-star exchange only if the triangle bounds a face.

The first auxiliary lemma is the analogue of Lemma (4.2.2).

(4.3.4) Lemma. *If a 2-connected graph or plane graph G is ΔY reducible, then every 2-connected minor H of G is ΔY reducible as well.*

Proof. We confine ourselves to the plane graph case. By omitting references to the embeddings of G, one obtains the proof for the general situation. We induct on the number of SP reductions and ΔY exchanges that reduce the planar graph G to the cycle with just two edges. We may suppose that H has no series or parallel edges. H is then a minor of G

as well as of any graph derived from G by any SP reductions. Thus, by induction, we only need to examine the case where the first step of a given reduction sequence for G is a ΔY exchange.

Consider a triangle to 3-star exchange. By assumption, the triangle, say $\{e, f, g\}$, bounds a face. Let G' be the plane graph resulting from the exchange. Suppose e, f, and g occur in H. Since H is 2-connected, these edges must form a triangle in H that also bounds a face of H. By assumption, H has no series or parallel edges, so we may replace the triangle of H by a 3-star, getting a 2-connected graph H'. Then H' is a 2-connected minor of G', and the conclusion follows by induction. Suppose at least one of the edges of the triangle $\{e, f, g\}$ does not occur in H. It is easily seen that we may delete e, f, or g from G while retaining H as a minor. Regardless of the specific situation, H is isomorphic to a minor of G', and once more induction can be invoked.

The case of a 3-star to triangle exchange is handled analogously. Indeed, in the plane graph case, we may invoke duality.   □

The next two lemmas involve *grid graphs*, which are plane graphs of the form

(4.3.5)

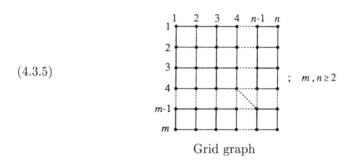

$;\quad m, n \geq 2$

Grid graph

(4.3.6) Lemma. *Every plane graph is a minor of some grid graph.*

Proof. (Sketch) We may assume that the given plane graph is 2-connected, since this can be achieved by the addition of edges. Split each vertex of that plane graph so that a 2-connected plane graph G results where the degree of each vertex is at most 3. By a suitable subdivision of edges, G can be embedded into a grid graph as follows. First embed any one face of G, but not the outer one. Then embed one face at a time so that each one of the successive subgraphs of G so embedded is 2-connected.   □

(4.3.7) Lemma. *Every grid graph is ΔY reducible.*

Proof. Two special ΔY reduction subsequences will be used repeatedly. For the first case, suppose we have a grid graph plus one edge e so that e and two edges of a degree 4 vertex of the grid graph form a triangle.

Consider the two ΔY exchanges depicted in (4.3.8) below.

(4.3.8)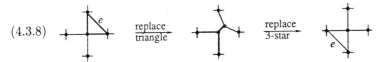

Moving edge e toward lower left-hand corner

Effectively, the two ΔY exchanges have moved the edge e one step closer toward the lower left-hand corner of the grid graph.

The second situation is even simpler. Suppose an edge e forms a triangle with two edges of a 3-star. Then that edge can be effectively eliminated via one ΔY exchange plus a series reduction as follows.

(4.3.9)

Removal of edge e

We are ready to describe a ΔY reduction sequence for grid graphs. By (4.3.5), the graph is obviously an SP graph if m or $n = 2$. Hence, suppose $m, n \geq 3$. First we reduce the two pairs of series edges in the upper right corner and lower left corner of the grid graph to one edge each. Thus, the upper right-hand corner has become

(4.3.10)

Upper right-hand corner after series reduction

By repeated application of (4.3.8), we move the edge e toward the left or bottom boundary of the graph. When that boundary is reached, e can be eliminated either by (4.3.9) or by the fact that it has become parallel to the lower left edge produced in one of the two initial series reductions.

We now have the upper right-hand portion as

(4.3.11)

Upper right-hand corner after removal of e

The edge g is in series with f and can be contracted. Then f can be eliminated analogously to the edge e. By repetition of this process and suitable adjustment in the last iteration, we effectively eliminate all nodes of the right-hand boundary of the grid graph. By induction, the lemma follows.   □

Proof of Theorem (4.3.3).   A given plane graph is by Lemma (4.3.6) a minor of some grid graph. By Lemmas (4.3.4) and (4.3.7), both graphs are ΔY reducible.   □

By Lemma (4.3.4), the class of ΔY graphs is closed under the taking of 2-connected minors. Thus, one is tempted to look for a characterization of the ΔY graphs by exclusion of minimal minors that are 2-connected and not ΔY reducible. As a first step toward finding these minors, we introduce the following equivalence relation on the class of 2-connected graphs that are not ΔY reducible and that are minimal with respect to that property. Two such graphs are defined to be related if one can be obtained from the other one by a sequence of ΔY exchanges. The above characterization problem is solved once one finds one member of each equivalence class. It is easy to see that the graph K_6 is one of the desired minimal graphs. By straightforward enumeration, the equivalence class represented by K_6 can be shown to consist of the following graphs.

(4.3.12)

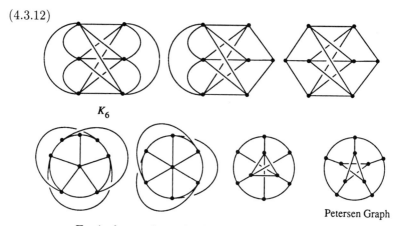

K_6

Petersen Graph

Equivalence class of minimal nonreducible graphs
represented by K_6

One might conjecture, and would not be the first person to do so, that the graphs of (4.3.12) constitute all minimal nonreducible graphs. The conjecture is appealing but false. A counterexample due to Robertson is the graph G constructed as follows. We start with the planar graph of (4.3.13) below. We add to that graph one vertex v, which is connected to the circled nodes of the planar graph by one edge each. The graph G is evidently 3-connected and has no triangles or 3-stars. Thus, it is not ΔY reducible. We claim that G does not have any one of the graphs of (4.3.12) as a minor. By the construction, G can be reduced to a planar graph by removal of the vertex v. But none of the graphs of (4.3.12) becomes planar when any one its vertices is removed, as is easily verified. Thus, G is a counterexample to the conjecture.

(4.3.13)

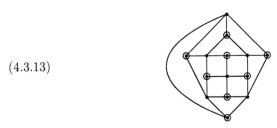

Planar graph for counterexample G

In the next section, we adapt the concept of ΔY extension and reduction sequences to binary matroids.

4.4 Delta-Wye Construction for Binary Matroids

In this section, we translate the definitions and conditions for ΔY graphs given in Section 4.3 to statements about binary matroids. Thus, we obtain ΔY extension steps, ΔY reduction steps, ΔY matroids, etc. Recall from Lemma (4.2.12) that every SP matroid is the graphic matroid of an SP graph, and vice versa. In contrast, the yet-to-be-defined ΔY matroids turn out to be not just the graphic matroids of ΔY graphs.

We start with the definitions. From Section 4.2, we already know how SP extensions are performed in binary matroids. So let us translate the ΔY exchange from graphs to binary matroids. To gain some intuitive insight, we perform a triangle to 3-star step in a 2-connected graph G with edge set E, then translate that step into matroid language for the graphic matroid $M = M(G)$ of G. Let the triangle of G be $C = \{e, f, g\}$. As in Section 4.3, we assume that C does not contain a cocycle.

To carry out the triangle to 3-star exchange in G, we first add the 3-star, say $C^* = \{x, y, z\}$, getting a graph G'. A partial drawing of G' with

C and C^* is shown below.

(4.4.1)

Triangle and 3-star

Then we delete the triangle C, getting the desired graph G''. Since C contains no cocycle, the graph G has a tree X and a cotree $Y = E - X$ so that $C \subseteq Y$. The representation matrix B of M for this X is assumed to be

(4.4.2)

$$B = \ X \ \boxed{\begin{array}{c|ccc} & e & f & g \\ \hline \overline{B} & a & b & c \end{array}}$$

Matrix B for $M = M(G)$

Clearly, $X \cup \{y\}$ is a tree of the graph G'. We claim that the representation matrix B' for $M' = M(G')$ corresponding to that tree must be

(4.4.3)

$$B' = \begin{array}{c} X \\ y \end{array} \boxed{\begin{array}{c|ccc|cc} & e & f & g & x & z \\ \hline \overline{B} & a & b & c & a & b \\ \hline & & 0 & & 1 & 1 \end{array}}$$

Matrix B' for $M' = M(G')$

The proof of this claim is as follows. First, the edges x and z of G' produce the only fundamental circuits with $X \cup \{y\}$ that contain y. This fact justifies the last row of B'. Second, contraction of y in G' makes the edge x (resp. z) parallel to e (resp. f). Correspondingly, upon deletion of row z from B', the columns x and e (resp. z and f) must be parallel. These facts uniquely determine B' as shown in (4.4.3).

Deletion of the columns e, f, and g from B' yields the desired representation matrix B'' for $M'' = M(G'')$, i.e.,

(4.4.4)

$$B'' = \begin{array}{c} X \\ y \end{array} \boxed{\begin{array}{c|cc} \overline{Y} & x & z \\ \hline \overline{B} & a & b \\ \hline 0 & 1 & 1 \end{array}} \quad ; \quad \overline{Y} = Y - \{e, f, g\}$$

Matrix B'' for $M'' = M(G'')$

Instead of $X \cup \{y\}$, we could also have chosen $X \cup \{x\}$ or $X \cup \{z\}$ as tree of G'. Correspondingly, the column vectors a and b explicitly shown in (4.4.4) would have been a and c, or c and b. Each one of the latter matrices may also be obtained by a GF(2)-pivot in row y of B''.

The above translation of a triangle to 3-star exchange in G to a triangle to triad exchange in M may be rephrased as follows. The latter exchange in M is allowed only if the triangle does not contain a cocycle. In the exchange, we first find a representation matrix B where the triangle elements are nonbasic. Next we delete one of the columns corresponding to the triangle. Then we add a row that has 1s in the remaining two columns of the triangle, and 0s elsewhere. Finally, we re-index the two columns formerly indexed by triangle elements. The diagram below summarizes this particular ΔY exchange process. For clarity, the diagram omits indices other than e, f, g and x, y, z.

(4.4.5)

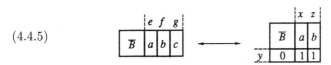

ΔY exchange, case 1

Note that $a + b + c = 0$ (in GF(2)). There are other ways to display the triangle to triad exchange. Specifically, when we select a B with exactly one element of the triangle, say f, basic, we get a case of the form

(4.4.6)

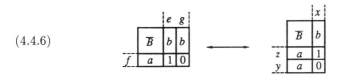

ΔY exchange, case 2

When we select a B with two elements of the triangle basic, say e and f, we get the third possible case

(4.4.7)

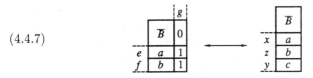

ΔY exchange, case 3

where $a + b + c = 0$ (in GF(2)). Simple checking, analogous to that proving (4.4.5), confirms these claims. We emphasize that in each situation, the symbols $\overline{B}$, a, b, and c refer to different matrices and vectors. However, the relationships between the triangle elements e, f, and g and the triad elements x, y, and z are displayed in agreement with (4.4.1).

The definition of a triangle to triad exchange of a binary but not necessarily graphic matroid M is as follows. We require that the triangle, say $\{e, f, g\}$, does not contain a cocycle. Let B be any representation matrix of M. Then B is one of the left-hand matrices of (4.4.5), (4.4.6), or (4.4.7). The matroid resulting from the exchange is given by the corresponding right-hand matrix. Suppose we select B of (4.4.7) to carry out the triangle to triad exchange, as we always may. Then the cocycle condition on the triangle is equivalent to the requirement that the row vectors a and b of B be nonzero and distinct. Hence, the row vectors a, b, and $c = a + b$ (in GF(2)) in the resulting matrix of (4.4.7) are distinct, and $\{x, y, z\}$ is indeed a triad in the corresponding matroid. The triad trivially does not contain a cycle since the resulting matrix indicates x, y, and z to be part of a basis.

For the definition of a triad to triangle exchange, we invert the above process. We demand that the triad not contain a cycle. The exchange is specified by (4.4.5), (4.4.6), and (4.4.7) when we start with the matrix on the right hand side and derive the one on the left hand side. By taking transposes of the matrices involved in (4.4.5), (4.4.6), and (4.4.7), we see that a triangle to triad exchange in M is precisely a triad to triangle exchange in M^*. Thus, these two operations are dual to each other.

We define ΔY *extension sequence*, ΔY *reduction sequence*, and ΔY *reducibility* for binary matroids by the obvious adaptation of the same terms for graphs. Later, we need the following lemma about ΔY exchanges.

(4.4.8) Lemma.   *Let M be a connected binary matroid. Then any ΔY exchange in M produces a connected binary matroid.*

Proof. By duality, we only need to consider the triangle to triad exchange depicted in (4.4.5). By Lemma (3.3.19), M is connected if and only if any representation matrix B of M is connected. The left-hand matrix of (4.4.5) is thus connected. Since the vectors a, b, c of that matrix are nonzero and $a + b + c = 0$ (in GF(2)), the right-hand matrix is easily verified to be connected as well. Thus, the related matroid is connected. $\square$

Suppose we start with the connected, binary, and graphic matroid having just two parallel elements, with representation matrix $B = [\,1\,]$. Let the ΔY *matroids* be the binary matroids that may be produced from that matroid by ΔY extension sequences. Due to Lemmas (4.2.16) and (4.4.8), SP extensions and ΔY exchanges maintain connectedness. Thus, the ΔY matroids are connected. By definition, the smallest ΔY matroid is graphic. Furthermore, all graphic matroids of ΔY graphs are ΔY matroids, as are the duals of these matroids. By (4.3.1), the nonplanar graph K_5 is ΔY reducible, as is, evidently, $K_{3,3}$. Thus, the graphic matroids $M(K_5)$ and

$M(K_{3,3})$ are ΔY-reducible, as well as $M(K_5)^*$ and $M(K_{3,3})^*$. By Lemma (3.2.48), the latter matroids are not graphic. Thus, ΔY matroids need not be graphic. Indeed, there are ΔY matroids that are not graphic and not cographic. For example, the nongraphic and noncographic matroid R_{12} introduced later in Chapter 10 via the representation matrix

$$(4.4.9) \qquad B^{12} = \begin{vmatrix} 1 & 0 & 1 & 1 & 0 & 0 \\ 0 & 1 & 1 & 1 & 0 & 0 \\ 1 & 0 & 1 & 0 & 1 & 1 \\ 0 & 1 & 0 & 1 & 1 & 1 \\ 1 & 0 & 1 & 0 & 1 & 0 \\ 0 & 1 & 0 & 1 & 0 & 1 \end{vmatrix}$$

Matrix B^{12} for R_{12}

is a ΔY matroid. It would take up too much space to describe the SP reduction and ΔY exchange steps that prove this claim. But with the machinery introduced in Chapter 10, we have deduced $B = [\,1\,]$ from B^{12} of (4.4.9) using eight ΔY exchange steps and ten SP reduction steps.

Analogously to Lemma (4.3.4), we now show that the class of ΔY-matroids is closed under the taking of connected minors. The proof mimics that of Lemma (4.3.4).

(4.4.10) Lemma. *If a connected binary matroid M is ΔY reducible, then every connected minor N of M is ΔY reducible as well.*

Proof. Select a representation matrix B for M that displays the minor N by a submatrix $\overline{\overline{B}}$. Thus,

$$(4.4.11)$$

Matrix B for M displaying minor N by $\overline{\overline{B}}$

We induct on the number of SP reductions and ΔY exchanges that reduce M to the matroid with just two parallel elements. We may assume that N has no series or parallel elements. N is then a minor of M as well as of any minor derived from M by SP reductions. Thus, by induction and duality, we only need to examine the case where the first step of a given reduction sequence for M is a triangle to triad exchange. Let the triangle of M be $\{e, f, g\}$.

Three cases are possible, depending on the number of elements of $\{e, f, g\}$ that index rows of B. In the first case, that number is 0. Thus, $e, f, g \in Y$. Suppose at most two of the columns e, f, g of B intersect

$\overline{\overline{\overline{B}}}$. Then according to (4.4.5), we drop from B one column that does not intersect $\overline{\overline{\overline{B}}}$, and add a row with two 1s. The matrix $\overline{\overline{\overline{B}}}$ is still a submatrix of the matrix so derived from B, and we are done by induction. Suppose all three columns e, f, g of B intersect $\overline{\overline{\overline{B}}}$. Then the change of B given by (4.4.5) may also be viewed as a triangle to triad change involving $\overline{\overline{\overline{B}}}$, provided we can prove that $\{e, f, g\}$ is a triangle of N that does not contain a cocycle. But the assumption that N is connected and has no series or parallel elements implies these two conditions, and once more we are done by induction. With similar ease one argues the case where one or two elements of $\{e, f, g\}$ index rows of B. □

In Chapter 11, it is shown that the class of ΔY matroids is a subset of the class of regular matroids, which we define in a moment. But otherwise that class is not well understood. Consider a slightly more complicated situation. This time, a class of binary matroids is produced by ΔY extension steps from a given connected binary matroid, or possibly from several such matroids. In the remainder of this section, we cover an interesting instance.

We need a few definitions concerning regular and almost regular matroids. These matroids are treated in depth in Chapters 9–12. We confine ourselves here to a rather terse introduction. A real $\{0, \pm 1\}$ matrix is *totally unimodular* if every square submatrix has real determinant equal to 0 or ± 1. A binary matrix is *regular* if its 1s can be signed to become ± 1s so that the resulting real matrix is totally unimodular. A binary matroid is *regular* if M has a regular representation matrix. It is not so difficult to see, and is also proved in Chapter 9, that for a given binary matroid either none or all representation matrices are regular.

A binary matroid M is *almost regular* if it is not regular, and if for any element z of the matroid, at least one of the minors M/z and $M \backslash z$ is regular. In addition, we demand the existence of a label for each element z so that we can identify at least one of the minors M/z or $M \backslash z$ as regular. The label must be "*con*" or "*del*." If the label for element z is "*con*" (resp. "*del*"), then M/z (resp. $M \backslash z$) must be regular. We still have a rather technical condition that must be satisfied by the labels. They must be so chosen that every circuit (resp. cocircuit) of M has an even number of elements with "*con*" (resp. "*del*") labels. Finally, there must be at least one "*con*" element and at least one "*del*" element.

The reader is likely to be puzzled by the strange parity condition and the existence condition. In Chapter 12, it is shown how they come about, and why one might want to define and investigate almost regular matroids. So for the time being, we suggest that the reader simply accept or at least tolerate these seemingly strange requirements.

One way to construct some almost regular matroids is as follows. We take any square $\{0, \pm 1\}$ real matrix A that is not totally unimodular but all of whose proper submatrices do have that property. Call any such matrix *minimal non-totally unimodular*. There is a rather simple subclass of these matrices where every row and every column has exactly two ±1s. We assume that A is not of this variety. Let B be the support matrix of A. View B as the binary representation matrix of a binary matroid M. To the elements of M corresponding to the rows (resp. columns) of B, assign "*con*" (resp. "*del*") labels. Then M is an almost regular matroid, a fact proved in Chapter 12. An example is the minimal non-totally unimodular matrix

(4.4.12)
$$A = \begin{bmatrix} 1 & 1 & 1 & 1 \\ 1 & 1 & 0 & 0 \\ 1 & 0 & 1 & 0 \\ 1 & 0 & 0 & 1 \end{bmatrix}$$

Minimal non-totally unimodular matrix

The binary support matrix B of A represents an almost regular matroid when labels are assigned as described above. The class of minimal non-totally unimodular matrices is quite rich. Thus, the class of almost regular matroids is interesting as well.

Here are the representation matrices of two almost regular matroids that cannot be produced by the above process. Instead of row and column indices, we record for each row and column the label of the corresponding element of the matroid.

(4.4.13)
$$B^7 = \begin{array}{c} \\ con \\ del \\ del \end{array} \begin{bmatrix} 0 & 1 & 1 & 1 \\ 1 & 1 & 0 & 1 \\ 1 & 0 & 1 & 1 \end{bmatrix} \; ; \qquad B^{11} = \begin{array}{c} \\ con \\ del \\ del \\ del \\ del \end{array} \begin{bmatrix} 0 & 1 & 1 & 1 & 1 & 1 \\ 1 & 1 & 0 & 0 & 1 & 1 \\ 1 & 1 & 1 & 0 & 0 & 1 \\ 1 & 0 & 1 & 1 & 0 & 1 \\ 1 & 0 & 0 & 1 & 1 & 1 \end{bmatrix}$$

Column labels for B^7: def, con, con, con. Column labels for B^{11}: def, con, con, con, con, con.

Labeled Matrices B^7 and B^{11}

The matrix B^7 should be familiar. It is, up to column/row permutations and labels, the representation matrix of (3.3.22) of the nonregular Fano matroid. The reader can easily check that the Fano matroid with the given labels does satisfy the above-mentioned parity and existence conditions on labels. The origin of B^{11} is explained in Chapter 12. At any rate, verification of the conditions on labels for B^{11} requires a moderate computational effort.

Due to the labels, we want to restrict SP extension steps and ΔY exchanges a bit. Specifically, we allow a parallel (resp. series) extension of M only if the involved element z of M has a "*con*" (resp. "*del*") label. The new element receives the same label as z. A triangle to triad exchange is permitted only if the triangle, say $\{e, f, g\}$, has exactly two "*con*" elements. The labels of the resulting triad, say $\{x, y, z\}$, can be deduced from the drawing below, which should be interpreted the same way the drawing of (4.4.1) is linked to (4.4.5), (4.4.6), and (4.4.7).

(4.4.14)

Triangle and 3-star with labels

Note that the resulting triad has two "*del*" labels, as demanded by the parity condition. A triad to triangle exchange is just the reverse of the above step. It is permitted only if the triad has exactly two "*del*" labels. A *restricted* ΔY *extension sequence* is a sequence of restricted SP extensions and of restricted ΔY exchanges. It is not difficult to prove that restricted ΔY extension sequences convert almost regular matroids to matroids with the same property. We show this in Chapter 12.

We include a short restricted ΔY extension sequence that starts with B^7 of (4.4.13), and that generates, with appropriate labels, the matrix B deduced earlier from the minimal non-totally unimodular A of (4.4.12).

(4.4.15)

		d	e	f	g					d	e	f	g	h		
		d con	e con	f con	g con					d con	e con	f con	g con	h con		
a	con	0	1	1	1	add h parallel to a		a	con	0	1	1	1	1	replace triangle $\{d,g,h\}$
b	del	1	1	0	1			b	del	1	1	0	1	0	
c	del	1	0	1	1			c	del	1	0	1	1	0	

		e	f	y	z					b	c	y	z	
		c con	c con	d del f	d del f					d del f	d del f	d del f	d del f	
a	con	1	1	1	1	GF(2)-pivot on two circled 1s		a	con	1	1	1	1
b	del	①	0	1	0			e	con	1	0	1	0
c	del	0	①	1	0			f	con	0	1	1	0
x	con	0	0	1	1			x	con	0	0	1	1

Example of restricted ΔY extension sequence

The next theorem states a rather surprising fact about the power of restricted ΔY extension sequences.

(4.4.16) Theorem. *The class of almost regular matroids has a partition into two subclasses. One of the subclasses consists of the almost regular* *matroids producible by* ΔY *extension sequences from the matroid represented by* B^7 *of* (4.4.13). *The other subclass is analogously generated by* B^{11} *of* (4.4.13). *There is a polynomial algorithm that obtains an appropriate* ΔY *extension sequence for creating any almost regular matroid from the matroid of* B^7 *or* B^{11}, *whichever applies.*

Unfortunately, the existing proof of Theorem (4.4.16) is so long that we cannot include it here. In this book, it is one of the few results about binary matroids that we state but do not prove. We outline a proof in Chapter 12, when we restate Theorem (4.4.16) as Theorem (12.4.8). At that time, we use the theorem to establish several matrix constructions.

In the final section, we cover applications, extensions, and references.

4.5 Applications, Extensions, and References

The series-parallel construction is a basic idea of electrical network theory. The characterization of SP graphs in Theorem (4.2.6) in terms of excluded K_4 minors is given in Dirac (1952). Another proof and basic results about SP graphs are provided in Duffin (1965). Theorem (4.2.9) is due to Chartrand and Harary (1967). The dual of that result, Theorem (4.2.11), is proved directly in Truemper and Soun (1979), and Soun and Truemper (1980). Decomposition results for SP graphs and outerplanar graphs are described in Wagner (1987).

One may attempt to generalize SP matroids by dropping the restriction that the matroids be binary. One still starts with the matroid having just two parallel elements. The SP extensions are defined via addition of parallel elements and expansion by series elements. These steps can be nicely displayed by abstract matrices. With that approach, one very easily proves that the supposedly more general procedure generates nothing but the graphic matroids of SP graphs. By Theorems (3.5.2) and (4.2.13), a connected matroid with at least two elements is thus an SP matroid if and only if it has no U_4^2 or $M(K_4)$ minors. Additional material on combinatorial aspects of series-parallel networks may be found in Brylawski (1971).

Akers (1960) and Lehman (1963) contain the following conjecture. Let G be a 2-connected graph that is a plane graph plus one edge called the *return edge*. Then G is conjectured to be ΔY reducible to a cycle with just two edges, one of which must be the return edge. Note that the return edge is not allowed to participate in any ΔY exchange. The conjecture was first proved in Epifanov (1966) by fairly complicated arguments. A simple proof is given in Truemper (1989a), which also contains the proof of Theorem (4.3.3) given here. An interesting but not simple proof of

Theorem (4.3.3) is provided in Grünbaum (1967). That reference relies on Theorem (4.3.3) to derive a very elementary proof of Steinitz's Theorem linking 3-connected graphs and 3-dimensional polytopes. Computational aspects of ΔY graphs are treated in Feo (1985), and Feo and Provan (1988). Gitler (1991) characterizes ΔY reducible planar graphs where k specified nodes, $k \geq 3$, may not be removed. That is, none of the k nodes may be eliminated by Y-to-Δ exchanges or series reductions. The return edge case discussed above is equivalent to the situation with $k = 2$.

The counterexample graph G defined via the planar graph of (4.3.13) is due to Robertson (1988). The class of ΔY graphs may be specialized by demanding that all ΔY exchanges are either Δ-to-Y exchanges or Y-to-Δ exchanges. The ΔY graphs so created, we call Δ-to-Y graphs and Y-to-Δ graphs. The two classes are completely characterized in Politof (1988a), (1988b).

The special structure of SP graphs, ΔY graphs, Δ-to-Y graphs, and Y-to-Δ graphs has been exploited for numerous applications, in particular for combinatorial optimization and reliability problems. Relevant references are Moore and Shannon (1956), Akers (1960), Lehman (1963), Nishizeki and Saito (1975), (1978), Rosenthal and Frisque (1977), Monma and Sidney (1979), Farley (1981), Farley and Proskurowski (1982), Takamizawa, Nishizeki, and Saito (1982), Wald and Colbourn (1983a), (1983b), Agrawal and Satyanarayana (1984), (1985), Satyanarayana and Wood (1985), Arnborg and Proskurowski (1986), Politof and Satyanarayana (1986), (1990), Colbourn (1987), and El-Mallah and Colbourn (1990).

The class of almost regular matroids is defined in Truemper (1991). Theorem (4.4.16) is taken from that reference.

Chapter 5

Path Shortening Technique

5.1 Overview

In this chapter, we introduce a matroid tool called the path shortening technique. It is an adaptation of an elementary graph method to matroids. In Section 5.2, we first motivate that technique, then use it to prove several results concerning the existence of certain separations and of certain minors in binary matroids. In subsequent chapters, we rely a number of times on these results.

In Section 5.3, we employ the technique to devise a polynomial algorithm that solves the following problem. Given are two matroids M_1 and M_2 on a common set E. One must find a maximum cardinality set $Z \subseteq E$ that is independent in both M_1 and M_2. This problem is called the cardinality intersection problem, for short, intersection problem. Correspondingly, the algorithm is called the intersection algorithm. That scheme also solves the following problem, which is called the partitioning problem. As before, two matroids M_1 and M_2 on a common set E are given, say with rank functions $r_1(\cdot)$ and $r_2(\cdot)$. One must partition E into two sets, say E_1 and E_2, such that $r_1(E_1) + r_2(E_2)$ is minimized. The intersection algorithm also provides a constructive proof of a max-min theorem that links the intersection problem with the partitioning problem. The results of Section 5.3 and many more on the intersection and partitioning of matroids are almost entirely due to Edmonds. They constitute some of the most profound results of matroid theory.

The results of Section 5.3 are related to the remaining chapters as follows. There we frequently must find certain separations of matroids.

The intersection algorithm may be employed to do this. But one may also locate the desired separations with the separation algorithm of the next chapter. Thus, in principle, we do not require the intersection algorithm for the remainder of the book. As a consequence, the material of Section 5.3 may be just scanned or even skipped on a first reading without loss of continuity.

The final section, 5.4, contains extensions and references. The chapter utilizes the material of Chapters 2 and 3.

5.2 Shortening of Paths

Consider the following simple graph problem. Given is a 2-connected graph G with at least two edges, among them edges e and f. We are asked to prove that G has a 2-connected minor consisting of just e and f. The solution is straightforward. As stated in Section 2.2, any two edges of a 2-connected graph lie on some cycle. In particular, the edges e and f lie on some cycle C of G. Evidently, C consists of e, f, and two node-disjoint paths, say P_1 and P_2. We obtain the desired minor by deleting all edges not in C and contracting all edges of P_1 and P_2. One could call the second step a shortening of the paths P_1 and P_2.

The path shortening operation has numerous uses in graph theory involving far more complicated situations than the trivial example treated above. The method also has an interesting translation to matroid theory, where we will refer to it as the *path shortening technique*. In this section, we describe that technique while proving the matroid analogue of the just-used fact that in a 2-connected graph any two edges lie on some cycle. Subsequently, we rely on the path shortening technique to establish other matroid results that will be repeatedly invoked in later chapters. We begin with the matroid version of the cited graph result.

(5.2.1) Lemma.   *A binary matroid M is connected if and only if for every two elements x and z of M, there is a circuit containing both x and z.*

Proof. Let B be a binary representation matrix of M. First we prove the "if" part by contradiction. If M is not connected, then by Lemma (3.3.19), B is not connected. Evidently, M then has elements x and z that cannot both be in any circuit.

We turn to the "only if" part. Thus, we assume M, and hence B, to be connected. If the rank of M is 0, then due to the connectedness assumption, M has at most one element, and the desired conclusion holds vacuously. So assume that the rank of M is at least 1. Correspondingly, B has at least one row. Let x and z be two elements of M. We must show that some circuit of M contains both x and z.

Since B is connected, by at most one GF(2)-pivot we can assure that x indexes a row of B. Suppose that z indexes a column of B. If the element B_{xz} of B is a 1, then x is in the fundamental circuit of M given by the column z of B, and we are done. If $B_{xz} = 0$, then any GF(2)-pivot in column z produces the case where z also indexes a row. Thus, from now on we may assume that both x and z index rows of B.

At this point, we switch from the connected matrix B to the connected bipartite graph $\mathrm{BG}(B)$. The analysis of $\mathrm{BG}(B)$ consists of the following elementary step. We examine a shortest path from row node x to row node z. Suppose that the row nodes of the path are $x, r_1, r_2, \ldots, r_m, z$, for some $m \geq 1$, and that the column nodes are $y, s_1, s_2, \ldots, s_m$. The nodes are encountered in the given order as one moves along the path from x to z. We claim that B can be partitioned as

(5.2.2)

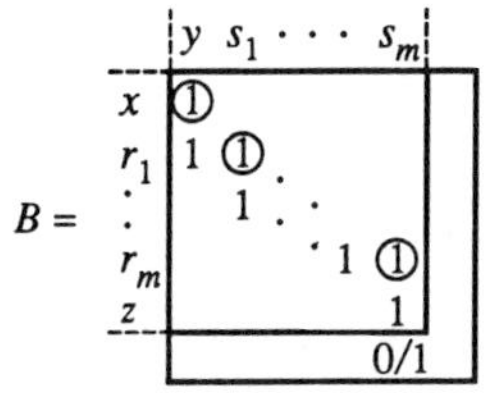

Shortest path from x to z displayed by B

In the submatrix $\overline{B}$ of B indexed by $x, r_1, r_2, \ldots, r_m, z$ and $y, s_1, s_2, \ldots,$ s_m, the 1s, circled or not, correspond to the edges of the path. Recall a convention of Section 2.3 about the display of matrices: If a submatrix or region of a matrix contains explicitly shown 1s while leaving the remaining entries unspecified, then the latter entries are to be taken as 0s. Accordingly, the display of the submatrix $\overline{B}$ implies that $\overline{B}$ does not contain any 1s beyond those corresponding to the edges of the path. We must prove, of course, that this is the case. So suppose there are additional 1s in $\overline{B}$. Then one readily confirms that the path has a chord, and thus is not shortest, a contradiction.

By GF(2)-pivots on the circled 1s of B of (5.2.2), we derive from B the following matrix B'.

(5.2.3)

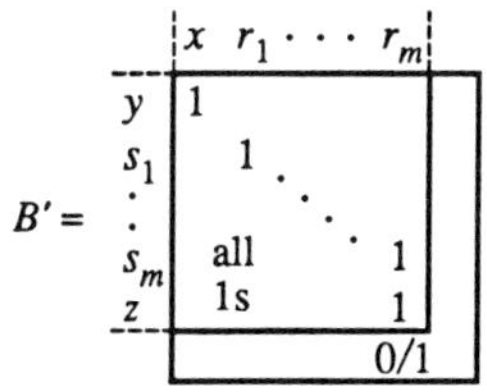

Matrix B' obtained by path shortening pivots

The matrix B' proves that z is contained in the fundamental circuit that x forms with the base producing B', and we have proved x and z to be in some circuit of M as desired.   □

Note that in $\mathrm{BG}(B)$, the path connecting x and z has at least two edges. In contrast, the graph $\mathrm{BG}(B')$ has an edge connecting x and z. Thus, the $\mathrm{GF}(2)$-pivots transforming B to B' have reduced the path of $\mathrm{BG}(B)$ to a shorter path in $\mathrm{BG}(B')$. For this reason, we call the above method the *path shortening technique*. Evidently, the proof procedure of Lemma (5.2.1) can be implemented in a polynomial algorithm that determines a circuit containing two given elements in any connected binary matroid.

We now prove additional matroid results using the path shortening technique. They concern the existence of certain separations or of certain minors. In each case, the proof procedure has an obvious translation to a polynomial algorithm that very effectively locates one of the claimed separations or minors.

The first case concerns the existence of connected 1-element extensions of a given minor in a binary matroid.

(5.2.4) Lemma.   *Let N be a connected minor of a connected binary matroid M. Define z to be an element of M not present in N. Then M has a connected minor N' that is a 1-element extension of N by z.*

Proof. By Lemma (3.3.12), M has a representation matrix B that displays N via a submatrix $\overline{B}$. Let the rows of $\overline{B}$ be indexed by $\overline{X}$ and the columns by $\overline{Y}$. Since M and N are connected matroids, both B and $\overline{B}$ are connected matrices. Suppose the element z indexes a column in B. We thus have B given by (5.2.5) below. If the subvector e of column z of B is nonzero, then the connected submatrix $[\overline{B} \mid e]$ represents the desired connected minor N' with z.

(5.2.5) $B =$

Matrix B displaying minor N by $\overline{B}$

So suppose $e = 0$. We know that the bipartite graph $\mathrm{BG}(B)$ is connected. Thus, that graph contains a path from $\overline{X} \cup \overline{Y}$ to z. Take a shortest path, say with row nodes $r_1, r_2, \ldots, r_m$ and column nodes $s_1, s_2, \ldots, s_n, z$, for

some $m, n \geq 1$. The increasing indices indicate the order in which the
nodes are encountered as one moves along the path from $\overline{X} \cup \overline{Y}$ to z. The
endpoint of the path in $\overline{X} \cup \overline{Y}$ is r_1 or s_1, whichever applies. Below, we
display B. The explicitly shown 1s correspond to the edges of the path.
Two cases are possible, depending on whether r_1 or s_1 is the endpoint of
the path in $\overline{X} \cup \overline{Y}$.

(5.2.6)

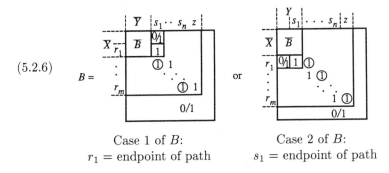

Case 1 of B: Case 2 of B:

r_1 = endpoint of path s_1 = endpoint of path

In each matrix of (5.2.6), the unspecified entries, which by our convention
are 0s, are justified as follows. Suppose such an entry is instead a 1, for
example in a row $x \in \overline{X}$ and column s_j, $j \geq 2$. Then the path does not
have minimal length since there is a path starting at node x that is shorter.
All other unspecified entries are argued similarly.

Assume case 1 of (5.2.6) applies. Then pivots on the circled 1s of B
produce the already-discussed instance with $e \neq 0$. For case 2 of (5.2.6),
pivots on the circled 1s produce a B' with a submatrix $[\overline{B}/d]$, where the
vector d is nonzero and indexed by z. That submatrix represents the desired
N'. The case where z indexes a row of B is handled by duality.   □

So far, we have explained each step of the path shortening technique
in detail. In the proofs to follow, we skip such details since the arguments
are identical or at least very similar to those above.

The next result concerns the presence of minors that are isomorphic
to the graphic matroids of wheels. Recall from Section 2.2 that for $n \geq 1$,
the wheel W_n is constructed from a cycle with n nodes as follows. Add one·
extra node and link it with one edge each to the nodes of the cycle. Small
wheels are as follows.

(5.2.7)

W_1 W_2 W_3 W_4

Small wheels

The spokes of each wheel W_n form a tree. With that tree as base of $M(W_n)$, we obtain from (5.2.7) the following representation matrices for $M(W_1), \ldots, M(W_4)$.

(5.2.8)

$$
[0] \qquad
\begin{bmatrix} 1 & 1 \\ 1 & 1 \end{bmatrix} \qquad
\begin{bmatrix} 1 & 0 & 1 \\ 1 & 1 & 0 \\ 0 & 1 & 1 \end{bmatrix} \qquad
\begin{bmatrix} 1 & 0 & 0 & 1 \\ 1 & 1 & 0 & 0 \\ 0 & 1 & 1 & 0 \\ 0 & 0 & 1 & 1 \end{bmatrix}
$$

$M(W_1) \quad M(W_2) \qquad M(W_3) \qquad M(W_4)$

Representation matrices

Indeed, for $n \geq 3$, $M(W_n)$ is represented by the $n \times n$ matrix

(5.2.9)

$$
\begin{bmatrix}
1 & & & & 1 \\
1 & 1 & & & \\
 & 1 & \ddots & & \\
 & & & 1 & \\
 & & & 1 & 1
\end{bmatrix}
$$

Representation matrix for $M(W_n)$, $n \geq 3$

Clearly for $n \geq 3$, W_n is 3-connected. Thus, by Theorem (3.2.36), W_n is the unique graph producing $M(W_n)$.

The graphic matroids of wheels are cited very often in the remainder of this book. We adopt an abbreviated terminology and simply call them *wheels*. The double meaning of "wheel" should not cause a problem. The context invariably makes it clear whether we mean a matroid or a graph. At any rate, we use *wheel graph* and *wheel matroid* if there is even a slight chance of confusion.

A basic result about wheels is as follows.

(5.2.10) Lemma. *Let M be a binary matroid with a binary representation matrix B. Suppose the graph $BG(B)$ contains at least one cycle. Then M has an $M(W_2)$ minor.*

Proof. Since the graph $BG(B)$ has at least one cycle, it has a cycle C without chords. Now $BG(B)$ is bipartite, so C has at least four edges. The submatrix $\overline{B}$ of B corresponding to C evidently is up to indices either the 2×2 matrix for $M(W_2)$ of (5.2.8), or for some $n \geq 3$, the $n \times n$ matrix of (5.2.9) for $M(W_n)$. Path shortening pivots on 1s of $\overline{B}$ convert the latter case to the former one. □

Lemma (5.2.10) and another application of the path shortening technique lead to a proof of the following result.

(5.2.11) Lemma.   *Let M be a connected binary matroid with at least four elements. Then M has a 2-separation or an $M(W_3)$ minor.*

Proof. Let B be a binary representation matrix of M. Since M is connected, the bipartite graph $BG(B)$ is connected. Suppose $BG(B)$ does not contain a cycle. Thus, that graph is a tree. Any tip node of the tree corresponds to a row or column unit vector in B. Thus, M has parallel or series elements. Since M has at least four elements, M has a 2-separation.

We turn to the remaining case where $BG(B)$ has a cycle. By Lemma (5.2.10), M has an $M(W_2)$ minor. We may assume that B displays that minor via the 2×2 matrix of (5.2.8), with four 1s. Enlarge that 2×2 matrix to a maximal submatrix of B containing only 1s. Let D be that submatrix, say with rows indexed by a set R and columns indexed by a set S. We use these sets to partition B as shown in (5.2.12) below. In B, each row of the submatrix U and each column of the submatrix V is assumed to be nonzero. By the maximality of D, at least one 0 must be contained in each row of U and in each column of V.

(5.2.12)

Partitioned matrix B

Let F be the graph obtained from $BG(B)$ by deletion of the edges corresponding to the 1s of the submatrix D. Suppose no path of F connects a node of R with one of S. Let X_1 (resp. Y_1) be the row (resp. column) nodes of F that are connected by some path with some node of R. Since $X_1 \supseteq R$, we have $|X_1 \cup Y_1| \geq 2$. Define $X_2 = X - X_1$ and $Y_2 = Y - Y_1$. Since $Y_2 \supseteq S$, we have $|X_2 \cup Y_2| \geq 2$. By derivation of F from $BG(B)$, any arc of $BG(B)$ connecting a node of $X_1 \cup Y_1$ with one of $X_2 \cup Y_2$ must correspond to a 1 of D. But $X_1 \supseteq R$ and $Y_2 \supseteq S$, so the partitioning of B according to X_1, X_2, Y_1, and Y_2 must result in

(5.2.13)

Matrix B with 2-separation $(X_1 \cup Y_1, X_2 \cup Y_2)$

Since GF(2)-rank $D = 1$, $(X_1 \cup Y_1, X_2 \cup Y_2)$ is a 2-separation of B, and we are done. So suppose a path in F does connect a node of R with one of S. Choose a path of minimal length. Evidently, the length must be odd and at least 3.

Assume the length to be 3. Then the center edge of the path corresponds in B of (5.2.12) to a 1 in some row $p \in P$ and some column $q \in Q$. Now row p of U contains a 0 and a 1, say in columns $s_1 \in S$ and $s_2 \in S$, respectively. Similarly, column q of V has a 0 and a 1, say in rows $r_1 \in R$ and $r_2 \in R$, respectively. Then in B of (5.2.12), the submatrix indexed by r_1, r_2, p and s_1, s_2, q is

(5.2.14)

$$
\begin{array}{c}
\quad\quad s_1\ s_2\ q \\
\begin{array}{c} r_1 \\ r_2 \\ p \end{array}
\left[\begin{array}{ccc}
1 & 1 & 0 \\
1 & 1 & 1 \\
0 & 1 & 1
\end{array} \right]
\end{array}
$$

Submatrix of B representing $M(W_3)$ minor

A GF(2)-pivot on the 1 in row p and column q produces up to indices the matrix for $M(W_3)$ of (5.2.8). Thus, M has an $M(W_3)$ minor.

Finally, suppose the shortest path in F from R to S has length greater than 3. With the path shortening technique, we reduce that path to one of length 3. The related GF(2)-pivots in B of (5.2.12) do not affect the submatrices D, U, or V. Thus, the pivots produce the earlier situation with length 3.   ☐

Lemma (5.2.11) supports the following conclusion about 3-connected binary matroids.

(5.2.15) Corollary.   *Every 3-connected binary matroid M with at least six elements has an $M(W_3)$ minor.*

Proof. By Lemma (5.2.11), M has a 2-separation or an $M(W_3)$ minor. Since M is 3-connected, the former case is not possible.   ☐

At times, one would like to claim that a given matroid M has a certain minor containing a specified element. A simple case is given in the following result.

(5.2.16) Lemma.   *Let M be a connected binary matroid with an $M(W_3)$ minor. Then for every element z of M, there is an $M(W_3)$ minor of M that contains z.*

Proof. Let N be the assumed $M(W_3)$ minor of M, and z be any element of M. If z is an element of N, we are done. Otherwise, by Lemma (5.2.4), M has a connected minor of the form $N+z$ or $N\&z$. Consider the first case. We may assume $N+z$ to be represented by the matrix of (5.2.8) for $M(W_3)$, plus one nonzero column indexed by z. A straightforward case analysis proves that for each possible column z, the matroid $N+z$ has an $M(W_3)$ minor with z. Thus, M has an $M(W_3)$ minor containing z as well. The situation for $N\&z$ is handled analogously, or by duality.   □

In the next section, we investigate the intersection and partitioning problems. The subsequent chapters do not rely on the material of the section, so it may be skipped without loss of continuity.

5.3 Intersection and Partitioning of Matroids

Let M_1 and M_2 be two arbitrary matroids defined on a common set E and with rank functions $r_1(\cdot)$ and $r_2(\cdot)$, respectively. The *cardinality intersection problem* demands that we find a maximum cardinality set $Z \subseteq E$ that is independent in M_1 and M_2. The *partitioning problem* requires that we locate a partition of E into E_1 and E_2 such that $r_1(E_1) + r_2(E_2)$ is minimized. In this section, we present an algorithm that simultaneously solves both problems. We call it the *intersection algorithm*. The method consists of repeated applications of the path shortening technique, though carried out in a rather unusual fashion. The algorithm also provides a constructive proof of a max-min theorem that links the two problems in an unexpected way.

We have elected to describe the intersection algorithm for general matroids instead of just binary ones, since in many, if not most, applications, at least one of the two matroids M_1, M_2 is nonbinary. Correspondingly, the algorithm makes use of abstract matrices. Thus, the reader should be familiar with the material on abstract matrices in Section 3.4, in particular with Lemma (3.4.11). That result says that any triangular submatrix of an abstract matrix is nonsingular if and only if the submatrix has only 1s on its diagonal.

We begin with the intersection problem. As stated above, we have two matroids M_1 and M_2 on a common set E. The algorithm must find a maximum cardinality set Z that is independent in both M_1 and M_2. The scheme begins with any set Z that is independent in both matroids. For example, $Z = \emptyset$ will do. The method iteratively replaces the given set Z by a larger one that is also independent in both matroids, until a set of maximum cardinality is found. It suffices that we describe one iteration. It consists of three steps, which we summarize next.

In step 1, we deduce two matrices from certain abstract matrices for M_1 and M_2. We combine the two matrices to a new matrix C that has a strange form, but that actually is a handy encoding of the initial abstract matrices of M_1 and M_2.

In step 2, we derive a graph from C and search in that graph for a certain path. Suppose a path of the desired kind can be located. We interpret that path in terms of the abstract matrices for M_1 and M_2 and deduce a set Z' that is larger than Z and independent in M_1 and M_2. With the set Z' in hand, we terminate the iteration. If a path of the desired kind cannot be found, we go to step 3.

In step 3, we conclude that the absence of paths of the desired kind implies a certain partition of the matrix C. We interpret that partition in terms of the abstract matrices for M_1 and M_2 and conclude that Z is optimal. Thus, we stop. The proof of optimality for Z also shows that the partition of C implies a partition of the set E into two sets, say E_1 and E_2, that solve the partitioning problem. In addition, the proof establishes the previously mentioned max-min theorem that connects the intersection problem with the partitioning problem.

We begin the detailed description. In step 1, we first find for $i = 1, 2$, a base X_i of M_i that contains the set Z. This is possible since Z is independent in M_1 and M_2. Let B^i be the abstract matrix of M_i corresponding to the base X_i. Thus, B^i has row index set X_i and column index set $E - X_i$. We adjoin an identity to B^i, getting $[I \mid B^i]$. In agreement with the indexing rules introduced in Section 2.3, we index the columns of the submatrix I of $[I \mid B^i]$ by X_i. Next we permute the columns of $[I \mid B^1]$ and $[I \mid B^2]$ such that the columns of the two matrices in same position have the same column index. Furthermore, the columns indexed by Z are to become the leftmost columns. Finally, we add zero rows if necessary, so that both matrices have the same number of rows. For $i = 1, 2$, let A^i be the matrix so obtained from $[I \mid B^i]$, and define $Y = E - Z$. Then the matrix A^i is of the form

(5.3.1)

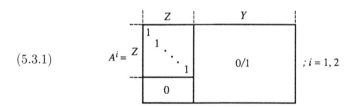

Matrix A^i obtained from $[I \mid B^i]$

The rows of A^i without index either are rows of $[I \mid B^i]$ indexed by $X_i - Z$, or are added zero rows.

Before going on, we would like to establish a simple lemma about the matrix A^i of (5.3.1). The result will allow easy verification that certain subsets of M_i are independent.

(5.3.2) Lemma.   *For some $k \geq 1$, let $\overline{Z}$ be a subset of E with k elements. If the column submatrix of A^i indexed by $\overline{Z}$ contains a $k \times k$ triangular*

submatrix that has only 1s on the diagonal, then $\overline{Z}$ *is independent in* M_i. □

Proof. Except possibly for column permutations and additional zero rows, A^i is the matrix $[I \mid B^i]$. Thus, the postulated $k \times k$ triangular submatrix of A^i has block triangular form, where one of the blocks is an identity submatrix of I, and where the other block is a triangular submatrix of B^i with only 1s on the diagonal. By Lemma (3.4.11), the set $\overline{Z}$ must be a subset of a base of M_i, and thus is independent in M_i. □

We continue with step 1. In the matrix A^1 (resp. A^2), we replace each 1 by α (resp. β), getting a matrix $\tilde{A}^1$ (resp. $\tilde{A}^2$). We compute a matrix $C = \tilde{A}^1 + \tilde{A}^2$ by adding entries termwise according to the following rule: $0 + 0 = 0$, $\alpha + 0 = \alpha$, $0 + \beta = \beta$, and $\alpha + \beta = \gamma$. By (5.3.1), the matrix C is of the form

(5.3.3)

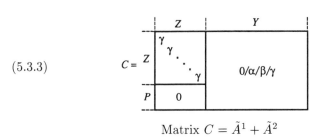

Matrix $C = \tilde{A}^1 + \tilde{A}^2$

Note the row index subset P in (5.3.3). The rows of P arise from the rows of A^1 and A^2 shown in (5.3.1) without index. We consider P to be a new index set that is disjoint from the index sets of A^1 and A^2. This concludes step 1.

In step 2, we first examine C of (5.3.3) for a trivial way of augmenting the set Z to a larger set Z' that is independent in M_1 and M_2. Specifically, assume that C contains a column z that in rows indexed by P has both an α and a β, or a γ. Then in the matrices A^1 and A^2, the column submatrices $\overline{A}^1$ and $\overline{A}^2$ indexed by $Z' = Z \cup \{z\}$ have triangular submatrices that via Lemma (5.3.2) prove Z' to be independent in M_1 and M_2. Thus, we can stop the iteration. So from now on, we assume that C has no such column z. Thus, each column of C contains in the rows indexed by P either just 0s and αs, or just 0s and βs, or just 0s. Let Q_1 (resp. Q_2) index the columns of the first (resp. second) kind. Using these two index sets, we partition C of (5.3.3) further as shown in (5.3.4) below. From the submatrix $\overline{C}$ defined by the row index set Z and the column index set Y of C of (5.3.4), we construct the following directed bipartite graph G. We start with the undirected bipartite graph $BG(\overline{C})$. Let (i, j) be an edge of $BG(\overline{C})$ where i is a row node of Z and j is a column node of Y. If the entry of $\overline{C}$ producing that edge is an α (resp. β), then we direct that edge from i to j (resp. j

to i). If that edge corresponds to a γ, then we replace it by two directed edges of opposite direction.

(5.3.4)

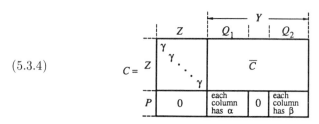

Matrix C partitioned by Q_1 and Q_2

Using any convenient shortest route algorithm, we locate a shortest path from Q_1 to Q_2, or determine that no such path exists. A case of the latter variety is on hand, for example, if Q_1 or Q_2 is empty.

If a shortest path does not exist, we go to step 3, to be covered shortly. Otherwise, let such a path connect a node $q_1 \in Q_1$ with a node $q_2 \in Q_2$. Let $\overline{\overline{C}}$ be the submatrix of $\overline{C}$ defined by the nodes of that path. We claim that $\overline{\overline{C}}$ either is the matrix

(5.3.5)

$$\overline{\overline{C}} = U \begin{array}{c} \\ \end{array}$$

Matrix $\overline{\overline{C}}$ defined by the nodes of the path

or is obtained from the matrix of (5.3.5) by replacing any number of the explicitly shown αs and βs by γs. In $\overline{\overline{C}}$ of (5.3.5), the statements "no α, γ" and "no β, γ" are valid since any violating entry would permit a shorter path from Q_1 to Q_2, a contradiction. For the same reason, we have for $i = 1, 2$, $Q_i \cap W = \{q_i\}$. We use the index sets U and W of $\overline{\overline{C}}$ to define $Z' = (Z - U) \cup W$. By (5.3.5), $|W| = |U| + 1$, so $|Z'| = |Z| + 1$. We claim that Z' is independent in M_1 and M_2. For a proof, we examine the column submatrix C' of C indexed by Z'. By (5.3.4) and (5.3.5), the corresponding column submatrices of A^1 and A^2 contain triangular matrices that confirm the claim. Thus, we have completed the iteration.

As an aside, consider one pivot in B^1 and B^2, each time on a particular 1 in column q_1. Specifically, in B^1 the 1 corresponds to an α entry of C in column q_1 and in a row of P, and in B^2 to the β entry of C explicitly shown in column q_1 and in the first row, say x, of $\overline{\overline{C}}$. Correspondingly, we drop the element x from Z and add the element q_1, getting, say, a set $\tilde{Z}$. The sets Z and $\tilde{Z}$ have the same cardinality, but if we repeat the above process for $\tilde{Z}$ instead of Z, we discover a shorter path. By suitable repetition of the above procedure, the path becomes ever shorter until we finally just add an element to obtain the set Z'.

Finally, we discuss step 3. We enter that step when a path from Q_1 to Q_2 does not exist. Let $Z_2 \subseteq Z$ and $Y_2 \subseteq Y$ be the nodes reachable from the nodes of Q_1, and define $Z_1 = Z - Z_2$ and $Y_1 = Y - Y_2$. The sets Z_1, Z_2, Y_1, Y_2 induce the following partition in C of (5.3.3).

(5.3.6)

$$C = $$

		Z		Y		
		Z_2	Z_1	Y_2		Y_1
Z	Z_2	γ (diag)	0	$0/\alpha/\beta/\gamma$		no α, γ
	Z_1	0	γ (diag)	no β, γ		$0/\alpha/\beta/\gamma$
	P	0	0	each column has α	0 0	each column has β

Matrix C when a path does not exist

In particular, the statements "no α, γ" and "no β, γ" in the submatrices indexed by Z_2, Y_1 and Z_1, Y_2, respectively, are correct since otherwise at least one additional node could be reached from Q_1. For $i = 1, 2$, define $E_i = Z_i \cup Y_i$. By definition, E_1 and E_2 partition E. From C of (5.3.6), it is obvious that E_i as subset of the matroid M_i has rank equal to $|Z_i|$. Furthermore, since $Z = Z_1 \cup Z_2$, we have

(5.3.7) $$|Z| = r_1(E_1) + r_2(E_2)$$

Now for any set Z independent in M_1 and M_2, and for any partition of E into any sets E_1 and E_2, we must have for $i = 1, 2$, $|Z \cap E_i| = r_i(Z \cap E_i) \leq r_i(E_i)$. Adding over $i = 1, 2$, we obtain

(5.3.8) $$|Z| = |Z \cap E_1| + |Z \cap E_2| \leq r_1(E_1) + r_2(E_2)$$

By (5.3.7) and (5.3.8), the set Z on hand in step 3 solves the intersection problem, and the sets E_1 and E_2 found at that time solve the partition problem. Thus, the algorithm has solved both problems simultaneously. Clearly, the algorithm is polynomial — indeed, very efficient.

The preceding arguments also prove the following theorem due to Edmonds.

(5.3.9) Theorem (Matroid Intersection Theorem). Let M_1 and M_2 be two matroids on a set E and with rank functions $r_1(\cdot)$ and $r_2(\cdot)$. Then

(5.3.10) $$\max |Z| = \min\{r_1(E_1) + r_2(E_2)\}$$

where the maximization is over all sets Z that are independent in both M_1 and M_2, and where the minimization is over all partitions of E into sets E_1 and E_2.

We describe two representative example applications. The first example involves an undirected bipartite graph G, say with edge set E connecting the nodes of a set V_1 with those of a set V_2. Define a *matching* to be a subset Z of E such that every node of G has at most one edge of Z incident. For $i = 1, 2$, let M_i be the matroid on E where a subset Z is independent if the nodes of V_i have at most one edge of Z incident. It is easily checked that M_i is the disjoint union of $|V_i|$ uniform matroids of rank 1, and thus is a matroid. Clearly, the matchings of G are precisely the subsets of E that are independent in both M_1 and M_2. Define a *node cover* to be a node subset such that every edge has at least one endpoint in that subset. We now reformulate the matroid intersection Theorem (5.3.9) to a basic result of graph theory due to König.

(5.3.11) Theorem.   *The cardinality of a maximum matching of a bipartite graph is equal to the cardinality of a minimum node cover.*

We leave it to the reader to prove that Theorem (5.3.11) follows from Theorem (5.3.9).

The second application concerns separations in matroids. Suppose we want to know whether a given matroid M on a set E has a k-separation with at least $k + l$ elements on each side, for some nonnegative integers k and l. How can we find such a separation or prove that none exists? Suppose we can efficiently solve the following, more restricted, problem. We are given two disjoint subsets F_1 and F_2 of E, each of cardinality $k + l$. We must decide whether M has a k-separation (E_1, E_2) such that $E_1 \supseteq F_1$ and $E_2 \supseteq F_2$. If we can solve the restricted problem, then we can solve the original one by enumerating all possible choices of the sets F_1 and F_2. The overall algorithm is polynomial if the possible values of $k + l$ can be uniformly bounded by some constant.

We analyze the restricted problem. Define $r(\cdot)$ to be the rank function of M. We need to find a pair (E_1, E_2) such that

$$(5.3.12) \qquad \begin{aligned} E_1 &\supseteq F_1;\ E_2 \supseteq F_2 \\ r(E_1) &+ r(E_2) \leq r(E) + k - 1 \end{aligned}$$

or prove that such a pair does not exist. An answer to the following problem,

$$(5.3.13) \qquad \min_{\substack{(E_1, E_2) \\ E_1 \supseteq F_1, E_2 \supseteq F_2}} \{r(E_1) + r(E_2)\}$$

obviously suffices. We want to eliminate the conditions $E_1 \supseteq F_1$ and $E_2 \supseteq F_2$ from (5.3.13) by some matroid construction. Let $M_1 = M / F_1 \backslash F_2$ and

$M_2 = M/F_2\backslash F_1$, with respective rank functions $r_1(\cdot)$ and $r_2(\cdot)$. Define $\overline{E} = E - (F_1 \cup F_2)$, and for $i = 1, 2$, $\overline{E}_i = E_i \cap \overline{E}$. Then for $i = 1, 2$, $r(E_i)$ is nothing but $r(F_i) + r_i(\overline{E}_i)$. Thus, $r(E_1) + r(E_2) = r(F_1) + r_1(\overline{E}_1) + r(F_2) + r_2(\overline{E}_2) = r_1(\overline{E}_1) + r_2(\overline{E}_2) +$ a constant. Thus, equivalent to (5.3.13) is the problem

$$(5.3.14) \qquad \min_{(\overline{E}_1, \overline{E}_2)} \{r_1(\overline{E}_1) + r_2(\overline{E}_2)\}$$

which can be solved by the intersection algorithm. For uniformly bounded $k + l$, the above scheme is polynomial. But from a practical standpoint, even small bounds on $k + l$, such as 4 or 5, are likely to make the scheme practically unusable. So one might want to explore other avenues to find k-separations. In the next chapter, we learn about a method for a particular case that turns out to be very important for several settings. Specifically, we will have $k = 3$ and $l \le 3$.

In the next section, we point out additional material about intersection and partitioning problems, and list references.

5.4 Extensions and References

The path shortening technique is introduced in Truemper (1984). As we have seen in Section 5.3, the technique fully applies to abstract representation matrices of general matroids. Thus, virtually all results of Section 5.2 can be translated to almost identical ones for general matroids. One additional class of nonbinary matroids must be introduced, though. It is the class of *whirls* $\mathcal{W}_n$. We define these matroids next. A *whirl* of rank $n \ge 1$ is derived from the wheel matroid with same rank by declaring the circuit containing the elements of the rim edges to be independent. Small whirls are represented over GF(3) by the matrices of (5.4.1) below. In general, for $n \ge 3$, $\mathcal{W}_n$ is represented over GF(3) by the matrix of (5.4.2) below, where $\alpha \in \{+1, -1\}$ is so chosen that the GF(3)-determinant of the matrix is -1. Equivalently, the real sum of the entries must be $2(\mathrm{mod}\,4)$. Note that $\mathcal{W}_2$ is also U_4^2, the uniform matroid of rank 2 on four elements.

(5.4.1)

$$\boxed{1} \qquad \begin{bmatrix} 1 & -1 \\ 1 & 1 \end{bmatrix} \qquad \begin{bmatrix} 1 & 0 & 1 \\ 1 & 1 & 0 \\ 0 & 1 & 1 \end{bmatrix} \qquad \begin{bmatrix} 1 & 0 & 0 & -1 \\ 1 & 1 & 0 & 0 \\ 0 & 1 & 1 & 0 \\ 0 & 0 & 1 & 1 \end{bmatrix}$$

$$\mathcal{W}_1 \qquad\qquad \mathcal{W}_2 \qquad\qquad \mathcal{W}_3 \qquad\qquad\qquad \mathcal{W}_4$$

Representation matrices over GF(3) for whirls $\mathcal{W}_1$–$\mathcal{W}_4$

(5.4.2)

$$\begin{bmatrix} 1 & & & & \alpha \\ 1 & 1 & & & \\ & 1 & \ddots & & \\ & & \ddots & 1 & \\ & & & 1 & 1 \end{bmatrix}$$

Representation matrix over GF(3) for whirl $\mathcal{W}_n$, $n \geq 3$

Lemmas (5.2.1) and (5.2.4) are valid for general matroids. Lemmas (5.2.10) and (5.2.11) and Corollary (5.2.15) are readily extended to the general case by allowing occurrence of the whirl $\mathcal{W}_2$ as possible minor besides the given wheel matroids. All these results, including the extensions to nonbinary matroids, are implicit in Whitney (1935) and Tutte (1958), (1965), (1971).

We have no reference for Lemma (5.2.16), but that result is well known. The nonbinary version of that lemma is proved in Bixby (1974). It involves U_4^2 instead of $M(W_3)$, and claims the existence of a U_4^2 minor with z, as follows.

(5.4.3) Lemma. *Let M be a connected matroid with a U_4^2 minor. Then for every element z of M, there is a U_4^2 minor of M that contains z.*

The proof is virtually identical to that of Lemma (5.2.16) except that abstract matrices are used here. The result motivated a long series of papers concerning the presence of specified elements in given minors (Seymour (1981e), (1985b), (1986a), (1986b), Oxley (1984), (1987a), (1990a), Kahn (1985), Coullard (1986), Coullard and Reid (1988), Oxley and Row (1989), Oxley and Reid (1990), and Reid (1990), (1991a)–(1991d)).

The matroid intersection and partitioning results of Section 5.3 are just a small sampling of a wealth of material. The roots of these problems can be traced back to several matching results of which Theorem (5.3.11), due to König (1936), is an example. Lovász and Plummer (1986) give a very complete account of these developments. Other early results related to matroid intersection and partitioning are the solution of the problem of partitioning a graph into forests in Nash-Williams (1961), (1964) and Tutte (1961), and the solution of the so-called optimum branching problem, first in Chu and Liu (1965), and later in Edmonds (1967a), Bock (1971), and Karp (1971).

Edmonds (1965a), (1970), (1979) introduced and solved the matroid intersection and partitioning problems, as well as generalizations to matroids with weighted elements and to so-called polymatroids. He proved that these problems can be converted to structurally simple linear programs since certain polytopes have only integer vertices. In the terminology of linear programming, the intersection problem is then the linear programming dual of the partitioning problem. This work and Edmonds's profound results for matching problems (for a complete coverage, see Lovász and Plummer (1986)) establish Edmonds as the founder of polyhedral combinatorics. For applications, generalizations, and other algorithms, see Lehman (1964), Edmonds (1965b), (1967b), Edmonds and Fulkerson (1965), Lawler (1975), (1976), Frank (1981), Hassin (1982), Lawler and Martel (1982a), (1982b), Grötschel, Lovász, and Schrijver (1988), and Fujishige (1989), (1991). Schrijver (1984) houses many of the problems treated in those references under one roof.

As far as we know, the graph formulation of the intersection algorithm described in Section 5.3 is due to Krogdahl (see Lawler (1976)), except for our use of abstract matrices to simplify the arguments. In Cunningham (1986) a considerably faster version of the algorithm is given. The improvement rests on Menger's Theorem and the observation that the path used to derive the set Z' from the given set Z should be chordless, but need not be shortest. Incidentally, Menger's Theorem may also be used to determine, for the given set Z, a set $R \subseteq (E - Z)$ of minimum cardinality so that Z becomes an independent set of maximum cardinality in the minors $M_1 \backslash R$ and $M_2 \backslash R$ of M_1 and M_2. The graph approach may also be used to improve the already very appealing algorithm of Frank (1981) for the intersection problem with weighted elements.

The second application cited at the end of Section 5.3, which involves certain k-separations of a matroid, is due to Cunningham (1973), and Cunningham and Edmonds (1978).

Finally, we should mention that the intersection case involving at least three matroids, is in general $\mathcal{NP}$-hard since it includes the $\mathcal{NP}$-complete Hamiltonian cycle problem (see Garey and Johnson (1979)).

Chapter 6

Separation Algorithm

6.1 Overview

So far, we have described two simple matroid tools: the series-parallel and delta-wye constructions of Chapter 4, and the path shortening technique of Chapter 5. In this chapter, we introduce a third tool called the separation algorithm. Before we summarize that method and some of its uses, let us recall from Section 3.3 some definitions and results concerning matroid separations. Let M be a binary matroid on a set E and with rank function $r(\cdot)$. Furthermore, let B be a binary representation matrix of M with row index set X and column index set Y. Suppose two sets E_1 and E_2 partition E. For $i = 1, 2$, define $X_i = E_i \cap X$ and $Y_i = E_i \cap Y$. Then (E_1, E_2) is a k-separation of M, provided the matrix B when partitioned as

(6.1.1)

$$B = \begin{array}{c|c|c}
 & Y_1 & Y_2 \\
\hline
X_1 & B^1 & D^2 \\
\hline
X_2 & D^1 & B^2 \\
\end{array}$$

Partitioned version of B

satisfies

(6.1.2)
$$|X_1 \cup Y_1|, |X_2 \cup Y_2| \geq k$$
$$\text{GF}(2)\text{-rank } D^1 + \text{GF}(2)\text{-rank } D^2 \leq k - 1$$

The separation is exact if the inequality of (6.1.2) involving the GF(2)-rank of D^1 and D^2 holds with equality. In terms of E_1, E_2, E, and the rank function $r(\cdot)$, the conditions of (6.1.2) are

(6.1.3)
$$|E_1|, |E_2| \geq k$$
$$r(E_1) + r(E_2) \leq r(E) + k - 1$$

In the case of an exact separation, the inequality involving the rank function holds with equality. Finally, for $k \geq 2$, the matroid M is k-connected if it does not have an l-separation for any $1 \leq l < k$.

We are ready to summarize the material of this chapter. In Section 6.2, we describe and validate the just-mentioned separation algorithm. The scheme solves the following problem. Given is a binary matroid M with a minor N. For some $k \geq 1$, an exact k-separation (F_1, F_2) is known for N. We want to decide whether or not M has a k-separation (E_1, E_2) where for $i = 1, 2$, $E_i \supseteq F_i$. In the affirmative case, we say that (F_1, F_2) *induces* the k-separation (E_1, E_2).

The problem of finding induced separations may seem rather technical. But several important matroid results can be derived from its solution. Two such results are included in Sections 6.3 and 6.4. The first result provides sufficient conditions for the existence of induced separations. In later chapters, we rely upon these conditions to prove the existence of a number of decompositions. The second result builds upon the first one. It concerns the existence of certain extensions of 3-connected minors in 3-connected binary matroids. We use that result in the next chapter to establish the so-called splitter theorem and the existence of some sequences of minors. Finally, in Section 6.5, we sketch extensions of the results to the nonbinary case and provide references.

The chapter relies on the material of Chapters 2, 3, and 5.

6.2 Separation Algorithm

Suppose we are given a binary matroid M on a set E. Let N be a minor of M on a set $F \subseteq E$. Assume that N has, for some $k \geq 1$, an exact k-separation (F_1, F_2). We want to know whether or not M has a k-separation (E_1, E_2) where for $i = 1, 2$, $E_i \supseteq F_i$. If such (E_1, E_2) exists, we declare it to be *induced* by (F_1, F_2). In this section, we describe a simple method called the separation algorithm for deciding the existence of (E_1, E_2).

We begin with an informal discussion that relies on a particular binary representation matrix B^N of N. Let X_2 be a maximal independent subset of N contained in F_2. Then select a subset X_1 from F_1 so that $X_1 \cup X_2$ is a basis of N. For $i = 1, 2$, let $Y_i = F_i - X_i$. The desired representation matrix B^N of N corresponds to the base $X_1 \cup X_2$ of N. By the derivation of that base, B^N is of the form

(6.2.1)

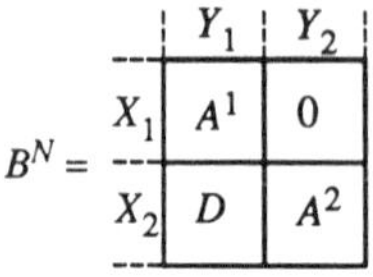

$$
B^N = \begin{array}{c|c|c}
 & Y_1 & Y_2 \\
\hline
X_1 & A^1 & 0 \\
\hline
X_2 & D & A^2
\end{array}
$$

Partitioned version of B^N

for some A^1, A^2, and D. Indeed, the zero submatrix indexed by X_1 and Y_2 is present since X_2 is a maximal independent subset of F_2, and since $Y_2 = F_2 - X_2$. By assumption, (F_1, F_2) is an exact k-separation of N, so we have by (6.1.2)

(6.2.2)
$$
|X_1 \cup Y_1|, |X_2 \cup Y_2| \geq k
$$
$$
\text{GF}(2)\text{-rank } D = k - 1
$$

We embed B^N into a representation matrix of M, and thus make the minor N visible, as follows. Since $X_1 \cup X_2$ is independent in N, that set is also independent in M. Thus, we can find a set $X_3 \subseteq E - F$ so that $X_1 \cup X_2 \cup X_3$ is a base of M. Let $Y_3 = (E - F) - X_3$. The representation matrix B of M for this base contains B^N as submatrix. We depict B below. For reasons to become clear shortly, we have placed A^1, A^2, D, and the 0 submatrix of B^N into the corners of B.

(6.2.3)

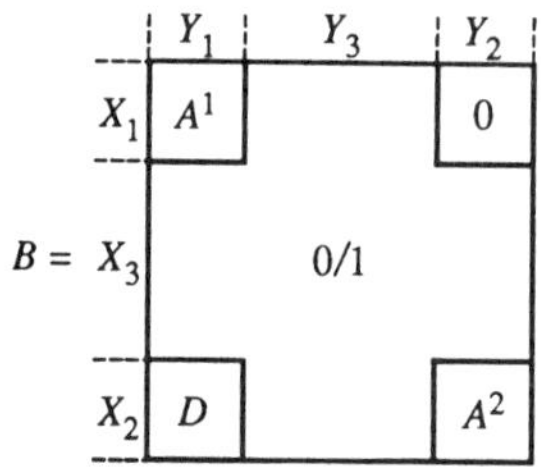

$$
B = \begin{array}{c|c|c|c}
 & Y_1 & Y_3 & Y_2 \\
\hline
X_1 & A^1 & & 0 \\
\hline
X_3 & & 0/1 & \\
\hline
X_2 & D & & A^2
\end{array}
$$

Matrix B for M displaying partitioned B^N

Recall that we want to find a k-separation (E_1, E_2) of M where for $i = 1, 2$, $E_i \supseteq F_i$, or to prove that no such k-separation exists. In terms of the index sets of (6.2.3), we want to partition the set X_3 into X_{31}, X_{32}, and the set Y_3 into Y_{31}, Y_{32} so that the correspondingly refined matrix B of (6.2.3) is

of the form

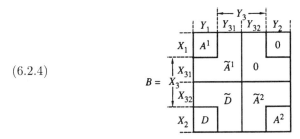

(6.2.4)

Partition of B induced by that of B^N

with GF(2)-rank $\tilde{D} \leq k - 1$, or we are to prove that such partitions of X_3 and Y_3 do not exist. In the affirmative case, D is a submatrix of $\tilde{D}$, and since GF(2)-rank $D = k - 1$, we must have GF(2)-rank $\tilde{D} = k - 1$. Put differently, if an induced k-separation of M exists at all, then it must be an exact induced k-separation.

We employ a recursive scheme to decide whether or not an induced k-separation exists. As the measure of problem size for the recursion, we use $|X_3 \cup Y_3|$. If $|X_3 \cup Y_3| = 0$, then $M = N$, and for $i = 1, 2$, $E_i = F_i$ gives the desired induced k-separation of M. Suppose $|X_3 \cup Y_3| \geq 0$. Redraw B of (6.2.3) so that an arbitrary row $x \in X_3$ and an arbitrary column $y \in Y_3$ are displayed as follows.

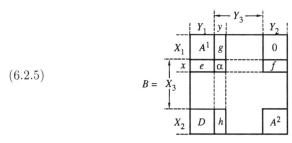

(6.2.5)

Matrix B for M with partitioned B^N,
row $x \in X_3$, and column $y \in Y_3$

The recursive method relies on the analysis of the following three cases of B of (6.2.5). Collectively, these cases cover all situations.

In the first case, we suppose that for some row $x \in X_3$, the subvector e is not spanned by the rows of D. We claim that in any induced k-separation, we must have $x \in X_{31}$. For a proof, take any such separation as depicted by (6.2.4). If $x \in X_{32}$, then the subvector e of row x occurs in $\tilde{D}$. Since e is not spanned by the rows of D, we have GF(2)-rank $\tilde{D} >$ GF(2)-rank D, which contradicts the condition GF(2)-rank $\tilde{D}$ = GF(2)-rank D. Thus, x must be in X_{31} as claimed. We now examine the subvector f of row x. Suppose that subvector is nonzero. Using (6.2.4) once more, we see that in any induced k-separation, the nonzero f forces x to be in X_{32}. But the latter requirement conflicts with the one determined earlier for x. Thus, an induced k-separation cannot exist, and we stop with that conclusion. So we now assume the subvector f of row x to be zero. We know already that x must be in X_{31} in any induced k-separation. Suppose in B of (6.2.5), we adjoin e to A^1 and f to the explicitly shown 0 submatrix, getting a new A^1 and a new 0 submatrix. Correspondingly, we extend N by x to $N\&x$ and redefine N to be the extended matroid. Evidently, $(X_1 \cup \{x\} \cup Y_1, X_2 \cup Y_2)$ is a k-separation of the new N, and that k-separation induces one in M if and only if this is so for the k-separation $(X_1 \cup Y_1, X_2 \cup Y_2)$ of the original N. Thus, we may replace the original problem by one involving the new N. By our measure of problem size, the new problem is smaller than the original one, and we may apply recursion.

In the second case, we suppose that for some column $y \in Y_3$, the subvector g is nonzero. Arguing analogously to the first case via (6.2.4), we conclude that y must be in Y_{31} in any induced k-separation. Furthermore, suppose that the column subvector h of column y is not spanned by the columns of D. Using (6.2.4) once more, we see that y must also be in Y_{32} in any induced k-separation. Thus, an induced k-separation cannot exist, and we stop with that conclusion. So suppose that h is spanned by the columns of D. Then we adjoin g to A^1, h to D, and correspondingly redefine N to become $N+y$. Then the k-separation $(X_1 \cup Y_1 \cup \{y\}, X_2 \cup Y_2)$ of the new N induces a k-separation of M if and only if this is so for the k-separation $(X_1 \cup Y_1, X_2 \cup Y_2)$ of the original N. Once more, we may replace the induced k-separation problem involving the original N by one with the new N. The latter problem is smaller, and we may invoke recursion.

For the discussion of the third and final case, we suppose that neither of the above cases applies. Equivalently, for all $x \in X_3$, the subvector e of row x is spanned by the rows of D, and for all $y \in Y_3$, the subvector g of column y is zero. By (6.2.5), $(X_1 \cup Y_1, X_2 \cup X_3 \cup Y_2 \cup Y_3)$ is a k-separation of M induced by the one of N, and we stop with that conclusion.

We call the above recursive method the *separation algorithm*. It clearly has a polynomial implementation. For later reference, we summarize the algorithm below.

Separation Algorithm

1. Suppose B of (6.2.5) has a row $x \in X_3$ with the indicated row subvectors e and f such that GF(2)-rank $[e/D]$ > GF(2)-rank D. Then x must be in X_{31}. Suppose, in addition, that f is nonzero. Then x must also be in X_{32}, i.e., B cannot be partitioned, and we stop with that declaration. On the other hand, suppose $f = 0$. Since x must be in

X_{31}, we adjoin e to A^1, and f to the explicitly shown 0 matrix. Then we start recursively again with the new B and the new B^N.

2. Suppose B of (6.2.5) has a column $y \in Y_3$ with the indicated column subvectors g and h such that g is nonzero. Then y must be in Y_{31}. Suppose, in addition, GF(2)-rank $[D \mid h] >$ GF(2)-rank D. Then y must also be in Y_{32}, i.e., B cannot be partitioned, and we stop with that declaration. On the other hand, suppose GF(2)-rank $[D \mid h] =$ GF(2)-rank D. Since y must be in Y_{31}, we adjoin g to A^1, and h to D. Then we start recursively again with the new B and the new B^N.

3. Finally, suppose that for all rows $x \in X_3$, the row subvector e satisfies GF(2)-rank $[e/D] =$ GF(2)-rank D, and that for all columns $y \in Y_3$, the column subvector g is 0. Then $X_{31} = Y_{31} = \emptyset$, $Y_{32} = Y_3$, $Y_{32} = Y_3$ gives the desired partition of B.

In the next two sections, we put the separation algorithm to good use. Preparatory to that discussion, we establish in the next lemma that certain extensions of binary matroids are 3-connected.

(6.2.6) Lemma. *Let N be a 3-connected binary matroid on at least six elements. Suppose a 1-, 2-, or 3-element binary extension of N, say M, has no loops, coloops, parallel elements, or series elements. Then M is 3-connected.*

Proof. Let C be a binary representation matrix of M that displays a representation matrix, say B, for N. By assumption, B is 3-connected. Suppose C is not connected. A straightforward case analysis proves C to contain a zero vector or unit vector. Thus, M has a loop, coloop, parallel elements, or series elements, a contradiction. Hence, C is connected. If C is not 3-connected, then by Lemma (3.3.20), there is a 2-separation of C with at least five rows/columns on each side. Then, necessarily, the matrix B has a 2-separation with at least two rows/columns on each side, another contradiction. We conclude that C, and hence M, are 3-connected. □

From Lemma (6.2.6), we deduce the following result for 1-edge extensions of 3-connected graphs.

(6.2.7) Lemma. *Let H be a 3-connected graph with at least six edges. Then a connected 1-edge extension of H is 3-connected if and only if it is producible as follows: Either two nonadjacent vertices of H are connected by a new edge, or a vertex of degree at least 4 is partitioned into two vertices, each of degree at least 2, and the two new vertices are connected by a new edge.*

Proof. The described extension steps are precisely the ways in which H can be extended by one edge to a larger connected graph that does not have loops, coloops, parallel edges, or series edges. Thus, the "only if" part is obvious. The "if" part follows from Lemma (6.2.6).   □

We are prepared for the next section, where we derive sufficient conditions for the existence of induced separations under various assumptions.

6.3 Sufficient Conditions for Induced Separations

In this section, we employ the separation algorithm to establish sufficient conditions under which induced separations can be guaranteed. Before we begin with the detailed discussion, we describe the general setting in which these conditions will be invoked. To this end, let $\mathcal{M}$ be a class of binary matroids. The class is assumed to be closed under minor-taking and isomorphism.

We select a matroid $N \in \mathcal{M}$, say on set F. Suppose by some method we find, for some $k \geq 2$, an exact k-separation (F_1, F_2) for N. At that point, we would like to claim the following.

(6.3.1) $\quad \begin{cases} \text{Suppose an } M \in \mathcal{M} \text{ has an } N \text{ minor, say } N' \text{ . Let } (F_1', F_2') \\ \text{be a } k\text{-separation of } N' \text{ that corresponds to } (F_1, F_2) \text{ under} \\ \text{one of the isomorphisms between } N' \text{ and } N. \text{ Then the } k\text{-} \\ \text{separation } (F_1', F_2') \text{ of } N' \text{ induces a } k\text{-separation of } M. \end{cases}$

Results of type (6.3.1) are valuable if an $M \in \mathcal{M}$ is known to have an N minor, and if the induced k-separation of M may be employed to effect a useful decomposition of M. In Chapters 10–13, we will see that these two assumptions are satisfied in a number of cases. Thus, nontrivial instances of (6.3.1) are indeed useful.

The technique for proving (6.3.1) is in principle straightforward. With machinery yet to be described, we compute all minimal binary matroids satisfying the assumptions of (6.3.1) but not its conclusion. If no such matroid is in $\mathcal{M}$, then (6.3.1) indeed holds.

Application of the technique entails two difficulties. First, $\mathcal{M}$, N, and (F_1, F_2) must be properly selected. Second, we need structural insight and computational tools to prove that $\mathcal{M}$ has no M for which (6.3.1) fails. In this section, we ignore the first aspect. It will be treated in depth in Chapters 10–13. Instead, we concentrate on the development of the structural insight and of the computational tools.

We break down that development into two phases. In the first one, we accomplish the following task, where N is the previously mentioned matroid with exact k-separation (F_1, F_2).

(6.3.2) $\left\{\begin{array}{l}\text{Find computationally tractable properties of minimal bi-}\\ \text{nary matroids } M \text{ that have } N \text{ as a minor, but that do}\\ \text{not have a } k\text{-separation induced by the exact } k\text{-separation}\\ (F_1, F_2) \text{ of } N.\end{array}\right.$

An M satisfying the conditions of (6.3.2) we simply call *minimal*. In the second phase, we expand the task of (6.3.2) to the task (6.3.3) below. It essentially says that the requirements of (6.3.2) are to hold for some minor isomorphic N, and that M is to be minimal with respect to that condition. The precise statement is as follows.

(6.3.3) $\left\{\begin{array}{l}\text{Find computationally tractable properties of binary ma-}\\ \text{troids } M \text{ satisfying the following conditions: } M \text{ must have}\\ \text{at least one } N \text{ minor. Some } k\text{-separation of at least one}\\ \text{such minor corresponding to } (F_1, F_2) \text{ of } N \text{ under one of the}\\ \text{isomorphisms must fail to induce a } k\text{-separation of } M. \text{ The}\\ \text{matroid } M \text{ is to be minimal with respect to those condi-}\\ \text{tions.}\end{array}\right.$

We say that an M satisfying the conditions of (6.3.3) is *minimal under isomorphism*. Evidently, minimality under isomorphism demands more than the previously defined minimality.

Answers to (6.3.3) give sufficient conditions so that (6.3.1) holds. That is, if none of the binary matroids M with the yet-to-be-determined properties of (6.3.3) is in $\mathcal{M}$, then necessarily (6.3.1) must hold. Thus, answers to (6.3.3) effectively supply sufficient conditions under which (6.3.1) is satisfied.

We begin with the task (6.3.2). We are given a binary matroid N on a set F. In the next lemma, we rely on rank functions instead of representation matrices. Thus, we let $r_N(\cdot)$ be the rank function of N. For some $k \geq 1$, we have an exact k-separation (F_1, F_2) of N. Thus, by (6.1.3),

(6.3.4)
$$|F_1|, |F_2| \geq k$$
$$r_N(F_1) + r_N(F_2) = r_N(F) + k - 1$$

Let M be any binary matroid on a set E and with rank function $r_M(\cdot)$. Assume that M has N as a minor, and that (F_1, F_2) does not induce a k-separation of M. Thus, the system

(6.3.5)
$$E_1 \supseteq F_1; \ E_2 \supseteq F_2$$
$$r_M(E_1) + r_M(E_2) \leq r_M(E) + k - 1$$

has no solution. Assume M to be minimal as defined above. Thus, every proper minor M' of M with N as a minor has a solution for an appropriately adapted (6.3.5). First, we show that there is a unique partition of the set $E - F$ into Z_1 and Z_2 so that $N = M/Z_1 \backslash Z_2$.

(6.3.6) Lemma.   Let M, E, N, F, and (F_1, F_2) be as defined above. Suppose (6.3.5) has no solution and that M is minimal. Then for all $z \in (E - F)$, M/z or $M \backslash z$ does not have N as a minor.

Proof. Suppose both M/z and $M \backslash z$ have N as a minor. By the minimality of M, they both have induced separations, say (U_1, U_2) for M/z and (W_1, W_2) for $M \backslash z$. Let $r_{M/z}(\cdot)$ and $r_{M \backslash z}(\cdot)$ be the rank functions of M/z and $M \backslash z$. According to (6.3.5),

$$U_1, W_1 \supseteq F_1; \ U_2, W_2 \supseteq F_2$$

(6.3.7)
$$r_{M/z}(U_1) + r_{M/z}(U_2) \leq r_{M/z}(E - \{z\}) + k - 1$$

$$r_{M \backslash z}(W_1) + r_{M \backslash z}(W_2) \leq r_{M \backslash z}(E - \{z\}) + k - 1$$

By the minimality of M, z cannot be a loop or coloop of M. Thus, the two inequalities of (6.3.7) imply the inequalities

(6.3.8)
$$r_M(U_1 \cup \{z\}) + r_M(U_2 \cup \{z\}) \leq r_M(E) + k$$
$$r_M(W_1) \qquad + r_M(W_2) \qquad \leq r_M(E) + k - 1$$

We add the two inequalities of (6.3.8) and apply submodularity to get

(6.3.9)
$$r_M(U_1 \cup \{z\} \cup W_1) + r_M((U_1 \cup \{z\}) \cap W_1)$$
$$+ \, r_M(U_2 \cup \{z\} \cup W_2) + r_M((U_2 \cup \{z\}) \cap W_2)$$
$$\leq 2r_M(E) + 2k - 1$$

But each one of $(U_1 \cup \{z\} \cup W_1, (U_2 \cup \{z\}) \cap W_2)$ and $(U_2 \cup \{z\} \cup W_2, (U_1 \cup \{z\}) \cap W_1)$ is a pair (E_1, E_2) satisfying $E_1 \supseteq F_1$ and $E_2 \supseteq F_2$. Since (6.3.5) cannot be satisfied, we have

(6.3.10)
$$r_M(U_1 \cup \{z\} \cup W_1) + r_M((U_2 \cup \{z\}) \cap W_2) \geq r_M(E) + k$$
$$r_M(U_2 \cup \{z\} \cup W_2) + r_M((U_1 \cup \{z\}) \cap W_2) \geq r_M(E) + k$$

Summing the latter two inequalities, we obtain a contradiction of (6.3.9).   □

For further insight into the structure of a minimal M, we employ the separation algorithm of Section 6.2. That is, we have the representation matrix B of (6.2.5) for M, repeated here for ease of reference.

(6.3.11)

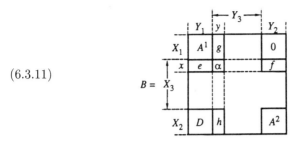

Matrix B for M with partitioned B^N,
row $x \in X_3$, and column $y \in Y_3$

The submatrix of B composed of A^1, A^2, D, and the explicitly shown 0 matrix is B^N of (6.2.1), which we also repeat here. The latter matrix represents N.

(6.3.12)

$$
B^N = \begin{array}{c|c|c}
 & Y_1 & Y_2 \\
\hline
X_1 & A^1 & 0 \\
\hline
X_2 & D & A^2
\end{array}
$$

Partitioned version of B^N

We apply the separation algorithm to search for a partition as given by (6.2.4). Since (6.3.5) cannot be satisfied, the algorithm terminates in step 1 or 2 announcing that no partition with the desired properties exist. Below, we list that algorithm again, with references adapted to the just-defined matrices.

Separation Algorithm

1. Suppose B of (6.3.11) has a row $x \in X_3$ with the indicated row subvectors e and f such that GF(2)-rank $[e/D]$ > GF(2)-rank D. Then x must be in X_{31}. Suppose, in addition, that f is nonzero. Then x must also be in X_{32}, i.e., B cannot be partitioned, and we stop with that declaration. On the other hand, suppose $f = 0$. Since x must be in X_{31}, we adjoin e to A^1, and f to the explicitly shown zero matrix. Then we start recursively again with the new B and the new B^N.

2. Suppose B of (6.3.11) has a column $y \in Y_3$ with the indicated column subvectors g and h such that g is nonzero. Then y must be in Y_{31}. Suppose, in addition, GF(2)-rank $[D \mid h]$ > GF(2)-rank D. Then y must also be in Y_{32}, i.e., B cannot be partitioned, and we stop with

that declaration. On the other hand, suppose GF(2)-rank $[D \mid h] =$ GF(2)-rank D. Since y must be in Y_{31}, we adjoin g to A^1, and h to D. Then we start recursively again with the new B and the new B^N.

3. Finally, suppose that for all rows $x \in X_3$, the row subvector e satisfies GF(2)-rank $[e/D] =$ GF(2)-rank D, and that for all columns $y \in Y_3$, the column subvector g is 0. Then $X_{31} = Y_{31} = \emptyset$, $Y_{32} = Y_3$, $Y_{32} = Y_3$ gives the desired partition of B.

By (6.3.11), for any $x \in X_3$ and any $y \in Y_3$, the minors M/x and $M \backslash y$ of M have N as a minor. Thus, by the minimality of M, the separation algorithm does find a partition if we delete any row $x \in X_3$ or any column $y \in Y_3$ from B.

We now prove some results about the structure of a matrix B produced by a minimal M. We start with the special case where B of a minimal M contains just one row or column beyond that of B^N. Consider the case of a single additional row x. In the notation of (6.3.11), $X_3 = \{x\}$ and $Y_3 = \emptyset$. By step 1 of the separation algorithm, the row subvectors e and f of row x satisfy

(6.3.13) e is not spanned by the rows of D, and

f is nonzero

Similarly, we deduce for the case of a single additional column y, i.e., when $X_3 = \emptyset$ and $Y_3 = \{y\}$,

(6.3.14) g is nonzero, and

h is not spanned by the columns of D

We now treat the remaining cases, where B has at least two additional rows or columns beyond those of B^N. Thus, $|X_3 \cup Y_3| \geq 2$. We want to show that both X_3 and Y_3 are nonempty, and that there exist $x \in X_3$ and $y \in Y_3$ so that the subvectors e, f of row x, and g, h of column y, as well as the scalar α, obey certain conditions. First we prove the following fact about the subvectors e of the rows $x \in X_3$ and about the subvectors g of the columns $y \in Y_3$.

(6.3.15) Lemma.   *Exactly one of the two cases (i) and (ii) below applies.*

(i) *There is exactly one row $x \in X_3$ such that the subvector e is not spanned by the rows of D. In that row x, the subvector f is zero. Furthermore, for all $y \in Y_3$, the subvector g of column y is zero.*

(ii) *There is exactly one column $y \in Y_3$ such that the subvector g is nonzero. In that column y, the subvector h is spanned by the columns of D. Furthermore, for all $x \in X_3$, the subvector e of row x is spanned by the rows of D.*

Proof. If the condition about f or h does not hold, then we have the smaller case of (6.3.13) or (6.3.14), a contradiction.

To prove the claims about e and g, we apply the separation algorithm to B. Consider each application of steps 1 and 2 except for the last application, when the algorithm stops. In each such application, a row $x \in X_3$ or column $y \in Y_3$ is moved to X_{31} or Y_{31}, and the related row subvector f satisfies $f = 0$, or the column subvector h satisfies GF(2)-rank $[D \mid h]$ = GF(2)-rank D. Exactly one of these conditions is violated in the last iteration. The rows and columns moved to X_{31} and Y_{31}, plus the row or column encountered in the last application, suffice to prove that B has no induced partition. Thus, by the minimality of M, these rows and columns comprise the rows and columns that B has beyond those of B^N. We conclude that in each row $x \in X_3$, the row subvector f is zero except for at most one such vector, and that in each column $y \in Y_3$, the column subvector h is spanned by D except for at most one such vector. Furthermore, if there is a nonzero f, then all vectors h are spanned by D, and if there is an h not spanned by D, then all vectors f are zero. To prove the claims about e and g, we use duality, or equivalently, we apply the above arguments to B^t. Then g plays the role of f above, and e that of h. The conditions just proved for f and h establish the statements about e and g of the lemma. $\qquad\Box$

We investigate the two cases of Lemma (6.3.15) further. We begin with the situation where a unique row $x \in X_3$ has a subvector e that is not spanned by the rows of D, and where for all $y \in Y_3$, the subvector g of column y is zero. The next lemma tells more about the columns $y \in Y_3$.

(6.3.16) Lemma. *Suppose case (i) of Lemma (6.3.15) applies. Then there exists a $y \in Y_3$ such that the scalar α of column y is 1 and the subvector h is nonzero.*

Proof. As in the proof of Lemma (6.3.15), we apply the separation algorithm to B. By the assumptions, in the first iteration the row x is moved to X_{31}. By Lemma (6.3.15), in the next recursive application of the separation algorithm, step 2 must apply. Thus, a column $y \in Y_3$ is moved to Y_{31}. By Lemma (6.3.15), the subvector g of that column is zero. Then $\alpha = 1$ since otherwise step 2 could not move column y to Y_{31}.

Suppose h is zero. Since g is also zero, we can pivot on $\alpha = 1$ without disturbing the submatrix B^N. This implies that both $M\backslash y$ and M/y have N as a minor, in violation of Lemma (6.3.6). Thus, h must be nonzero. $\qquad\Box$

Consider now the second case of Lemma (6.3.15). Thus, B has a unique column $y \in Y_3$ with nonzero subvector g, and the subvector h is spanned by the columns of D. Furthermore, for all $x \in X_3$, the subvector

e of row x is spanned by the rows of D. Analogously to Lemma (6.3.16), we have the following result.

(6.3.17) Lemma.   *Suppose case* (ii) *of Lemma* (6.3.15) *applies. Then there exists an* $x \in X_3$ *such that the subvector* e *is spanned by the rows of* D. *If the subvector* f *is zero, then the subvector* e *is nonzero. Furthermore, the subvector* $[e \mid \alpha]$ *is not spanned by the rows of* $[D \mid h]$.

Proof. Apply the separation algorithm to B. In the first iteration, the column y is moved to Y_{31}. In the next recursive application of the separation algorithm, step 1 must apply. Thus, a row $x \in X_3$ is moved to X_{31}. The subvector e must be spanned by the rows of D, but the subvector $[e \mid \alpha]$ of row x is not spanned by the rows of $[D \mid h]$.

Suppose f is zero. If e is also zero, then necessarily $\alpha = 1$. A pivot on α does not disturb the submatrix B^N. Thus, both M/x and $M \backslash x$ contain N as a minor, a contradiction of Lemma (6.3.6). Thus, e is nonzero.   □

With (6.3.13), (6.3.14), and Lemmas (6.3.16) and (6.3.17), we assemble the following theorem.

(6.3.18) Theorem.   *Let* M *be minimal, and let* B *be the representation matrix of* (6.3.11) *for* M.

(a) *Suppose* $X_3 = \{x\}$ *and* $Y_3 = \emptyset$. *Then the subvector* e *of row* x *is not spanned by the rows of* D, *and* f *is nonzero.*

(b) *Suppose* $X_3 = \emptyset$ *and* $Y_3 = \{y\}$. *Then the subvector* g *of column* y *is nonzero, and* h *is not spanned by the columns of* D.

(c) *Suppose* $|X_3 \cup Y_3| \geq 2$. *Then either* (c.1) *or* (c.2) *below applies for some* $x \in X_3$ *and* $y \in Y_3$.

    (c.1)   *The subvector* e *of row* x *is not spanned by the rows of* D, *and* f *is zero. The subvector* g *of column* y *is zero,* $\alpha = 1$, *and the subvector* h *is nonzero.*

    (c.2)   *The subvector* g *of column* y *is nonzero, and the subvector* h *is spanned by the columns of* D. *The subvector* e *is spanned by the rows of* D. *If the subvector* f *is zero, then the subvector* e *is nonzero. The subvector* $[e \mid \alpha]$ *is not spanned by the rows of* $[D \mid h]$.

Proof. Statements (6.3.13) and (6.3.14) establish (a) and (b). Lemmas (6.3.15), (6.3.16), and (6.3.17) prove parts (c.1) and (c.2).   □

For our purposes, Theorem (6.3.18) suffices as answer for the task (6.3.2). Thus, we turn to the task (6.3.3). That problem demands that we find computationally tractable properties of binary matroids M that are minimal under isomorphism. That is, any such M has a minor isomorphic to N. For at least one such minor the following holds. Some k-separation of that minor corresponds to (F_1, F_2) of N under one of the isomorphisms, and fails to induce a k-separation of M. We want M to be minimal with respect to these conditions.

Let B of (6.3.11) be the representation matrix of an M that is minimal under isomorphism. Since minimality under isomorphism implies the minimality defined for (6.3.2), B observes the conditions of Theorem (6.3.18). Evidently, minimality under isomorphism is a stronger requirement than minimality. Thus, we expect e, f, g, h, and α of Theorem (6.3.18) to obey additional conditions. The next lemma supplies computationally tractable ones. The notation is that of Theorem (6.3.18).

(6.3.19) Lemma.

($\bar{c}$.1) *If case (c.1) of Theorem (6.3.18) applies, then the following holds.*

    ($\bar{c}$.1.1)   *The subvector e of row x of B is not parallel to a row of the submatrix A^1.*

    ($\bar{c}$.1.2)   *Suppose column $z \in Y_1$ of A^1 is nonzero. Then the subvector e of row x of B is not a unit vector with 1 in column z of B.*

($\bar{c}$.2) *If case (c.2) of Theorem (6.3.18) applies, then the following holds.*

    ($\bar{c}$.2.1)   *Suppose $\overline{D}$, the matrix obtained from D by deletion of a column $z \in Y_1$ of D, has the same GF(2)-rank as D. Then the subvector $[g/h]$ of column y of B is not parallel to column z of $[A^1/D]$.*

    ($\bar{c}$.2.2)   *Suppose the rows of D do not span a row $z \in X_1$ of A^1. Then $[g/h]$ is not a unit vector with 1 in row z.*

Proof. ($\bar{c}$.1.1): For a proof by contradiction, suppose the subvector e of row x of B is parallel to a row $z \in X_1$ of A^1. In B, we exchange the rows x and z, and appropriately adjust X_1 to $X_1' = (X_1 - \{z\}) \cup \{x\}$ and X_3 to $X_3' = (X_3 - \{x\}) \cup \{z\}$. By (c.1) of Theorem (6.3.18), f is zero. Thus, the submatrix of B indexed by X_1', X_2, Y_1, and Y_2 is B^N except for the change of the index z to x. Let N' be the corresponding minor of M. We know that $N\&x$ does not induce a k-separation of M. Thus, $N'\&z = N\&x$ does not induce one either. The same conclusion applies to N', since row z contains e. Now M is minimal under isomorphism, so M must be minimal with respect to N'. We show that the latter conclusion leads to a contradiction. For the proof, let us examine the effect of the exchange of rows x and z on column $y \in Y_3$. By that exchange, the role of the zero subvector g indexed by X_1 is taken on by a vector g' indexed by X_1'. By (c.1) of Theorem (6.3.18), the entry α in row x is nonzero. Thus, the vector g' is nonzero. Apply Lemma (6.3.15) to N' and the subvectors e and g'. Since e is not spanned by the rows of D and since g' is nonzero, these subvectors violate the conclusions of that lemma, and thus provide the desired contradiction.

($\bar{c}$.1.2): Suppose that column $z \in Y_1$ of A^1 is nonzero, and that e is a unit vector with 1 in column z of B. Perform a pivot in column $z \in Y_1$ of A^1.

Then the subvector e becomes parallel to a row of A^1, and the above case ($\bar{c}.1.1$) applies.

($\bar{c}.2.1$): Suppose $\overline{D}$, the matrix obtained from D' by deletion of a column $z \in Y_1$ of D, has the same GF(2)-rank as D. Further assume that the subvector $[g/h]$ of column y of B is parallel to column z of $[A^1/D]$. Exchange columns y and z of B. Adjust Y_1 to $Y_1' = (Y_1 - \{z\}) \cup \{y\}$ and Y_3 to $Y_3' = (Y_3 - \{y\}) \cup \{z\}$. The swap of columns effectively replaces N by an isomorphic minor N' and modifies D to D' and e to e'. A simple rank calculation confirms that under the assumption on $\overline{D}$, the rows of D' do not span e'. Arguing analogously to the case ($\bar{c}.1.1$), M is not minimal under isomorphism.

($\bar{c}.2.2$): By a pivot in row z of A^1, this case becomes ($\bar{c}.2.1$), as is readily checked.   □

We summarize the preceding conclusions in the next theorem, which finishes the task (6.3.3). The statement of the theorem is rather detailed to simplify its application.

(6.3.20) Theorem.   *Let M be minimal under isomorphism. Then one of* (a), (b), *or* (c) *below holds.*

(a) *M is represented by*

(6.3.21)

$$B = \begin{array}{c|c|c} & Y_1 & Y_2 \\ \hline X_1 & A^1 & 0 \\ \hline x & e & f \\ \hline X_2 & D & A^2 \end{array}$$

Matrix B for M minimal
under isomorphism, case (a)

In row x, e is not spanned by the rows of D, and f is nonzero.

(b) *M is represented by*

(6.3.22)

$$B = \begin{array}{c|c|c|c} & Y_1 & y & Y_2 \\ \hline X_1 & A^1 & g & 0 \\ \hline X_2 & D & h & A^2 \end{array}$$

Matrix B for M minimal
under isomorphism, case (b)

In column y, g is nonzero, and h is not spanned by the columns of D.

(c) M has a minor $\overline{M}$ with representation matrix

(6.3.23)

$$
\overline{B} = \begin{array}{c|c|c|c}
 & Y_1 & y & Y_2 \\
\hline
X_1 & A^1 & g & 0 \\
\hline
x & e & \alpha & f \\
\hline
X_2 & D & h & A^2
\end{array}
$$

Matrix $\overline{B}$ for minor $\overline{M}$
of M minimal under isomorphism

Either (c.1) or (c.2) below holds for e, f, g, h, and α.

(c.1) *e is not spanned by the rows of D; $f = 0$; $g = 0$; $h \neq 0$; $\alpha = 1$; e is not parallel to a row of A^1. If column $z \in Y_1$ of A^1 is nonzero, then e is not a unit vector with 1 in column z of $\overline{B}$.*

(c.2) *$g \neq 0$; h is spanned by the columns of D; e is spanned by the rows of D; $f = 0$ implies $e \neq 0$; $[e \mid \alpha]$ is not spanned by the rows of $[D \mid h]$. If $\overline{D}$, the matrix obtained from D by deletion of a column $z \in Y_1$, has the same GF(2)-rank as D, then $[g/h]$ is not parallel to column z of $[A^1/D]$. If the rows of D do not span a row $z \in X_1$ of A^1, then $[g/h]$ is not a unit vector with 1 in row z.*

Proof. The statements follow directly from Theorem (6.3.18) and Lemma (6.3.19). $\qquad\square$

Recall our main goal for this section: We want to determine sufficient conditions for induced separations. In the next corollary, we deduce such conditions from Theorem (6.3.20).

(6.3.24) Corollary. *Let $\mathcal{M}$ be a class of binary matroids that is closed under isomorphism and under the taking of minors. Suppose that N given by B^N of (6.3.12) is in $\mathcal{M}$, but that the 1- and 2-element extensions of N given by (6.3.21), (6.3.22), (6.3.23), and by the accompanying conditions are not in $\mathcal{M}$. Assume that a matroid $M \in \mathcal{M}$ has an N minor. Then any k-separation of any such minor that corresponds to $(X_1 \cup Y_1, X_2 \cup Y_2)$ of N under one of the isomorphisms induces a k-separation of M.*

Proof. Take $M \in \mathcal{M}$ satisfying the assumptions. We know $\mathcal{M}$ to be closed under isomorphism. Thus, we may suppose that the N minor of M is N itself. Suppose the k-separation of N does not induce one in M. Then M, or a minor of M containing N, is minimal under isomorphism. By Theorem (6.3.20), M has a minor represented by one of the matrices of (6.3.21), (6.3.22), (6.3.23). Since $\mathcal{M}$ is closed under minor-taking, any such minor of M is in $\mathcal{M}$. But presence of such a minor in $\mathcal{M}$ is ruled out by assumption, a contradiction. $\qquad\square$

Sometimes an abbreviated version of Corollary (6.3.24) suffices to produce the desired conclusion of induced separations. The following result is one such version.

(6.3.25) Corollary. *Let $\mathcal{M}$ be a class of binary matroids that is closed under isomorphism and under the taking of minors. Suppose a 3-connected N given by B^N of (6.3.12) is in $\mathcal{M}$. Assume that $N/(X_2 \cup Y_2)$ has no loops and that $N \backslash (X_2 \cup Y_2)$ has no coloops. Furthermore, assume for every 3-connected 1-element extension of N in $\mathcal{M}$, say by an element z, that the pair $(X_1 \cup Y_1, X_2 \cup Y_2 \cup \{z\})$ is a k-separation of that extension. Then for any 3-connected matroid $M \in \mathcal{M}$ with an N minor, the following holds. Any k-separation of any such minor that corresponds to $(X_1 \cup Y_1, X_2 \cup Y_2)$ of N under one of the isomorphisms induces a k-separation of M.*

Proof. Suppose the conclusion is false. By Corollary (6.3.24), $\mathcal{M}$ contains a matroid M represented by one of the matrices (6.3.21), (6.3.22), or (6.3.23). We first dispose of the cases (6.3.21) and (6.3.22). The assumed 3-connectedness of B^N and the conditions of Theorem (6.3.20) on the matrices of (6.3.21) and (6.3.22) imply that these matrices do not contain zero vectors, unit vectors, or parallel vectors. Then by Lemma (6.2.6), these matrices represent 3-connected 1-element extensions of N. By assumption, any 3-connected 1-element extension of N does have an induced k-separation. Hence, the cases (6.3.21) and (6.3.22) cannot occur.

Consider (6.3.23), case (c.1). Delete column y from that matrix. The reduced matrix represents $N \& x$. We claim that $N \& x$ is 3-connected. By (c.1), the subvector e of row x is not spanned by the rows of D. It also is not parallel to a row of A^1. Now $N/(X_2 \cup Y_2)$ has no loop, so A^1 has no zero columns. Then by (c.1), e is not a unit vector. By Lemma (6.2.6), $N \& x$ is 3-connected as claimed. Since e is not spanned by the rows of D, $(X_1 \cup Y_1, X_2 \cup Y_2 \cup \{x\})$ is not a k-separation of $N \& x$. By assumption, $N \& x$ cannot be in $\mathcal{M}$. Yet $N \& x$ is a minor of $M \in \mathcal{M}$, a contradiction.

Consider (6.3.23), case (c.2). We first establish an auxiliary result. Suppose that for some $z \in Y_1$, deletion of column z from D reduces the GF(2)-rank, or that for some $z \in X_1$, the rows of D span row z of A^1. We claim that z is a coloop of $N \backslash (X_2 \cup Y_2)$, contrary to assumption. For a proof, we delete from $[I \mid B^N]$ the columns indexed by $X_2 \cup Y_2$. By a simple rank calculation, every basis of the reduced matrix contains column z. This establishes the claim.

By (c.2) and the auxiliary result, g and h of (6.3.23) satisfy the following conditions: $g \neq 0$, and $[g/h]$ is not parallel to a column of $[A^1/D]$ and is not a unit vector. Then by Lemma (6.2.6) and (6.3.23), $N+y$ is 3-connected, and $(X_1 \cup Y_1, X_2 \cup Y_2 \cup \{y\})$ is not a k-separation of $N+y$. Thus, $N+y$ cannot be in $\mathcal{M}$. Yet $N+y$ is a minor of $M \in \mathcal{M}$, a contradiction. $\qquad\square$

In Chapter 10, we require the graph version of Corollary (6.3.25) for

$k = 3$. For that situation, we adapt the above matroid language as follows. Suppose we have 3-connected graphs G and H. On hand is a 3-separation (F_1, F_2) for H. Then that 3-separation of H *induces* one for G if the latter graph has a 3-separation (E_1, E_2) where $E_1 \supseteq F_1$ and $E_2 \supseteq F_2$. Here is the special graph version of Corollary (6.3.25) for $k = 3$.

(6.3.26) Corollary. *Let $\mathcal{G}$ be a class of connected graphs that is closed under isomorphism and under the taking of minors. Let a 3-connected graph $H \in \mathcal{G}$ have a 3-separation (F_1, F_2) with $|F_1|, |F_2| \geq 4$. Assume that H/F_2 has no loops and $H \backslash F_2$ has no coloops. Furthermore, assume that for every 3-connected 1-edge extension of H in $\mathcal{G}$, say by an edge z, the pair $(F_1, F_2 \cup \{z\})$ is a 3-separation of that extension. Then for any 3-connected graph $G \in \mathcal{G}$ with an H minor, the following holds. Any 3-separation of any such minor that corresponds to (F_1, F_2) of H under one of the isomorphisms induces a 3-separation of G.*

Proof. By the assumptions and Corollary (6.3.25), (F_1, F_2) is a 3-separation of $M(H)$, and that 3-separation induces (in the matroid sense) a 3-separation (E_1, E_2) in $M(G)$. For $i = 1, 2$, $E_i \supseteq F_i$, and thus $|E_i| \geq |F_i| \geq 4$. By Theorem (3.2.25), part (c), (E_1, E_2) is a 3-separation of G as desired. $\square$

We touch upon the complexity of finding a minor that prevents an induced k-separation. We consider this problem in the following setting. We are given a binary matroid M, a minor N of M, and a k-separation of N. We would like to obtain a k-separation of M induced by that of N. If that is not possible, we want to find a minor M represented up to indices by one of the matrices of (6.3.21)–(6.3.23). The next theorem says that this problem can be solved in polynomial time.

(6.3.27) Theorem. *There is a polynomial algorithm for the following problem. The input consists of a binary matroid M, a minor N of M, and a k-separation of N. The output is to be either a k-separation of M induced by that of N, or a minor of M that is isomorphic to one of the matroids represented by the matrices of (6.3.21)–(6.3.23).*

Proof. If an induced k-separation does exist, then one such k-separation is found by the separation algorithm. Suppose there is no such k-separation of M. Then we use a polynomial implementation of the constructive proofs of Lemmas (6.3.15)–(6.3.17) and (6.3.19) to locate a minor of M represented up to indices by one of the matrices of (6.3.21)–(6.3.23). $\square$

Sometimes a class $\mathcal{M}$ of matroids under investigation is only closed under restricted isomorphism and under special minor-taking. We want sufficient conditions under which the above results for induced decompositions remain valid in the new setting. We state one such instance following a definition.

Let L be a set of elements. Assume that two binary matroids contain the set L. We say that the two matroids are *L-isomorphic* if there is an isomorphism that is an identity on the set L. The conditions on $\mathcal{M}$ are as follows. Each matroid of $\mathcal{M}$ contains L. Furthermore, $\mathcal{M}$ is closed under L-isomorphism and under the taking of minors, provided the minors are connected and contain L. Analogous definitions apply to graphs, or to the term "minimal under L-isomorphism."

As before, let N be a binary matroid with a k-separation (F_1, F_2). We also assume that N is in $\mathcal{M}$, and that $L \subseteq F_2$. The next theorem says that Corollaries (6.3.24) and (6.3.25) remain valid under the additional conditions. Under a suitable change to graph terminology, the same conclusion applies to Corollary (6.3.26). Finally, Theorem (6.3.27) remains valid when L-isomorphisms replace isomorphisms.

(6.3.28) Theorem.   *Corollaries (6.3.24) and (6.3.25) remain valid when $\mathcal{M}$ and N satisfy the following two conditions for some set L contained in the set F_2 of N. First, each matroid of $\mathcal{M}$ contains the set L. Second, $\mathcal{M}$ is closed under L-isomorphism and under the taking of minors, provided the minors are connected and contain L. Corollary (6.3.26) remains valid when the above conditions on $\mathcal{M}$ and N are applied to the class $\mathcal{G}$ and to the graph H. Theorem (6.3.27) remains valid when L-isomorphisms are claimed instead of isomorphisms.*

Proof. The cited results rely on Theorem (6.3.20), which is nothing but Theorem (6.3.18) plus Lemma (6.3.19). Now Theorem (6.3.18) is a statement about a minimal M, and thus does not involve any isomorphism. But in the proof of Lemma (6.3.19), the matroid N is replaced by an isomorphic matroid N'. However, N' can be derived from N by a relabeling of some elements of $F_1 = X_1 \cup Y_1$. Thus, the elements of F_2 are not affected, and N' is F_2-isomorphic to N. Since $L \subseteq F_2$, N' is also L-isomorphic to N. We apply these observations to rewrite Theorem (6.3.20) so that it becomes a statement about a matroid minimal under L-isomorphism. Indeed, we only need to change the claims about the matrices of (6.3.21), (6.3.22), and (6.3.23) by allowing for a relabeling of indices other than those of L to get the desired theorem. It is now an easy matter to verify that the theorem so derived from Theorem (6.3.20) implies the claimed results for Corollaries (6.3.24), (6.3.25), and (6.3.26), and Theorem (6.3.27).   □

An example application of Theorem (6.3.20) is covered in the next section. There we prove the existence of certain extensions of 3-connected binary minors in 3-connected binary matroids. In Chapters 10, 11, and 13, we use Corollaries (6.3.24)–(6.3.26) and Theorem (6.3.28).

6.4 Extensions of 3-Connected Minors

An important matroid problem is as follows. We are given a 3-connected binary matroid M with a 3-connected minor N. The minor has at least six elements. We want to obtain a 3-connected minor N' of M that, for some small $k \geq 1$, is a k-element extension of an N minor of M. In this section, we show that an N' with $k = 1$ or 2 can always be found. Our main tool for establishing this result is Theorem (6.3.20) of the preceding section. In Chapter 7, we refine the conclusion proved here to obtain the so-called splitter theorem.

The precise statement of the above claim about N' and k is as follows.

(6.4.1) Theorem. *Let M be a 3-connected binary matroid with a 3-connected proper minor N. Suppose N has at least six elements. Then M has a 3-connected minor N' that is a 1- or 2-element extension of some N minor of M. In the 2-element case, N' is derived from the N minor by one addition and one expansion.*

Proof. Let z be any element of M that is not in N. Lemma (5.2.4) says that the connected M has a connected minor N' that is a 1-element extension of N by z. Now Theorem (6.4.1) holds for M and N if and only if it holds for M^* and N^*. Hence, by duality, we may assume that the extension is an addition. Let N be represented by a matrix $\overline{B}$. Thus, for some vector a, the minor N' of M is represented by the matrix

(6.4.2)

$$\begin{array}{c|c|c} & Y & z \\ \hline X & \overline{B} & a \end{array}$$

Matrix for 1-element extension N' of N

Since N' is connected, the vector a must be nonzero. If a is not a unit vector and is not parallel to a column of $\overline{B}$, then by Lemma (6.2.6), N' is 3-connected and we are done. Otherwise, due to at most one GF(2)-pivot in $\overline{B}$, we may assume a to be a unit vector, say with 1 in row $u \in Y$. Let d be the row vector of $\overline{B}$ indexed by u. Partition $\overline{B}$ into d and $\overline{\overline{B}}$, and also partition $\overline{X}$ into $\{u\}$ and $\overline{\overline{X}} = \overline{X} - \{u\}$. We thus can rewrite $[\overline{B} \mid a]$ of (6.4.2) as

(6.4.3)

$$\begin{array}{c|c|c} & Y & z \\ \hline \overline{\overline{X}} & \overline{\overline{B}} & 0 \\ \hline u & d & 1 \end{array}$$

Partitioned version of matrix of (6.4.2) for N'

The partition in (6.4.3) corresponds to the 2-separation $(\overline{\overline{X}} \cup \overline{Y}, \{u, z\})$ of N'. Since M is 3-connected, that 2-separation of N' does not induce one in M. There must be a minor M' of M that proves this fact and that is minimal under isomorphism. The matroid M' has an N' minor. If necessary, we change the element labels of M' so that N' itself is that N' minor.

We apply Theorem (6.3.20). The just-defined M' plays the role of M of the theorem. The submatrices $\overline{\overline{B}}$, d, and $[1]$ of (6.4.3) correspond to A^1, D, and A^2, respectively, of the theorem. We list enough conditions of parts (a), (b), and (c) of Theorem (6.3.20) to derive the desired conclusion. At the same time, we substitute $\overline{\overline{B}}$, d, and $[1]$ for A^1, D, and A^2. On the other hand, the indices x and y, the subvectors e, f, g, h, and the scalar α employed below should be interpreted exactly as in Theorem (6.3.20).

(a) M' is represented by

(6.4.4)

	Y	z
$\overline{\overline{X}}$	$\overline{\overline{B}}$	0
x	e	f
u	d	1

Case (a) of Theorem (6.3.20)

In row x, the subvector e is not spanned by d, and $f = 1$.

We evaluate this condition. Evidently, e is nonzero and not parallel to d. Indeed, one easily verifies that the matrix of (6.4.4) has no zero vectors, unit vectors, or parallel vectors. By Lemma (6.2.6), the matroid M' is therefore 3-connected, and hence is a 3-connected 2-element extension of N produced by one addition and one expansion.

(b) M' is represented by

(6.4.5)

	Y	y	z
$\overline{\overline{X}}$	$\overline{\overline{B}}$	g	0
u	d	h	1

Case (b) of Theorem (6.3.20)

In column y, the subvector h is not spanned by the columns of d.

This condition is clearly incompatible with the fact that d is a nonzero vector. Thus, this case cannot occur.

(c) M' has a minor $\overline{M}$ with representation matrix

(6.4.6)

	Y	y	z
$\overline{\overline{X}}$	$\overline{\overline{B}}$	g	0
x	e	α	f
u	d	h	1

Case (c) of Theorem (6.3.20)

From the conditions (c.1) and (c.2) of Theorem (6.3.20), we extract the following.

(c.1) The vector e is not spanned by d and is not parallel to a row of $\overline{\overline{B}}$. Furthermore, e is not a unit vector with 1 in a column $t \in \overline{Y}$ for which column t of $\overline{\overline{B}}$ is nonzero.

We evaluate these conditions. Since the matrix $\overline{B} = [\overline{\overline{B}}/d]$ for N is 3-connected, each column of $\overline{\overline{B}}$ is nonzero. Thus, the above conditions on e imply that the matrix composed $\overline{\overline{B}}$, d, and e has no zero vectors, unit vectors, or parallel vectors. By Lemma (6.2.6), that matrix represents a 3-connected 1-element extension of N.

(c.2) The vector g is nonzero. If $\overline{d}$, the subvector obtained from d by deletion of an element $t \in \overline{Y}$, has the same GF(2)-rank as d, then $[g/h]$ is not parallel to the column t of $[\overline{\overline{B}}/d]$. If d does not span a row $t \in \overline{X}$ of $\overline{\overline{B}}$, then $[g/h]$ is not a unit vector with 1 in row t.

We interpret these conditions. By the 3-connectedness of N, the vector d has at least two 1s and does not span any row of $\overline{\overline{B}}$. Thus, the above conclusions about $[g/h]$ hold for all $t \in \overline{Y}$ and all $t \in \overline{X}$. Put differently, $[g/h]$ must be nonzero, cannot be a unit vector, and cannot be parallel to a column of $[\overline{\overline{B}}/d]$. By Lemma (6.2.6), the matrix composed of $\overline{\overline{B}}$, d, g, and h represents a 3-connected 1-element extension of N.   □

The minor N' of Theorem (6.4.1) may be efficiently found. Indeed, one only needs to implement the preceding constructive proof. The precise complexity claim is as follows.

(6.4.7) Theorem.   *There is a polynomial algorithm for the following problem. The input is a connected binary matroid M, and a 3-connected proper minor N of M on at least six elements. The output is either a 2-separation of M, or a minor N' of M that is a 3-connected 1- or 2-element extension of an N minor of M. In the 2-element case, N' is derived from the N minor by one addition and one expansion.*

Proof. We implement the proof of Theorem (6.4.1), using as subroutine the polynomial algorithm of Theorem (6.3.27) to produce either a

2-separation of M or one of the matrices (6.3.21)–(6.3.23). It is easy to see that a polynomial algorithm can thus be assembled for the stated problem.   □

In the next section, we discuss extensions of the results of this chapter and include references.

6.5 Extensions and References

The results of Section 6.3 overlap significantly with results of Seymour (1980b) for general matroids, even though the terminology is quite different. The precise relationships are as follows. Lemma (6.3.6) is the binary version of a result taken from Seymour (1980b). That reference proceeds to describe a number of properties of matroids called minimal here. These results are then used to deduce a version of Corollary (6.3.24). In contrast, the approach taken here is based on properties that can be efficiently verified for matroids that are minimal or minimal under isomorphism. Such properties are investigated via abstract matrices for general matroids in Truemper (1986). The sufficiency conditions and testing algorithms so obtained are substantially stronger than those relying on Corollary (6.3.24) or on the even weaker Corollary (6.3.25). The simpler results given here suffice for the proofs in the chapters to come. Truemper (1986) contains additional material about induced decompositions. For example, it is shown that the number of non-isomorphic minimal matroids is finite for a given N, provided all matroids under consideration are representable over a given finite field.

Truemper (1988) treats in detail the case of graphs, which we have skipped here entirely except for the specialized Corollary (6.3.26).

Theorem (6.4.1) may be viewed as a weak version of the splitter theorem of Seymour (1980b). We use Theorem (6.4.1) in the next chapter to prove that result. Upon slight modification, the approach of Section 6.4 yields other important theorems about 3-connected extensions of 3-connected matroids. We sketch the main ideas and provide related references in Section 7.5 of the next chapter.

Chapter 7

Splitter Theorem and Sequences of Nested Minors

7.1 Overview

Chapters 4, 5, and 6 cover three basic matroid tools: the series-parallel and delta-wye constructions, the path shortening technique, and the separation algorithm. The chapters also include a number of basic matroid results whose proofs rely on these tools. With this foundation, we derive in this chapter and the next one several fundamental results about the decomposition and composition of matroids. Specifically in this chapter, we define matroid splitters, characterize them, and deduce consequences of that characterization.

The concept of splitters and their characterization is due to Seymour. The idea can be summarized as follows. Let $\mathcal{M}$ be a class of binary matroids that is closed under isomorphism and under the taking of minors. Then a 3-connected matroid $N \in \mathcal{M}$ on at least six elements is declared to be a *splitter* of $\mathcal{M}$ if every matroid $M \in \mathcal{M}$ with a proper N minor has a 2-separation. Some researchers define graph or matroid 2-separations to be splits. The term "splitter" is in agreement with that notion.

The concept of splitters may seem rather abstract. But in subsequent chapters, we rely on it a number of times, and without doubt it is one of the central ideas for the decomposition of matroids. We characterize splitters in Section 7.2 in the so-called splitter theorem.

Define two minors of a graph or matroid to be *nested* if one of them is a minor of the other one. In Section 7.3, we derive from the splitter theorem several existence theorems about sequences of nested minors, among them

151

Tutte's wheel theorem for graphs. A special case of nested minor sequences is used in Section 7.4 to prove Kuratowski's characterization of planar graphs. According to that result, a graph is planar if and only if it does not have $K_{3,3}$ or K_5 minors. In the final section, 7.5, we point out a number of extensions and list references.

The chapter requires familiarity with the material of Chapters 2, 3, 5, and 6.

7.2 Splitter Theorem

Let $\mathcal{M}$ be a class of binary matroids that is closed under isomorphism and under the taking of minors. Recall that a *splitter* of $\mathcal{M}$ is a 3-connected matroid $N \in \mathcal{M}$ such that every matroid $M \in \mathcal{M}$ with a proper N minor is 2-separable. We employ the same terminology for graphs. For example, if $\mathcal{G}$ is a class of graphs that is closed under isomorphism and under the taking of minors, then a *splitter* of $\mathcal{G}$ is a 3-connected graph $G \in \mathcal{G}$ such that every graph of $\mathcal{G}$ with a proper G minor is 2-separable.

In this section, we derive surprisingly simple necessary and sufficient conditions for a given $N \in \mathcal{M}$ to be a splitter. The next theorem stating these conditions is the splitter theorem due to Seymour.

(7.2.1) Theorem (Splitter Theorem). *Let $\mathcal{M}$ be a class of binary matroids that is closed under isomorphism and under the taking of minors. Let N be a 3-connected matroid of $\mathcal{M}$ on at least six elements.*

(a) *If N is not a wheel, then N is splitter of $\mathcal{M}$ if and only if $\mathcal{M}$ does not contain a 3-connected 1-element extension of N.*

(b) *If N is a wheel, then N is a splitter of $\mathcal{M}$ if and only if $\mathcal{M}$ does not contain a 3-connected 1-element extension of N and does not contain the next larger wheel.*

Proof. If N is a splitter of $\mathcal{M}$, then the 3-connected extensions cited in (a) or (b) obviously cannot occur in $\mathcal{M}$. We prove the converse by contradiction. Thus, we suppose that $\mathcal{M}$ does not contain the 3-connected extensions cited in (a) or (b), whichever applies, and that nevertheless N is not a splitter of $\mathcal{M}$. Thus, $\mathcal{M}$ contains a 3-connected matroid M with a proper N minor, and M is not one of the cases excluded under (a) or (b). Since $\mathcal{M}$ is closed under isomorphism, we may assume N itself to be that N minor. To M and N we apply Theorem (6.4.1). According to that theorem, M has a 3-connected minor N' that is a 3-connected 1- or 2-element extension of an N minor. In the 2-element extension case, N' is derived from the N minor by one addition and one expansion. Again, since $\mathcal{M}$ is closed under isomorphism and minor taking, we may take N itself to be that N minor. The 1-element extension case has been ruled out by (a) and (b). Thus, N' is derived from N by one addition and one expansion. Suppose a binary matrix $\overline{B}$ with row index set $\overline{X}$ and column index set $\overline{Y}$ represents N. Then N' can be represented by a binary matrix C that displays $\overline{B}$, and thus N, as follows.

(7.2.2)

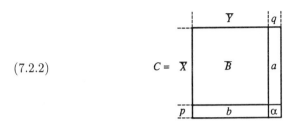

Matrix C representing N'

We now show either that N' contains a 3-connected 1-element extension of an N minor, a case ruled out by both (a) and (b), or that N is a wheel and N' is the next larger wheel, a case ruled out by (b). We accomplish this by the following investigation into the structure of C of (7.2.2).

Since N' is 3-connected, the matrix C does not contain zero vectors, unit vectors, or parallel vectors. In particular, the subvectors a and b of C must be nonzero. Furthermore, the submatrices $[\overline{B} \mid a]$ and $[\overline{B}/b]$ of C, which represent 1-element extensions of N, cannot be 3-connected since otherwise we have an eliminated case of (a) and (b). Thus, the subvector a (resp. b) is a unit vector or is parallel to a column (resp. row) of $\overline{B}$.

Because of pivots in $\overline{B}$ and row exchanges in C, we may assume that a is a unit vector with 1 in the topmost position, and that b is parallel to a row of $\overline{B}$. Since C is 3-connected, we necessarily have $\alpha = 1$. We then may partition C as follows.

(7.2.3)

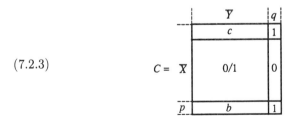

Initial partition of C with $a = $ unit vector

In the subsequent processing of C of (7.2.3), we introduce a number of row and column exchanges and pivots that affect the index sets substantially. Since $\mathcal{M}$ is closed under isomorphism, we do not have to keep track of such index changes. So, instead, we just make sure that the matrix obtained from the current C by deletion of the rightmost column and bottom row does represent N up to a relabeling of the elements. We always refer to that matrix as the current $\overline{B}$. With this convention, we can freely introduce new indices or reuse old ones.

Suppose during the subsequent processing of C of (7.2.3), we detect that the current C contains the current $\overline{B}$ plus a nonzero row or column that is not parallel to a row or column of $\overline{B}$ and that is not a unit vector. Then by Lemma (6.2.6), $\overline{B}$ plus that row or column up to indices represents a 3-connected 1-element extension of N, which has been ruled out. Thus, we assume below that this case does not occur. The proof of the theorem is complete once we show N to be a wheel and N' to be the next larger wheel. We are now ready to process C of (7.2.3).

We know that the vector b of C of (7.2.3) is parallel to a row of $\overline{B}$. That vector of $\overline{B}$ cannot be c, for otherwise, C is 2-separable. So assume b is parallel to the second row of $\overline{B}$, say row v. An exchange of rows v and p of C produces

(7.2.4)

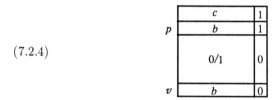

Matrix C after exchange of rows p and v

Except for the replacement of the row index p by v, that row exchange does not affect the submatrix $\overline{B}$. The column vector to the right of the current $\overline{B}$ must be parallel to, say, the first column of $\overline{B}$. Exchange the first column and the last column of C. By the 3-connectedness of N', we must have

(7.2.5)

1	$\overline{c}$	1
①	$\overline{b}$	1
0	0/1	0
0	$\overline{b}$	1

Matrix C after exchange of first and last column

Pivot on the circled 1 of (7.2.5). That pivot produces the matrix of (7.2.6) below.

Inductively, assume that the current C is given by (7.2.7) below. Suppose that the subvector b of row x is a unit vector, say with 1 in column $z \in Y_2$, and that column z of the submatrix $\overline{\overline{B}}$ is zero. Then we have $(X_1 \cup Y_1 \cup \{x, y, z\}, (X_2 \cup Y_2) - \{z\})$ as a 2-separation of C unless $|(X_2 \cup Y_2) - \{z\}| \le 1$. If $|(X_2 \cup Y_2) - \{z\}| = 1$, then C contains a zero row indexed by X_2, or C has a zero or unit vector column indexed by $Y_2 - \{z\}$. Either case is a contradiction of the 3-connectedness of N'.

(7.2.6)

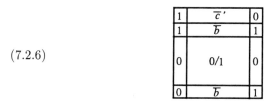

Matrix C after pivot

(7.2.7)

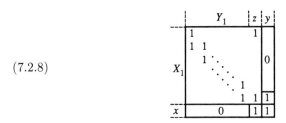

Matrix C for inductive proof

Thus, $|(X_2 \cup Y_2) - \{z\}| = 0$, i.e., $X_2 = \emptyset$ and $Y_2 = \{z\}$, which implies $b = [1]$. Since the columns z and y of C must be distinct, we also have $c = [1]$. Then C is

(7.2.8)

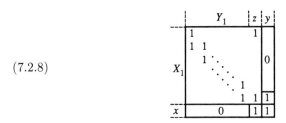

Matrix C displaying wheel case

Evidently, the current $\overline{B}$ is a matrix of type (5.2.9). Accordingly, N is wheel. A pivot in C on the 1 in the bottom right corner confirms that N' is a wheel as well.

Once more, assume that in the matrix C of (7.2.7), the row subvector b is a unit vector with 1 in column z. But this time, suppose that column

z of $\overline{\overline{B}}$ is nonzero. By a pivot in column z of $\overline{\overline{B}}$, we convert the situation to the third possible case, where b is not a unit vector. In that third case, the vector b in row x of C must be parallel to a row of $\overline{\overline{B}}$, say row p. Exchange rows p and x of C. We get

(7.2.9)

$$
\begin{array}{c|ccc|c}
 & & & & y \\
\hline
1 & 1 & & & c & 0 \\
 & & 1 \cdot \cdot \cdot & & 0 & 0 \\
 & & & \cdot \cdot 1 & b & 1 \\
\hline
x & & 0 & & b & 1 \\
 & & & & 0/1 & 0 \\
\hline
p & & 0 & & b & 0 \\
\end{array}
$$

Matrix C after exchange of rows p and x

The remaining arguments are analogous to those for (7.2.4)–(7.2.6). They produce an instance of (7.2.7) where $|Y_1|$ has been increased by 1. By induction, the case already discussed must eventually be encountered where N is a wheel and N' is the next larger wheel.   □

When specialized to graphs, the splitter Theorem (7.2.1) becomes the following result.

(7.2.10) Corollary. *Let $\mathcal{G}$ be a class of connected graphs that is closed under isomorphism and under the taking of minors. Let H be a 3-connected graph of $\mathcal{G}$ with at least six edges.*

(a) *If H is not a wheel, then H is a splitter of $\mathcal{G}$ if and only if $\mathcal{G}$ does not contain any graph derived from H by one of the following two extension steps:*

 (1) *Connect two nonadjacent nodes by a new edge.*

 (2) *Partition a vertex of degree at least 4 into two vertices, each of degree at least 2, then connect these two vertices by a new edge.*

(b) *If H is a wheel, then H is a splitter of $\mathcal{G}$ if and only if $\mathcal{G}$ does not contain any of the extensions of H described under (a) and does not contain the next larger wheel.*

Proof. Let $\mathcal{M}$ be the collection of graphic matroids produced by the graphs of $\mathcal{G}$. Define N to be the graphic matroid of the graph H. Lemma (6.2.7) says that the extensions of H described under (a) are precisely the 3-connected 1-edge extensions of H. Thus, these extensions correspond to the 3-connected 1-element graphic extensions of N. The result then follows from the splitter Theorem (7.2.1).   □

Typically, we will specify $\mathcal{M}$ or $\mathcal{G}$ by exclusion of certain minors and of all their isomorphic copies. Clearly, any collection of matroids or graphs so specified is closed under isomorphism and under the taking of minors.

Two graph examples of splitters are given in the next theorem. Recall that W_n is the wheel graph with n spokes.

(7.2.11) Theorem. W_3 *is a splitter of the graphs without W_4 minors, and K_5 is a splitter of the graphs without $K_{3,3}$ minors.*

Proof. There is no 3-connected 1-edge extension of W_3. Thus, by part (b) of Corollary (7.2.10), W_3 is a splitter of the graphs without W_4 minors. For the second part, we note that up to isomorphism there is just one 3-connected 1-edge extension of K_5. To obtain it, one partitions one vertex of K_5 into two vertices of degree 2 and connects the two vertices by a new edge. The resulting graph is readily seen to have a $K_{3,3}$ minor. Thus, by part (a) of Corollary (7.2.10), K_5 is a splitter of the graphs without $K_{3,3}$ minors. $\square$

We will see a number of other splitter examples in subsequent chapters. In the next section, we deduce from the splitter Theorem (7.2.1) certain sequences of nested minors and Tutte's wheel theorem.

7.3 Sequences of Nested Minors and Wheel Theorem

Recall from Section 7.1 that two matroids are *nested* if one of them is a minor of the other one. In this section, we prove the existence of certain sequences of nested minors of binary matroids. As a special case, we establish Tutte's wheel theorem. Main tools are the splitter Theorem (7.2.1) and results proved in Chapter 5 with the path shortening technique.

For a given binary matroid, an arbitrary sequence of nested minors is easy to find. The task becomes difficult and interesting when one imposes conditions on the sequence. We need a few definitions to express such conditions.

Suppose two matroids are nested. Define the *rank gap* between the two matroids to be the absolute difference in rank between them. Analogously, define the *corank gap*. Finally, let the *gap* be the sum of the rank gap and the corank gap. Evidently, the gap is the number of elements that occur only in the larger matroid.

Consider a sequence of nested minors of a given binary matroid. Then the *rank gap* of the sequence is the maximum rank gap among the pairs of successive minors of the sequence. Analogously, define the *corank gap* and the *gap* of the sequence.

We now state the conditions under which we want to find nested minor sequences. Suppose for some $k \geq 2$, we have a k-connected binary matroid M with a k-connected minor N. In the typical situation, we want to find a sequence of nested k-connected minors $M_0, M_1, M_2, \ldots, M_t = M$, where M_0 is demanded to be isomorphic to N. Furthermore, the rank gap, or the corank gap, or the gap is to be bounded by some given constant. Other variants are possible. For example, one may require that a given element of M that does not occur in M_0 be present in M_1, that a given element not occurring in M_1 be present in M_2, and so on.

Sequences of the desired kind are readily determined when $k = 2$. The main ingredient for their construction is a recursive application of Lemma (5.2.4). We skip the details of that simple case, and instead turn immediately to the much more complicated case with $k = 3$. Specifically, we prove the existence of three types of sequences for $k = 3$, then point out extensions in Section 7.5.

In each of the cases treated here, we are given a 3-connected binary matroid M with a 3-connected proper minor N on at least six elements. In the first case, we desire a sequence of nested 3-connected minors M_0, $M_1, \ldots, M_t = M$, where M_0 is isomorphic to N and where the gap is small. The next theorem shows that the gap can be held to 1 or 2.

(7.3.1) Theorem.   *Let M be a 3-connected binary matroid having a 3-connected proper minor N on at least six elements.*

(a) *Assume N is not a wheel. Then for some $t \geq 1$, there is a sequence of nested 3-connected minors $M_0, M_1, \ldots, M_t = M$, where M_0 is isomorphic to N and where the gap is 1.*
(b) *Assume N is a wheel. Then for some $t \geq 1$, there is sequence of nested 3-connected minors $M_0, M_1, \ldots, M_t = M$ with the following features. M_0 is isomorphic to N. For some $0 \leq s \leq t$, the subsequence M_0, $M_1, \ldots, M_s$ consists of wheels and has gap 2, and the subsequence $M_s, M_{s+1}, \ldots, M_t = M$ has gap 1.*

Proof. We first establish part (a). Thus, we assume that N is not a wheel. Indeed, inductively we assume for some $i \geq 0$, the existence of a sequence of nested 3-connected minors $M_0, M_1, \ldots, M_i$ of M, where M_0 is isomorphic to N, where M_i is not a wheel, and where the gap is 1. If $M_i = M$, we are done. So assume that M_i is a proper minor of M.

We rely on the contrapositive statement of part (a) of the splitter Theorem (7.2.1) to find a larger sequence. To this end, we define $\mathcal{M}$ to be the matroid collection containing M, all minors of M, and all matroids isomorphic to these matroids. By this definition, $\mathcal{M}$ is closed under isomorphism and under the taking of minors. Since M_i is a 3-connected proper minor of the 3-connected $M \in \mathcal{M}$, it cannot be a splitter of $\mathcal{M}$. Thus, by part (a) of Theorem (7.2.1), $\mathcal{M}$ contains a matroid M_{i+1} that is a 3-connected 1-element extension of a matroid isomorphic to M_i. Now every 1-element reduction of a wheel with at least six elements is 2-separable. Thus, if M_{i+1} is a wheel, then M_i is 2-separable, a contradiction. We conclude that M_{i+1} is not a wheel.

If necessary, we relabel M_0, $M_1, \ldots$, M_i so that they constitute a sequence of nested minors of M_{i+1}. These matroids plus M_{i+1} satisfy the induction hypothesis for $i + 1$. By induction, the claimed sequence exists for M.

The proof of part (b) is essentially the same, except that we establish M_{i+1} using part (b) of Theorem (7.2.1) when M_i is a wheel. $\Box$

We have used the splitter Theorem (7.2.1) for a simple proof of Theorem (7.3.1). Indeed, the two theorems are essentially equivalent, since one may deduce the splitter Theorem (7.2.1) from Theorem (7.3.1) just as easily. We sketch the proof.

Let $\mathcal{M}$ and N be as specified in the splitter Theorem (7.2.1). Suppose N is not a wheel. We must show that N is a splitter of $\mathcal{M}$ if and only if $\mathcal{M}$ does not contain any 3-connected 1-element extension of N. We prove the nontrivial "if" part by contradiction. So let M be a 3-connected matroid of $\mathcal{M}$ with N as proper minor. By Theorem (7.3.1), there is a sequence of nested 3-connected minors M_0, $M_1, \ldots$, $M_t = M$, where M_0 is isomorphic to N, and where the gap is 1. Since $\mathcal{M}$ is closed under isomorphism, we may assume M to be so chosen that M_0 is equal to N. Then M_1 is a 3-connected 1-element extension of N and $M_1 \in \mathcal{M}$, which contradicts the assumed absence of such extensions. The case where N is a wheel is treated analogously.

A direct translation of Theorem (7.3.1) into graph language results in the following corollary.

(7.3.2) Corollary.   *Let G be a 3-connected graph having a 3-connected proper minor H with at least six edges.*

(a) *Assume H is not a wheel. Then for some $t \geq 1$, there is a sequence of nested 3-connected minors G_0, $G_1, \ldots$, $G_t = G$, where G_0 is isomorphic to H, and where each G_{i+1} has exactly one edge beyond those of G_i.*

(b) *Assume H is a wheel. Then for some $t \geq 1$, there is a sequence of nested 3-connected minors G_0, $G_1, \ldots$, $G_t = G$ with the following features. G_0 is isomorphic to H. For some $0 \leq s \leq t$, the subsequence G_0, $G_1, \ldots$, G_s consists of wheels, where each G_{i+1} has exactly one additional spoke beyond those of G_i. Furthermore, in the subsequence G_s, $G_{s+1}, \ldots$, $G_t = G$, each G_{i+1} has exactly one edge beyond those of G_i.*

One may combine Corollary (7.3.2) with Corollary (5.2.15) to obtain Tutte's wheel theorem, which is listed next.

(7.3.3) Theorem (Wheel Theorem).   *Let G be a 3-connected graph on at least six edges. If G is not a wheel, then G has some edge z such that at least one of the minors G/z and $G \backslash z$ is 3-connected.*

Proof. Corollary (5.2.15) says that a 3-connected graph with at least six edges, in particular G specified here, has a W_3 minor. Thus, G has a largest wheel minor, say H. Since G is not a wheel, H is a proper minor of G. We apply Corollary (7.3.2) to G and H. Accordingly, G has a sequence of nested 3-connected minors $G_0, G_1, \ldots, G_t = G$, where G_0 is isomorphic to H. Since H is a largest wheel minor of G and since G is not a wheel, the index s of part (b) of Corollary (7.3.2) must be zero, and $t \geq 1$. We also conclude from that part that $G = G_t$ has exactly one edge beyond those of G_{t-1}. Put differently, the 3-connected minor G_{t-1} is for some edge z equal to G/z or $G \backslash z$, which proves the theorem. □

Theorem (7.3.3) can obviously be rewritten so that it becomes a wheel theorem for binary matroids instead of graphs. The proof relies on Theorem (7.3.1) instead of Corollary (7.3.2).

From the sequence of nested 3-connected minors $M_0, M_1, \ldots, M_t = M$ of Theorem (7.3.1), one can derive a number of other interesting sequences. A particular construction utilizes the representation matrices of these minors. We present details following some observations.

By Lemma (3.3.12), a binary matroid M with a given minor $\overline{M}$ has a representation matrix that displays $\overline{M}$. We apply this result inductively to the sequence of nested minors $M_0, M_1, \ldots, M_t = M$, and conclude that M has a representation matrix B that simultaneously displays M_0, $M_1, \ldots, M_t = M$, say by nested matrices $B^0, B^1, \ldots, B^t = B$. Let C^i be the column submatrix of B that has the same column index set as B^i. Let b^i be a row vector of C^i, say with row index x. Assume that b^i is not a row of B^i. Indeed, assume that b^i is nonzero, is not a unit vector, and is not parallel to a row of B^i. Lemma (6.2.6) shows that under these assumptions, B^i plus b^i represent a 3-connected minor $M_i\&x$ of M. There are other minors M_j without x for which $M_j\&x$ is 3-connected. Specifically, let k be the largest index, $i \leq k \leq t$, such that M_j does not contain x. We claim that for each $i < j \leq k$, $M_j\&x$ is 3-connected. The proof consists of the following observation. Let b^j be the row vector of C^j indexed by x. Since $j \leq k$, b^j is not part of B^j. Evidently, B^i is a submatrix of B^j, and b^i is a subvector of b^j. We know that b^i is nonzero, is not a unit vector, and is not parallel to a row of B^i. Then the latter statement must also hold when we use j as superscript instead of i. Accordingly, $M_j\&x$ is 3-connected.

We use these observations as follows. In the sequence $M_0, M_1, \ldots, M_t = M$, we redefine each M_j, $i \leq j \leq k$, to be $M_j\&x$. Correspondingly, we redefine the B^j matrices by adjoining the b^j row vectors. The result is a new sequence of nested 3-connected minors. The rank gap of that sequence is larger than that for the original one. But the corank gap is still the same. Indeed, since the gap of the original sequence was 1, the corank gap of the new sequence is 0 or 1.

We repeat the above process, using all possible indices i and x, until no further changes are possible. This occurs when each final B^i cannot be extended to a larger 3-connected matrix within C^i. The final sequence may contain duplicate minors. In that case, we delete just enough minors to eliminate all such duplicates. Then we redefine the indices so that M_0, $M_1, \ldots, M_t = M$ is now the sequence resulting from the above process. Correspondingly, we redefine the indices of the B^i and C^i matrices.

If M_0 and M have same corank, then $t = 0$, and the outcome is $M_0 = M_t = M$, an uninteresting case. If M_0 and M have different corank, we must have $t \geq 1$. Consider the latter case. Evidently, each B^{i+1} is deduced from B^i by first adjoining any number of row vectors b^i to B^i, and then adjoining a column vector a. We claim that each one of these vectors b^i is a unit vector or is parallel to a row of B^i. Indeed, in any other case of a nonzero b^i, the previously described process would have adjoined b^i to B^i and would have redefined M_i prior to termination, a contradiction. In the case of a zero vector b^i, B^{i+1} would have a zero or unit vector, and thus would be 2-separable, again a contradiction.

In terms of the minors of the final sequence, we thus have proved that each M_{i+1} may be derived from M_i by extensions involving any number of series elements, possibly none, followed by a 1-element addition.

The above construction is the main ingredient in the proof of the following variant of Theorem (7.3.1).

(7.3.4) Theorem.   *Let M be a 3-connected binary matroid with a 3-connected proper minor N on at least six elements. If M does not contain a 3-connected 1-element expansion (resp. addition) of any N minor, then M has a sequence of nested 3-connected minors $M_0, M_1, \ldots, M_t = M$, where M_0 is an N minor of M, and where each M_{i+1} is obtained from M_i by expansions (resp. additions) involving some series (resp. parallel) elements, possibly none, followed by a 1-element addition (resp. expansion).*

Proof. Clearly, the parenthetic case is dual to the stated one. Thus, we only consider the case where M does not contain any 3-connected 1-element expansion of any N minor. We prove the existence of the claimed sequence of nested minors as follows.

Apply the above construction process to the sequence of nested 3-connected minors $M_0, M_1, \ldots, M_t = M$ of Theorem (7.3.1). The result is a new sequence $M_0, M_1, \ldots, M_t = M$ where each M_{i+1} may be derived from M_i by expansions involving any number of series elements, possibly none, followed by a 1-element addition. Since M does not contain a 3-connected 1-element expansion of any N minor, the construction must have left M_0 unchanged. The sequence so produced is the desired one.   □

Theorem (7.3.4) may be specialized to graphs as follows.

(7.3.5) Corollary.   *Let G be a 3-connected graph with a 3-connected minor H on at least six edges.*

(a) *Suppose no H minor of G can be extended to another 3-connected minor of G by the following process: Some node of H of degree at least 4 is partitioned into two nodes, each with degree at least 2, and then the two nodes are connected by a new edge. Then G has a sequence of nested 3-connected minors G_0, $G_1, \ldots, G_t = G$ with the following properties. G_0 is an H minor of G. Each G_{i+1} may be obtained from G_i as follows. First, at most two edges of G_i are replaced by paths of length 2. Next, an edge is added so that the new graph has no degree 2 nodes and no parallel edges.*

(b) *Suppose no H minor of G can be extended to another 3-connected minor of G by connecting two nonadjacent nodes by a new edge. Then G has a sequence of nested 3-connected minors G_0, $G_1, \ldots, G_t = G$ with the following properties. G_0 is an H minor of G. Each G_{i+1} may be obtained from G_i as follows. First, some edges, possibly none, incident at some vertex are replaced by two parallel edges each. The new vertex must have degree of at least 4. Next, that vertex is partitioned into two vertices, each with degree at least 2, such that no two edges remain parallel. Finally, the two vertices just created are joined by a new edge.*

Proof. Part (a) is a routine translation of the non-parenthetic part of Theorem (7.3.4) into graph language, in the same sense that Corollary (7.3.2) is the graph version of Theorem (7.3.1). Part (b) is a translation of the parenthetic part of Theorem (7.3.4).   □

We conclude this section by proving that the sequences of nested 3-connected minors of the above theorems and corollaries can be readily found. Additional material about nested sequences is contained in Section 7.5.

(7.3.6) Theorem.   *There is a polynomial algorithm for the following problem. The input consists of M and N of Theorem (7.3.1) or (7.3.4), or of G and H of Corollary (7.3.2) or (7.3.5). The output is a 2-separation of M or G, or a sequence of nested 3-connected minors with properties as specified in the respective theorem or corollary.*

Proof. The arguments that prove the cited theorems and corollaries can be summarized as follows. Theorem (6.4.1) implies the splitter Theorem (7.2.1), which in turn implies Theorem (7.3.1) and Corollary (7.3.2). The latter results imply Theorem (7.3.4) and Corollary (7.3.5). The polynomial algorithm claimed for the stated problems is essentially an efficient implementation of the constructive proofs of these implications. We first sketch the details for the matroid case with given M and N. With the polynomial algorithm of Theorem (6.4.7), we locate a 2-separation of M or produce the 3-connected 1- or 2-element extensions claimed by Theorem (6.4.1). We use these extensions plus the proof procedure of the splitter Theorem

(7.2.1) to derive the sequence of nested minors of Theorem (7.3.1). From that sequence we construct the sequence of Theorem (7.3.4).

The graph case may be handled by appropriate translation of each step of the above algorithm into graph language. Alternately, one could represent the given G and H by graphic matroids M and N, apply the polynomial algorithm already described, and extract from G the minors $G_0, G_1, \ldots, G_t = G$ corresponding to the sequence $M_0, M_1, \ldots, M_t = M$ found by the algorithm.   □

The remainder of this book includes several demonstrations of the power and utility of sequences of nested 3-connected minors. The discussion in the next section may be viewed to be one such demonstration.

7.4 Characterization of Planar Graphs

In this section, we prove Kuratowski's characterization of graph planarity in terms of excluded $K_{3,3}$ and K_5 minors. We state that theorem and describe a proof of beautiful simplicity due to Thomassen.

(7.4.1) Theorem.   *A graph is planar if and only if it has no $K_{3,3}$ or K_5 minors.*

Proof. The "only if" part follows from the fact that planarity is maintained under the taking of minors, and that by Lemma (3.2.48) both $K_{3,3}$ and K_5 are not planar. For a proof of the nontrivial "if" part, let G be a connected nonplanar graph each of whose proper minors is planar. We need to show that G is isomorphic to $K_{3,3}$ or K_5.

We first prove that G cannot be 1- or 2-separable. The case of 1-separability is trivial. Suppose G is 2-separable. Let G be produced by identifying nodes k and l of a graph G_1 with nodes m and n, respectively, of a graph G_2. Now G_1 plus an edge connecting nodes k and l is planar, since that graph, say G_1', is isomorphic to a proper minor of G. The graph G_2' similarly defined from G_2, is also planar. It is easy to combine planar drawings of G_1' and G_2' to one of G.

Thus, G is 3-connected. According to Lemma (5.2.15), G must have a W_3 minor, say H. Since W_3 is also the complete graph K_4, no H minor of G can be extended to another minor of G by addition of an edge that connects two nonadjacent nodes. Under the latter condition, part (b) of Corollary (7.3.5) states that G has a sequence of nested 3-connected minors $G_0, G_1, \ldots, G_t = G$ with the following properties. G_0 is an H minor of G. Each G_{i+1} may be obtained from G_i as follows. First, some edges, possibly none, incident at some vertex are replaced by two parallel edges each. The new vertex must have degree of at least 4. Next, that vertex is partitioned into two vertices, each with degree at least 2, such that no two edges remain parallel. Finally, the two vertices just created are joined by a new edge.

By the minimality of G, the graph G_{t-1} of the sequence is planar, while $G = G_t$ is not. Consider a realization of G_{t-1} in the plane. Note that we do not rely on the fact, not proved here, that the drawing of G_{t-1} is essentially unique. As just described, G may be derived from G_{t-1} by the replacement of some edges incident at a vertex by some parallel edges, followed by a partitioning of that new vertex, etc. Define G'_{t-1} to be the graph on hand when the first step has been carried out — that is, when in G_{t-1} some edges incident at a vertex have been replaced by two parallel edges each. Let v be the vertex of G'_{t-1} to be partitioned, say into vertices v_1 and v_2. By the connectivity conditions, a partial drawing of G'_{t-1} that emphasizes vertex v is either

(7.4.2)

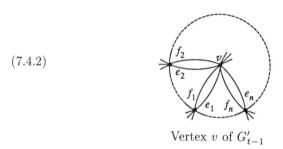

Vertex v of G'_{t-1}

or is deduced from that drawing by deletion of any number of the parallel edges $f_1, f_2, \ldots, f_n$. The dashed segments represent internally node-disjoint paths. The vertex v is so partitioned into v_1 and v_2 that the resulting graph G is nonplanar. This implies that we cannot replace v of the assumed drawing for G'_{t-1} by v_1 and v_2, and then join these two nodes by an edge, while retaining a planar drawing. There are two possible causes for the nonplanarity. First, some of the edges $e_1, e_2, \ldots, e_n$ may cross when they are attached to v_1 and v_2. In fact, no matter which particular situation occurs, just four edges can always be selected to produce the graph of (7.4.3) below.

(7.4.3)

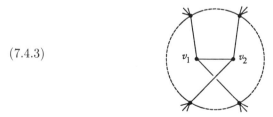

Partition of vertex v producing $K_{3,3}$ minor

The graph contains a subdivision of $K_{3,3}$, so we are done. In the second case, the e_i edges can be attached to v_1 and v_2 so that they do not cross. However, crossing edges are encountered when we attach the f_i edges. We may suppose this to be so even when we relabel edges such that some e_i become f_i, since otherwise we can produce the earlier situation.

This leaves just one case, where $n = 3$, and where for $i = 1, 2, 3$, both e_i and f_i are present. We then have

(7.4.4)

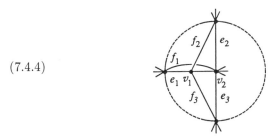

Partition of vertex v producing K_5 minor

which contains a subdivision of K_5.   □

Actually, Kuratowski proved for a nonplanar graph G the presence of a subgraph that is a subdivision of $K_{3,3}$ or K_5 and not just a $K_{3,3}$ or K_5 minor. Now a $K_{3,3}$ minor induces a subdivision of $K_{3,3}$. For a proof, carry out the expansion steps that convert the $K_{3,3}$ minor to a subgraph of G. The same argument applies to a K_5 minor, except when an expansion step splits a vertex of degree 4 into two vertices each of which has degree 3 upon insertion of the new edge. The graph so produced has a $K_{3,3}$ minor. Thus, Theorem (7.4.1) is equivalent to Kuratowski's original formulation.

In the final section, we cover extensions and references.

7.5 Extensions and References

The splitter Theorem (7.2.1) and its extension to nonbinary matroids is due to Seymour (1980b); see also Tan (1981) and Truemper (1984). The extension to the nonbinary case requires the introduction of the whirls given by (5.4.1) and (5.4.2). The latter matroids constitute a second special case besides that of wheels. The graph version of Theorem (7.3.1), and thus effectively Corollary (7.2.10), were independently proved in Negami (1982). An early splitter example is implicit in Wagner (1937a). A number of splitters are given in Seymour (1980b), Oxley (1987b), (1987c), (1989a), (1989b), (1990a), and Truemper (1988).

Tutte's wheel Theorem (7.3.3) appeared in Tutte (1966a). The extension of that result to general matroids is also due to Tutte, and is known as the wheels and whirls theorem (Tutte (1966b)). Strengthened versions are in Halin (1969), Oxley (1981b), and Coullard and Oxley (1991).

Theorem (7.3.4) is a binary and slightly weaker version of results in Seymour (1980b) and Truemper (1984). The related graph result appeared for the first time in Barnette and Grünbaum (1969). That graph result has been repeatedly rediscovered.

Kuratowski's original characterization of planar graphs in terms of forbidden subgraphs appeared in Kuratowski (1930). The equivalence of that result to the excluded minor version given by Theorem (7.4.1) is due to Wagner (1937b). The amazingly short proof of Theorem (7.4.1) is from Thomassen (1980). The matroid version of Theorem (7.4.1) is proved in Bixby (1977).

Substantially stronger results about sequences of nested 3-connected minors exist. Bixby and Coullard (1987) contains the most recent and strongest one: Let N be a 3-connected proper minor of a 3-connected matroid M. Then for any element z of M that is not in N, there is a 3-connected minor N' of M that contains z, that has N as a minor, and that has at most four elements beyond those of N. Note that isomorphisms are not involved in this result. When the reference to the element z is dropped, then N' can be guaranteed to have at most three elements beyond those of N instead of four (Truemper (1984)). The latter result can be strengthened when N has no triangles and no triads (= dual triangles). In that situation, N' need to have at most two elements beyond those of N (Bixby and Coullard (1984)). The sequences of nested 3-connected minors that are implied by these theorems may be modified by the construction given in Section 7.3 prior to Theorem (7.3.4). Some examples are worked out in Truemper (1984). For a recursive characterization of 3-connectivity using so-called separating cocircuits, see Bixby and Cunningham (1979).

The original proofs of the results cited in the preceding paragraph are by no means simple. But there is a unified way in which they can be obtained. We sketch the main idea. Recall that the splitter Theorem (7.2.1) is based on the notion of induced 2-separations and on Theorem (6.3.20), which deals with minors called minimal under isomorphism in Section 6.3. Similarly, the theorems cited in the preceding paragraph may be derived using the notion of induced 2-separations plus the following result of Truemper (1986) about minors called minimal in Section 6.3: Suppose that a 2-separation of a matroid N on at least four elements does not induce a 2-separation of a matroid M containing N, and that M is minimal with respect to that condition. Then M has at most five elements beyond those of N.

For $k \geq 4$, characterizations of sequences of nested k-connected minors seem to become very complex. Rajan (1986) and Robertson (1984) contain results about the graph case for various kinds of 4-connectivity.

Work cited in Section 5.4 for minors with specified elements (Bixby (1974), Seymour (1981e), (1985b), (1986a), (1986b), Oxley (1984), (1987a), (1990a), Kahn (1985), Coullard (1986), Coullard and Reid (1988), Oxley and Row (1989), Oxley and Reid (1990), and Reid (1990), (1991a)–(1991d)) is closely related to the material of this chapter. Several of these results are readily proved with the earlier mentioned generalization of Theorem (7.3.1) due to Bixby and Coullard (1987).

Related to the above results are theorems claiming the following process to be possible. First, a minor N of a matroid M is replaced by an isomorphic copy that still contains a specified subset Z of the elements of N. Second, that isomorphic copy is extended to another minor of M with certain attractive properties. A theorem of this type is given in Tseng and Truemper (1986).

Chapter 8

Matroid Sums

8.1 Overview

In this chapter, we describe ways of decomposing or composing binary matroids, using a class of constructs called k-sums, where k ranges over the positive integers. Thus, there are 1-sums, 2-sums, 3-sums, etc. In subsequent chapters, we use k-sums frequently, in particular to analyze or construct certain graphs or matroids, for example the graphs without K_5 minors, the regular matroids, and the max-flow min-cut matroids. The setting in which k-sums are then invoked is as follows. One wants to understand or construct a given class of graphs or matroids. Already available is some insight into several proper subclasses. One conjectures that each graph or matroid not in any one of those subclasses can be recursively constructed by composition steps where the elementary building blocks are taken from the subclasses. It turns out that the k-sums defined in this chapter are well suited for such a composition process, as well as for the inverse decomposition process.

Generally, the structural complexity of k-sums grows as k increases. Thus, 1-sums represent the simplest, indeed trivial, case of decomposition or composition. We cover that case in Section 8.2. In the same section, we also discuss the more interesting but still elementary case of 2-sums.

For 3-sums, or generally for k-sums with $k \geq 3$, the simplicity of 1- and 2-sums gives way to a setting of rich structure that permits many interesting conclusions. In Section 8.3, we explore these k-sums, especially 3-sums. In Section 8.4, we acquire an efficient method for finding 1-, 2-, and 3-sums. Two alternatives of 3-sums, called Δ-sum and Y-sum, are covered in Section 8.5.

In the final section, 8.6, we summarize applications of k-sums, list extensions of the k-sum concept to general matroids via abstract matrices, mention other ways to decompose or compose matroids, and provide references.

We close this section with a review of the exact k-separations defined in Section 3.3. They turn out to be the key ingredient for k-sums. Suppose a binary matroid M on a set E has rank function $r(\cdot)$. Then a pair (E_1, E_2) partitioning E is an exact k-separation if $|E_1|, |E_2| \geq k$ and $r(E_1) + r(E_2) = r(E) + k - 1$. Given such a separation, let X_2 be a maximal independent subset of E_2, then enlarge X_2 by a subset X_1 of E_1 to a base of M. Define for $i = 1, 2$, $Y_i = E_i - X_i$. The representation matrix B of M corresponding to the base $X_1 \cup X_2$ must be of the form

(8.1.1)

$$
B = \begin{array}{c|c|c|}
 & Y_1 & Y_2 \\
\hline
X_1 & A^1 & 0 \\
\hline
X_2 & D & A^2 \\
\hline
\end{array}
$$

Matrix B with exact k-separation

where $|X_1 \cup Y_1|, |X_2 \cup Y_2| \geq k$ and GF(2)-rank $D = k - 1$. Below, we repeatedly utilize B of (8.1.1) when we work with exact k-separations.

The chapter requires knowledge of Chapters 2, 3, 5, 6, and 7. We also use the process of ΔY exchanges of Chapter 4.

8.2 1- and 2-Sums

In this section, we learn how to deduce and manipulate 1- and 2-sum decompositions and compositions. We start with the 1-sum case. Let M be a binary matroid on a set E and with a representation matrix B. Lemma (3.3.19) says that M has a 1-separation if and only if B is not connected. Assume the latter case. Then B can clearly be partitioned as shown in (8.2.1) below, with $|X_1 \cup Y_1|, |X_2 \cup Y_2| \geq 1$. The latter condition is equivalent to demanding that the submatrices A^1 and A^2 of B are nonempty, i.e., they are not 0×0 matrices. We declare that the binary matroids represented by A^1 and A^2, say M_1 and M_2, are the two *components* of a *1-sum decomposition* of M. The decomposition is reversed in the obvious way, giving a *1-sum composition* of M_1 and M_2 to M. We mean either process when we say that M is a 1-*sum* of M_1 and M_2, denoted by $M = M_1 \oplus_1 M_2$.

(8.2.1)

$$B = \begin{array}{c|c|c} & Y_1 & Y_2 \\ \hline X_1 & A^1 & 0 \\ \hline X_2 & 0 & A^2 \end{array}$$

Matrix B of 1-separation

Under the assumption of graphicness, the 1-sum has a straightforward graph interpretation given by the next lemma. We omit the elementary proof via Theorem (3.2.25), part (a).

(8.2.2) Lemma. *Let M be a binary matroid. Assume M to be a 1-sum of two matroids M_1 and M_2.*

(a) *If M is graphic, then there exist graphs G, G_1, and G_2 for M, M_1, and M_2, respectively, such that identification of a node of G_1 with one of G_2 creates G.*

(b) *If M_1 and M_2 are graphic (resp. planar), then M is graphic (resp. planar).*

Analogously to the matroid case, we call the graph G of Lemma (8.2.2) a 1-*sum* of G_1 and G_2, and denote this by $G = G_1 \oplus_1 G_2$.

We move to the more interesting case of 2-sums. We assume that the given binary matroid M is connected and has a 2-separation (E_1, E_2). Since M is connected, the 2-separation must be exact. Thus, in any matrix B of (8.1.1) corresponding to that exact 2-separation, the submatrix D must have $\mathrm{GF}(2)$-rank 1. Hence, B is the following matrix, where for the moment the indices $x \in X_2$ and $y \in Y_1$ are to be ignored.

(8.2.3)

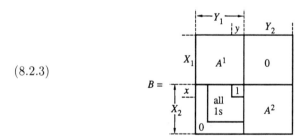

Matrix B of exact 2-separation

We refer to the submatrix of B of (8.2.3) indexed by X_2 and Y_1 as D, in agreement with (8.1.1). We want to extract from B two submatrices B^1 and B^2 that contain A^1 and A^2, respectively, and that also contain enough information to reconstruct B. Evidently, the latter requirement is equivalent to the condition that we must be able to compute D from B^1 and B^2. Now D has GF(2)-rank 1. Thus, knowledge of one nonzero row of D and of one nonzero column of D suffices to compute D.

With this insight, we choose B^1 to be A^1 plus one nonzero row of D, say row x, and B^2 to be A^2 plus one nonzero column of D, say column y. The two indices x and y are shown in (8.2.3). Below, we display B^1 and B^2 so selected.

(8.2.4)

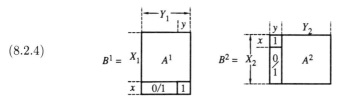

Matrices B^1 and B^2 of 2-sum

We reconstruct D, and thus implicitly B, from B^1 and B^2 by computing

(8.2.5) $$D = (\text{column } y \text{ of } B^2) \cdot (\text{row } x \text{ of } B^1)$$

Let M_1 and M_2 be the minors of M represented by B^1 and B^2. We call these minors the *components* of a *2-sum decomposition* of M. The reverse process, which corresponds to a reconstruction of B from B^1 and B^2, is a *2-sum composition* of M_1 and M_2 to M. Both cases are handled by saying that M is a *2-sum* of M_1 and M_2, denoted by $M = M_1 \oplus_2 M_2$. For future reference, we record in the next lemma that 2-separations of connected binary matroids produce 2-sums with connected components.

(8.2.6) Lemma.   *Any 2-separation of a connected binary matroid M produces a 2-sum with connected components M_1 and M_2. Conversely, any 2-sum of two connected binary matroids M_1 and M_2 is a connected binary matroid M.*

Proof. The above definitions establish the lemma except for the connectedness claims. It is easily verified that connectedness of B of (8.2.3) implies connectedness of B^1 and B^2 of (8.2.4), and vice versa. By Lemma (3.3.19), connectedness of representation matrices is equivalent to connectedness of the corresponding matroids. Thus, connectedness of B, B^1, and B^2 is equivalent to connectedness of M, M_1, and M_2, respectively.   □

The 2-sum decomposition or composition has the following appealing graph interpretation.

(8.2.7) Lemma.   *Let M be a connected binary matroid that is a 2-sum of M_1 and M_2, as given via B, B^1, and B^2 of (8.2.3) and (8.2.4).*

(a) If M is graphic, then there exist 2-connected graphs G, G_1, and G_2 for M, M_1, and M_2, respectively, with the following feature. The graph G is produced when one identifies the edge x of G_1 with the edge y of G_2, and when subsequently the edge so created is deleted. The drawing below depicts the general case.

(8.2.8)

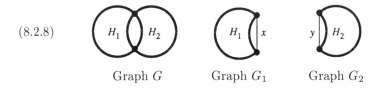

Graph G Graph G_1 Graph G_2

(b) If M_1 and M_2 are graphic (resp. planar), then M is graphic (resp. planar).

Proof. (a) Given B of (8.2.3) for M, let $E_1 = X_1 \cup Y_1$ and $E_2 = X_2 \cup Y_2$. We know that the pair (E_1, E_2) is a 2-separation of M. Select any graph G for M. Since M is connected, G is 2-connected. Let H_1 and H_2 be the subgraphs of G with edge sets E_1 and E_2, respectively. By Theorem (3.2.25) part (b), the 2-separation (E_1, E_2) of M implies that the connected components of H_1 and H_2 are connected in cycle fashion. By the switching operation of Section 3.2, we may rearrange G so that both H_1 and H_2 are connected. We may assume that G already is that graph. Thus, G is as given by (8.2.8), with connected H_1 and H_2. By Lemma (8.2.6), the matrices B^1 and B^2 of (8.2.4) are connected. By (8.2.3) and the connectedness of B^1, contraction in G of the edges of $X_2 - \{x\}$ and deletion of the edges of Y_2 must produce the graph G_1 of (8.2.8). Similarly, contraction of the edges of X_1 and deletion of the edges of $Y_1 - \{y\}$ must produce the graph G_2 of (8.2.8). Thus, G_1 and G_2 are graphs for the graphic matroids given by B^1 and B^2 of (8.2.4). The two graphs are 2-connected since the corresponding matrices are connected.

(b) Let G_1 and G_2 be 2-connected graphs for M_1 and M_2. For some H_1 and H_2, the drawings of (8.2.8) correctly depict G_1 and G_2. Identify the edge x of G_1 with the edge y of G_2, then delete the so-created edge. There are two ways to accomplish the identification, but either way is acceptable and produces a graph G as depicted by (8.2.8). Elementary checking of fundamental cycles of G versus fundamental circuits of M confirms that G represents the matroid M defined by B of (8.2.3). Thus, M is graphic. If G_1 and G_2 are plane graphs, we can carry out the edge identification so that G becomes a plane graph. Thus, M is planar if M_1 and M_2 are planar.   □

We turn to the more challenging case of general k-sums, with $k \geq 3$.

8.3 General k-Sums

We are given a 3-connected binary matroid M on a set E. For some $k \geq 3$, we also know an exact k-separation (E_1, E_2). We want to decompose M in some useful way. We investigate this problem using the matrix B of (8.1.1). Slightly enlarged, we repeat that matrix below.

(8.3.1)

Matrix B with exact k-separation

Recall that the submatrix D of B has GF(2)-rank $k - 1$. We want to decompose M into two matroids M_1 and M_2 that correspond to two submatrices B^1 and B^2 of B. As in the 2-sum case, we postulate that B^1 and B^2 include A^1 and A^2, respectively. Furthermore, B^1 and B^2 must permit a reconstruction of B. The latter requirement can be satisfied by including in B^1 (resp. B^2) a row (resp. column) submatrix of D with the same rank as D, i.e., with GF(2)-rank $k - 1$. Indeed, the submatrix D of B can be computed from these row and column submatrices. We provide the relevant formulas in a moment. Last but not least, we want B^1 and B^2 to be proper submatrices of B.

There are numerous ways to satisfy these requirements. In the most general case, both B^1 and B^2 intersect all four submatrices A^1, A^2, D, and 0 of B, and thus induce the following rather complicated-looking partition of B. We explain the partition momentarily.

(8.3.2)

Partition of B displaying k-sum

In the notation of (8.3.2), the submatrix B^1 of B, which is not explicitly indicated, is indexed by $X_1 \cup \overline{X}_2$ and $Y_1 \cup \overline{Y}_2$. Furthermore, the submatrix B^2 is indexed by $\overline{X}_1 \cup X_2$ and $\overline{Y}_1 \cup Y_2$. Hence, B^1 contains A^1, intersects A^2 in C^2, and intersects D in $[D^1 \mid \overline{D}]$. The submatrix B^2 contains A^2, intersects A^1 in C^1, and intersects D in $[\overline{D}/D^2]$. We assume that C^1 (resp. C^2) is a proper submatrix of A^1 (resp. A^2). This implies that both B^1 and B^2 are proper submatrices of B. Observe that $\overline{D}$ is the submatrix of D contained in both B^1 and B^2. We assume that both submatrices $[D^1 \mid \overline{D}]$ and $[\overline{D}/D^2]$ of D have GF(2)-rank equal to $k-1$. By Lemma (2.3.14), this implies that $\overline{D}$ has GF(2)-rank $k-1$. We display the matrices B^1 and B^2 below.

(8.3.3)

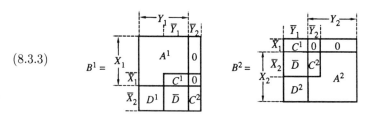

Matrices B^1 and B^2 of k-sum

The decomposition of B into B^1 and B^2 corresponds to a decomposition of M into two matroids, say M_1 and M_2, that are represented by B^1 and B^2. We call that decomposition of M a *k-sum decomposition* and declare M_1 and M_2 to be the *components* of the k-sum. The decomposition process is readily reversed. All submatrices of B except for the submatrix D^{12}, which is indexed by $X_2 - \overline{X}_2$ and $Y_1 - \overline{Y}_1$, are present in B^1 and B^2, and thus are already known. For the computation of D^{12} from B^1 and B^2, we first depict D and its submatrices.

(8.3.4)

$$D = \begin{array}{c|cc} & Y_1 \\ & \overline{Y}_1 \\ \hline \overline{X}_2 & D^1 & \overline{D} \\ X_2 & & \\ & D^{12} & D^2 \end{array}$$

Partitioned version of D

Since GF(2)-rank D = GF(2)-rank $\overline{D}$, there is a matrix F so that $[D^{12} \mid D^2] = F \cdot [D^1 \mid \overline{D}]$. Thus,

(8.3.5) $$D^2 = F \cdot \overline{D}$$

and

(8.3.6) $$D^{12} = F \cdot D^1$$

We first solve (8.3.5) for F. The solution may not be unique, but any solution is acceptable. Then we use F in (8.3.6) to obtain D^{12}.

Suppose $\overline{D}$ is square and nonsingular. Then from (8.3.5), we deduce $F = D^2 \cdot (\overline{D})^{-1}$. With that solution, we compute D^{12} according to (8.3.6) as

$$(8.3.7) \qquad\qquad D^{12} = D^2 \cdot (\overline{D})^{-1} \cdot D^1.$$

The reconstruction of B from B^1 and B^2 corresponds to a k-*sum composition* of M_1 and M_2 to M. Both the k-sum decomposition of M and the k-sum composition of M_1 and M_2 to M, we call a k-*sum* and denote it by $M = M_1 \oplus_k M_2$.

It is a simple exercise to prove that the 1- and 2-sums of Section 8.2 are special instances of the above situation. In the 1-sum case, D is a zero matrix. Thus, we may select C^1, C^2, D^1, D^2, and $\overline{D}$ to be trivial or empty, whichever applies. Then by (8.3.3), we have $B^1 = A^1$ and $B^2 = A^2$, which is the 1-sum of Section 8.2. In the 2-sum situation, $\overline{D}$ is the 1×1 matrix $[\,1\,]$, C^1 and C^2 are trivial matrices, $[D^1 \mid \overline{D}]$ is row x of D, and $[\overline{D}/D^2]$ is column y of D. The formula (8.2.5) for D is nothing but (8.3.7) with the just-defined D^1, D^2, and $\overline{D}$.

The reader is well justified to wonder why we consider such complex k-sums. We argue in favor of our approach as follows. Suppose we intend to investigate a certain matroid property that is inherited under minor-taking, say graphicness. To be even more specific, let us assume that we want to obtain the minor-minimal matroids that are not graphic. Suppose the above general conditions for k-sums admit a particular k-sum case where graphicness of the components implies graphicness of the k-sum. Then no minor-minimal nongraphic matroid can be such a k-sum. For if such a k-sum M exists, then its components, smaller as they are, are graphic. But by the just assumed feature of the k-sum, M is graphic as well, a contradiction. Thus, we can restrict our search for the minimal nongraphic matroids to instances that are not k-sums, a possibly very attractive insight. A second situation is as follows. We want to find a construction for matroids having a certain property. Indeed, we are willing to allow composition steps to be part of the construction. In that case, we need composition rules that preserve the property of interest.

The k-sum compositions described here, complex as they may be, do preserve a number of interesting matroid properties. Thus, for investigations into these properties, the k-sums are very useful. The evidence supporting this claim will be presented in subsequent chapters.

So far we have expressed the k-sum decomposition or composition in terms of representation matrices. It is interesting to consider k-sums in terms of matroid minors as well. By the very derivation of B^1 and B^2, the components M_1 and M_2 of the k-sum are minors of M. It is easy to see from (8.3.3) that B^1 and B^2 share precisely the submatrix $\overline{B}$ given by (8.3.8) below. Define $\overline{M}$ to be the matroid represented by $\overline{B}$. Clearly, $\overline{M}$ is a minor of M, but more importantly, of M_1 and M_2 as well. Indeed, one may view the composition of M_1 and M_2 to be an identification of the minor $\overline{M}$ of M_1 with the minor $\overline{M}$ of M_2. One might also say that $\overline{M}$ as minor of M forms the connection between the components M_1 and M_2 in M. In agreement with the latter terminology, we call $\overline{M}$ the *connecting matroid* or *connecting minor* of the k-sum. By (8.3.8) and the fact that D and $\overline{D}$ have the same rank, the pair $(\overline{X}_1 \cup \overline{Y}_1, \overline{X}_2 \cup \overline{Y}_2)$ is an exact k-separation of $\overline{M}$, provided that $|\overline{X}_1 \cup \overline{Y}_1|, |\overline{X}_2 \cup \overline{Y}_2| \geq k$. The latter condition is satisfied, for example, if both C^1 and C^2 are nonempty and nontrivial.

(8.3.8)

$$
\overline{B} = \begin{array}{c|c|c}
 & \overline{Y}_1 & \overline{Y}_2 \\
\hline
\overline{X}_1 & C^1 & 0 \\
\hline
\overline{X}_2 & \overline{D} & C^2
\end{array}
$$

Submatrix $\overline{B}$ representing the connecting minor $\overline{M}$
of the k-sum

Recall that we want important matroid properties to be preserved under k-sum composition. Evidently, $\overline{M}$ is the minor of M that decides whether or not we achieve that goal. Thus, the selection of $\overline{M}$ requires great care when one desires k-sums suitable for the investigation of matroid properties. Note that a particular choice of $\overline{M}$ imposes constraints on the exact k-separations that are to be converted to a k-sum. For example, any k-separation (E_1, E_2) capable of producing a k-sum with a given $\overline{M}$ minor must satisfy for $i = 1, 2, |E_i| \geq |\overline{X}_i \cup \overline{Y}_i| + 1$, since otherwise the submatrices C^1 and C^2 of B are not proper submatrices of A^1 and A^2. The selection of $\overline{M}$ becomes even more complex when computational aspects are considered. For example, one might demand that the k-sums defined by $\overline{M}$ can be located in polynomial time.

For this book, 3-sums are of considerable importance. A good choice for the matrix $\overline{B}$ representing $\overline{M}$ turns out to be

(8.3.9)

$$
\overline{B} = \begin{array}{|c|c|}
\hline
C^1 & 0 \\
\hline
\overline{D} & C^2 \\
\hline
\end{array} = \begin{array}{|cc|c|}
\hline
1 & 1 & 0 \\
\hline
\overline{D} & & 1 \\
 & & 1 \\
\hline
\end{array}
$$

Matrix $\overline{B}$ of 3-sum

where $\overline{D}$ is any 2×2 GF(2)-nonsingular matrix. If $\overline{D}$ is the 2×2 identity matrix, then by (5.2.8), $\overline{B}$ represents up to indices $M(W_3)$, which is the graphic matroid of the wheel with three spokes. If $\overline{D}$ contains exactly three

1s, the only other choice, then by one GF(2)-pivot, say in $C^1 = [1\ 1]$, we obtain the former case. Thus, in all instances, $\overline{M}$ is an $M(W_3)$ minor of M. With $\overline{B}$ of (8.3.9), the matrix B of (8.3.2) becomes the following matrix.

(8.3.10)

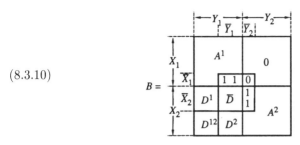

Matrix B of 3-sum

The representation matrices B^1 and B^2 of the components M_1 and M_2 are

(8.3.11)

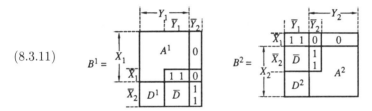

Matrices B^1 and B^2 of 3-sum

When does an exact 3-separation (E_1, E_2) of a binary matroid M lead to a 3-sum with this $\overline{M}$? Appealing sufficient conditions turn out to be 3-connectedness of M and the restriction that $|E_1|, |E_2| \geq 4$. We state this result in the next lemma.

(8.3.12) Lemma. *Let M be a 3-connected binary matroid on a set E. Then any 3-separation (E_1, E_2) of M with $|E_1|, |E_2| \geq 4$ produces a 3-sum, and vice versa.*

Proof. Let (E_1, E_2) be a 3-separation of M with $|E_1|, |E_2| \geq 4$. Since M is 3-connected, the 3-separation must be exact. Hence, by (8.3.1), M has a binary representation matrix B given by (8.3.13) below, with GF(2)-rank $D = 2$. In particular, any column submatrix of D containing four or more columns must contain a zero column or two parallel columns. We claim that the row index subset X_1 of B of (8.3.13) is nonempty. Assume otherwise. By $|X_1 \cup Y_1| \geq 4$, we have $|Y_1| \geq 4$. But then D has a zero column or two parallel columns, and M is not 3-connected, a contradiction. Thus, $X_1 \neq \emptyset$, and trivially $Y_1 \neq \emptyset$.

(8.3.13)

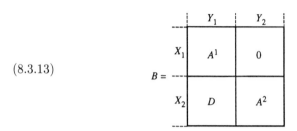

Matrix B with 3-separation

Let A^{11} be a connected block of A^1. There must be at least one such block, since otherwise M has coloops. To exhibit A^{11}, we partition B of (8.3.13) further as follows.

(8.3.14)

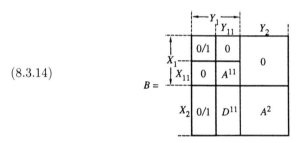

Matrix B of (8.3.13) with partitioned submatrix A^1

Note the column submatrix D^{11} of D corresponding to A^{11}. We claim that GF(2)-rank $D^{11} = 2$. If this is not the case, then the pair $(X_{11} \cup Y_{11}, E - (X_{11} \cup Y_{11}))$ constitutes a 1- or 2-separation of M, a contradiction.

Let us examine the submatrix $[A^{11}/D^{11}]$ of B of (8.3.14) more closely. Consider the paths in the bipartite graph $\mathrm{BG}(A^{11})$ between all pairs of nodes y and z in Y_{11} for which the columns y and z of D^{11} are GF(2)-independent. Since A^{11} is connected and GF(2)-rank $D^{11} = 2$, there is at least one path. In a shortest path, all intermediate nodes in Y_{11} correspond to zero columns of D^{11}. Suppose a shortest path has at least four arcs. With the path shortening technique of Chapter 5, we reduce that path by pivots in A^{11} to one with exactly two arcs. Thus, we may assume that a shortest path of length 2 exists, say connecting nodes y and z in $\mathrm{BG}(A^{11})$. Put differently, we may assume that A^{11} contains a row with two 1s in columns indexed by $y, z \in Y_{11}$ such that the columns of D^{11} indexed by y and z are GF(2)-independent. The path shortening pivots, if any, can also be carried out in B of (8.3.14) without affecting the entries of D and A^2. Thus, we may assume A^1 to contain a row with two 1s in columns y and z such that the columns y and z of D are GF(2)-independent.

By duality, we may suppose that A^2 contains a column having two 1s in rows u and v such that the rows u and v of D are GF(2)-independent. By Lemma (2.3.14), GF(2)-rank $D = 2$ implies that the rows u and v of D and the columns y and z of D must intersect in a 2×2 GF(2)-nonsingular submatrix $\overline{D}$ of D. When we partition B of (8.3.13) to exhibit the two 1s of A^1, the two 1s of A^2, and $\overline{D}$, we get an instance of (8.3.10), with the connecting minor given by $\overline{B}$ of (8.3.9). Thus, M is a 3-sum.

The converse is obvious, since any 3-sum given by (8.3.10) produces the 3-separation $(X_1 \cup Y_1, X_2 \cup Y_2)$ with $|X_1 \cup Y_1|, |X_2 \cup Y_2| \geq 4$.   □

Let us translate the 3-sum operation of binary matroids to graphs. For simplicity, we confine ourselves to the case where a given graph G with edge set E is 3-connected. In agreement with Lemma (8.3.12), we assume that G has a 3-separation (E_1, E_2) where $|E_1|, |E_2| \geq 4$. By that lemma, the graphic matroid $M = M(G)$ has a 3-sum decomposition induced by the 3-separation (E_1, E_2). The previously discussed results for the connecting minor $\overline{M}$ of that 3-sum can be restated for the case at hand as follows: The graph G has a minor $\overline{G}$ on edge set $\overline{E}$ and with a 3-separation $(\overline{E}_1, \overline{E}_2)$ such that $\overline{E}_1 \subseteq E_1$ and $\overline{E}_2 \subseteq E_2$. Up to indices, $\overline{G}$ is the wheel W_3.

The deletions and contractions reducing G to $\overline{G}$ evidently involve the edges of $E_1 - \overline{E}_1$ and $E_2 - \overline{E}_2$. Suppose in G we carry out these deletions and contractions just for the edges of $E_2 - \overline{E}_2$. Declare G_1 to be the resulting graph. The process producing G_1 corresponds to the reduction of the 3-sum M to the component M_1. Thus, $M_1 = M(G_1)$. By analogous reductions, this time confined to the edges of $E_1 - \overline{E}_1$, we obtain a graph G_2 for which $M_2 = M(G_2)$. Examine the representation matrices B^1 and B^2 of (8.3.11) for M_1 and M_2. Clearly, the set $\overline{X}_2 \cup \overline{Y}_2$ is a triangle in M_1, and the set $\overline{X}_1 \cup \overline{Y}_1$ is a triad of M_2. Correspondingly, $\overline{E}_2$ must be a triangle in G_1, and $\overline{E}_1$ must be a cocircuit of G_2 of size 3. Indeed, by the structure of G and G_2, that cocircuit of G_2 must be a 3-star. The drawing below summarizes these conclusions for G, G_1, and G_2.

(8.3.15)

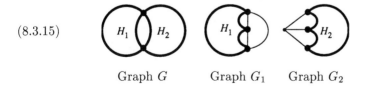

Graph G \qquad Graph G_1 \qquad Graph G_2

So far we have described the 3-sum decomposition of G into component graphs G_1 and G_2. The composition is even easier. We identify the three nodes of G_1 explicitly shown in (8.3.15) with the three nodes of G_2 so that the triangle of G_1 and the 3-star of G_2 form a copy of $\overline{G}$. The resulting graph is G plus that copy of $\overline{G}$. We delete the edges of that copy of $\overline{G}$ and have the desired G.

We use the term 3-*sum* to describe the decomposition of G into G_1 and G_2, as well as the composition of the latter graphs to G. In agreement with the matroid case, we denote this by $G = G_1 \oplus_3 G_2$ and call $\overline{G}$ the *connecting graph* or *connecting minor* of the 3-sum of G_1 and G_2.

We saw earlier that certain separations can be converted to 1-, 2-, or 3-sums. In the next section, we discuss how one may locate such separations and thus 1-, 2-, and 3-sums.

8.4 Finding 1-, 2-, and 3-Sums

Suppose that for a given binary matroid M, we want to either find a 1-, 2-, or 3-sum decomposition, or conclude that there is no such decomposition. In this section, we solve this seemingly difficult problem. Main tools are the separation algorithm of Section 6.2, some results about sequences of nested minors of Section 7.3, and the lemmas of Section 8.3 that assure us of a 1-, 2-, or 3-sum decomposition when certain separations are present.

We start with the simplest case, where the given binary matroid M is not connected. By Lemma (3.3.19), every representation matrix of M is disconnected. Thus, we easily detect this situation using an arbitrarily selected representation matrix B of M. From that B, we may deduce a 1-sum decomposition of M as described in Section 8.2. Thus, from now on, we may assume M to be connected. We also suppose that M has at least four elements.

To find a 2- or 3-sum decomposition, if it exists, we first rely on Lemma (5.2.11). That result says that under the above assumptions, M has a 2-separation or an $M(W_3)$ minor. Implicit in the proof of the lemma is a polynomial algorithm that decides which case applies. Suppose a 2-separation is determined. Then Lemma (8.2.6) tells us that this 2-separation can be easily converted to a 2-sum decomposition of M. So we now assume that we detect an $M(W_3)$ minor, say N.

To M and N we apply Theorem (7.3.6), which in turn cites Theorem (7.3.1). Accordingly, there is a polynomial algorithm that finds a 2-separation of M, or that establishes a sequence of nested 3-connected minors $M_0, M_1, \ldots, M_t = M$ with the following features. M_0 is isomorphic to N. For some s, $1 \leq s \leq t$, the subsequence $M_0, M_1, \ldots, M_s$ consists of wheel matroids and has gap 2, and the subsequence $M_s, M_{s+1}, \ldots, M_t$ has gap 1. We know already how to handle the 2-separation case. Thus, we may assume that the prescribed sequence of nested 3-connected minors is found. The description of that sequence can be simplified by saying that the sequence starts with an $M(W_3)$ minor and has gap 1 or 2. We use that sequence to find a 3-sum decomposition for M or to prove that there is no such decomposition. The details are as follows.

First, we observe that by Lemma (8.3.12), M has a 3-sum decomposition if and only if it has a 3-separation (E_1, E_2) with $|E_1|, |E_2| \geq 4$. Indeed, the decomposition can be efficiently determined from such a 3-separation using the algorithm implicit in the proof of that lemma. Thus, our task is finding such a 3-separation or proving that none exists.

From Section 5.3, we know that the intersection algorithm can solve the latter problem in polynomial time. However, the order of the algorithm is so large that the scheme is practically unusable unless the matroid M is quite small. In addition, it presently is not known how one might reduce the order to an acceptable level. For this reason, we outline below another algorithm that by suitable refinements and appropriate implementation does become very efficient. The method is based on the following observations, where for the moment we assume that M does have a 3-separation (E_1, E_2) with the desired property. For $j = 0, 1, \ldots, t$, let E^j be the groundset of M_j. For each j, and for $i = 1, 2$, define $E_i^j = E^j \cap E_i$. Clearly, each pair (E_1^j, E_2^j) satisfying $|E_1^j|, |E_2^j| \geq 4$ is a 3-separation of M_j. By its very derivation, any such 3-separation of M_j induces a 3-separation of M with at least four elements on each side, for example (E_1, E_2).

For the moment, let j be the smallest index so that $|E_1^j|, |E_2^j| \geq 4$. There is such an index since $(E_1^t, E_2^t) = (E_1, E_2)$ satisfies $|E_1^t|, |E_2^t| \geq 4$. Note that M_j has necessarily at least eight elements. Since M_0 has only six elements, we conclude that $j \geq 1$. Because of the minimality of j, we must have $|E_1^{j-1}|$ or $|E_2^{j-1}| \leq 3$. Since the gap of the sequence is at most 2, we deduce from the latter fact that $|E_1^j|$ or $|E_2^j| \leq 5$. There are only polynomially many pairs (E_1^j, E_2^j) for M_j where $|E_1^j|, |E_2^j| \geq 4$ and $|E_1^j|$ or $|E_2^j| \leq 5$. Indeed, there are only polynomially many pairs satisfying that condition when we allow j to range over $1, 2, \ldots, t$.

We use these observations as follows. Note that we still assume M to have a 3-separation (E_1, E_2) where $|E_1|, |E_2| \geq 4$. To find such a 3-separation, we first locate for $j = 1, 2, \ldots, t$ all pairs (E_1^j, E_2^j) satisfying $|E_1^j|, |E_2^j| \geq 4$ and $|E_1^j|$ or $|E_2^j| \leq 5$. For each pair, we test whether it is a 3-separation of M_j. We discard the cases where the answer is "no." For each remaining pair, say (E_1^j, E_2^j) of M_j, we check with the separation algorithm of Section 6.2 whether it induces a 3-separation of M. For at least one such pair, say (E_1^j, E_2^j) of M_j, we must obtain an affirmative answer. The 3-separation of M so found, say (E_1', E_2'), satisfies for $i = 1, 2$, $E_i' \supseteq E_i^j$ and thus $|E_i'| \geq |E_i^j| \geq 4$. Hence, (E_1', E_2') is the sought-after 3-separation of M.

So far, we have assumed that M has a 3-separation (E_1, E_2) where $|E_1|, |E_2| \geq 4$. If that is not the case, we can still carry out the above polynomial process, without success of course. But the lack of success proves that M has no 3-separation (E_1, E_2) with $|E_1|, |E_2| \geq 4$.

The next theorem summarizes the above discussion.

(8.4.1) Theorem. *There is a polynomial algorithm that accepts any binary matroid M as input, and that outputs the conclusion that M is not connected, or that M is connected but not 3-connected, or that M is 3-connected. In the first case, the algorithm also supplies a 1-sum decomposition, in the second case, a 2-sum decomposition, and in the third case, a 3-sum decomposition if it exists.*

In the next section, we meet close relatives of 3-sums, called Δ-sums and Y-sums.

8.5 Delta-Sum and Wye-Sum

There are two variations of the 3-sum. We call them delta-sum and wye-sum, for short Δ-sum and Y-sum. The relationships among 3-, Δ-, and Y-sums are very elementary. That does not imply that they may be employed interchangeably. Indeed, for some applications one definitely prefers one type over another. Both Δ-sum and Y-sum are derived from the 3-sum, so for convenient reference we repeat to related matrices of (8.3.10) and (8.3.11) below.

(8.5.1)

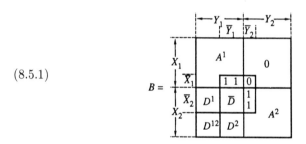

Matrix B of 3-sum

(8.5.2)

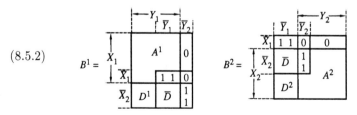

Matrices B^1 and B^2 of 3-sum

We proceed as follows. We first define Δ- and Y-sums. Then we show that the components of these sums are isomorphic to some minors of the sum, provided the sum is 3-connected. Finally, we indicate why one would want to use one type of these sums over another in certain applications.

We derive the Δ-sum from the 3-sum as follows. Let M be a 3-connected binary matroid with a 3-sum decomposition into M_1 and M_2, with representation matrices B, B^1, and B^2 given by (8.5.1) and (8.5.2). Suppose in M_2 we perform a ΔY exchange that replaces the triad $\overline{X}_1 \cup \overline{Y}_1$ by a triangle Z_2 as specified in (4.4.5). The new matroid, say $M_{2\Delta}$, is represented by

(8.5.3)

$$B^{2\Delta} = \quad X_2 \begin{array}{c|c|c} & Z_2 & Y_2 \\ \hline & \begin{array}{c|c} \overline{D} & \begin{array}{c} 1 \\ 1 \end{array} \end{array} & A^2 \\ d & \hline D^2 & \end{array}$$

Matrix $B^{2\Delta}$ for $M_{2\Delta}$

where the three columns of $B^{2\Delta}$ indexed by Z_2 sum to 0 (in GF(2)).

We say that M has been decomposed in a Δ-*sum decomposition* into *components* M_1 and $M_{2\Delta}$. The reverse process we call a Δ-*sum composition* of M_1 and $M_{2\Delta}$ to M. We refer to both cases as a Δ-*sum* and denote it by $M = M_1 \oplus_\Delta M_{2\Delta}$. The relevant triangle in M_1 or $M_{2\Delta}$, i.e., $\overline{X}_2 \cup \overline{Y}_2$ or Z_2, is the *connecting triangle* of M_1 or $M_{2\Delta}$. It is not difficult to check, say using the circuits of M, that the $\oplus_\Delta$ operator is commutative. Thus, $M_1 \oplus_\Delta M_{2\Delta} = M_{2\Delta} \oplus_\Delta M_1$. We omit the details of the proof, since we will make no use of this result.

The graph interpretation of the Δ-sum is as follows. Recall that by (8.3.15), the 3-sum decomposition of a graph G into component graphs G_1 and G_2 can be depicted as

(8.5.4)

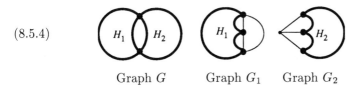

Graph G Graph G_1 Graph G_2

The ΔY exchange transforming M_2 to $M_{2\Delta}$ becomes a replacement of the explicitly shown 3-star of G_2 by a triangle. Let $G_{2\Delta}$ be the resulting graph. We may view the composition of G_1 and $G_{2\Delta}$ to G to be an identification of the triangle of G_1 with that of G_2, followed by a deletion of the so-created triangle from the resulting graph. Borrowing from the terminology of the matroid case, we call G a Δ-*sum* of G_1 and $G_{2\Delta}$, and denote this by $G = G_1 \oplus_\Delta G_{2\Delta}$.

The Y-sum of binary matroids is defined in an analogous fashion. As for the Δ-sum, we start with a 3-sum $M = M_1 \oplus_3 M_2$. But now we convert the triangle $\overline{X}_2 \cup \overline{Y}_2$ of M_1 by a ΔY exchange to a triad Z_1, getting a matroid M_{1Y}. By B^1 of (8.5.2) and the ΔY exchange of (4.4.7), the following matrix B^{1Y} represents M_{1Y}.

(8.5.5)

Matrix B^{1Y} for M_{1Y}

Here, the three rows of B^{1Y} indexed by Z_1 sum to 0 (in GF(2)). We call M a Y-*sum* with *components* M_{1Y} and M_2, and write $M = M_{1Y} \oplus_Y M_2$. The relevant triad in M_{1Y} or M_2 is the *connecting triad* of M_{1Y} or M_2. The Y-sum operator can be verified to be commutative, as one would expect.

In the graph case of a Y-sum, we replace the explicitly shown triangle of G_1 of (8.5.4) by a 3-star, getting a graph G_{1Y}. G is obtained from G_{1Y} and G_2 as follows. Let i, j, k (resp. p, q, r) be the three nodes of attachment of the 3-star in G_{1Y} (resp. G_2). Assume that i (resp. j, k) corresponds to p (resp. q, r). Now connect the node i (resp. j, k) of G_{1Y} minus its 3-star with the node p (resp. q, r) of G_2 minus its 3-star, say using an edge e (resp. f, g). The edges e, f, and g form a cutset in the new graph, and contraction of that cutset produces the graph G. We call G a Y-*sum* of G_{1Y} and G_2, and denote this by $G = G_{1Y} \oplus_Y G_2$.

The above Y-sum operation may seem cumbersome. In fact, the reader most likely has an alternate Y-sum composition in mind where the 3-star of G_{1Y} is identified with that of G_2, and where the edges of the resulting 3-star are deleted. Both Y-sum processes produce the same outcome, but they do differ in the way in which the composition is carried out. We chose the above, more complicated description because then the Δ-sum and Y-sum composition steps are dual operations. The reader may want to try to dualize the alternate Y-sum process to a new Δ-sum process for graphs. This should turn out to be a difficult task. Indeed, one can show that such a dual composition process cannot exist for graphs. At any rate, we know that if M is the Δ-sum $M_1 \oplus_\Delta M_{2\Delta}$, then M^* is the Y-sum $M_1^* \oplus_Y M_{2\Delta}^*$.

The matrices B^1 and B^2 of (8.5.2) are submatrices of B of (8.5.1). Thus, by definition, M_1 and M_2 are minors of M. One cannot expect M_{1Y} or $M_{2\Delta}$ also to be minors of M due to the assigned index sets Z_1 and Z_2. However, one would hope that M_{1Y} and $M_{2\Delta}$ are isomorphic to minors of M. The next result supports this notion.

(8.5.6) Lemma. *Let M be the 3-connected matroid represented by the* *binary matrix B of (8.5.1). Then M has minors isomorphic to M_{1Y} and $M_{2\Delta}$.*

Proof. By duality, it suffices that we prove the claim for $M_{2\Delta}$. For this, we may assume that the nonsingular submatrix $\overline{D}$ of B of (8.5.1) is a 2×2 identity matrix, since this can always be achieved by at most one GF(2)-pivot in B on one of the 1s indexed by $\overline{X}_1$ and $\overline{Y}_1$. Given $\overline{D}$ as a 2×2 identity matrix, we first perform in B two GF(2)-pivots on the 1s of $\overline{D}$. These pivots convert B to the matrix B' below. We explain the structure of B' momentarily.

(8.5.7)

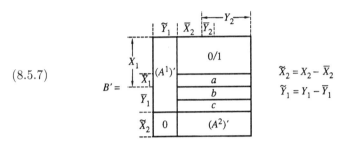

Matrix B' derived from B of (8.5.1) by two pivots

Note that the pivots in B are made within the submatrix B^2. Thus, the submatrix of B' indexed by the rows of $\overline{X}_1 \cup \overline{Y}_1 \cup \tilde{X}_2$ and the columns of $\overline{X}_2 \cup Y_2$ represents M_2. Since $(X_1 \cup Y_1, X_2 \cup Y_2)$ is a 3-separation of M, the submatrix D' of B' indexed by $X_1 \cup \overline{Y}_1$ and $\overline{X}_2 \cup Y_2$ has GF(2)-rank 2. Indeed, since $\overline{X}_1 \cup \overline{Y}_1$ is a triad of M_2, the row subvectors a, b, c are nonzero and add (in GF(2)) to the zero vector. By Lemma (8.3.12), pivots in $(A^1)'$ of B' are possible so that the new $(A^1)'$ of the new B' has two 1s in some column w and in rows u and v for which the rows u and v of D' are GF(2)-independent. The submatrix of the new B' indexed by $\tilde{X}_2 \cup \{u, v\}$ and $\overline{X}_2 \cup Y_2 \cup \{w\}$ then is isomorphic to $M_{2\Delta}$ by the rule (4.4.7) for ΔY exchanges. □

The simple relationships among 3-sum, Δ-sum, and Y-sum may deceive one into thinking that each such sum can be substituted for another one in all applications. A first warning of the incorrectness of this conclusion comes from applications involving planar graphs. That is, a Y-sum of two planar graphs must be planar, while a Δ-sum of two planar graphs need not be planar. Planarity of the Y-sum is most easily seen when one embeds each component on a sphere and combines the two drawings to one drawing on another sphere. The Δ-sum need not be planar. For example, the nonplanar graph $K_{4,3}$ can be created from two copies of the planar graph given below.

(8.5.8)

Planar component of nonplanar Δ-sum

The triangle $\{e, f, g\}$ of one copy is identified with that of the other copy. Subsequently, the triangle created by the identification process is deleted, and $K_{4,3}$ results. Other significant applications where caution is in order involve the max-flow min-cut matroids and the convex hull of the disjoint unions of circuits of a binary matroid (the so-called cycle polytope). On the other hand, when one examines matroid regularity, one may freely switch among the three types of sums without penalty. Validity of these claims is proved in Chapters 11 and 13.

In the final section, we link the material of this chapter to related results and provide references.

8.6   Extensions and References

Basic aspects of the composition or decomposition of matroids have been explored in a number of references, e.g. in Edmonds and Fulkerson (1965), Edmonds (1965a), (1979), Nash-Williams (1961), (1964), (1966), Bixby (1972), (1975), Brylawski (1972), (1975), Cunningham (1973), (1979), (1982a), (1982b), Cunningham and Edmonds (1978), (1980), Iri (1979), Nakamura and Iri (1979), Seymour (1980b), Tomizawa and Fujishige (1982), Fujishige (1983), (1985), and Conforti and Laurent (1988), (1989). Decomposing highly connected matroids and composing them again has not been treated extensively. Related to the approach taken here are the decomposition and composition of the graphs without K_5 minors in Wagner (1937a), the modular constructions of Brylawski (1975), and the decomposition and composition of the regular matroids in Seymour (1980b). The concept of the connecting minor is taken from Truemper (1985a), where it is developed for general matroids using abstract matrices, instead of just for binary matroids as done here. The material of Sections 8.2 and 8.3 is also derived from general matroid results of that reference. The methods of Section 8.4 for finding 1-, 2-, and 3-sums are from Truemper (1985a), (1990).

Locating k-sums for general $k \geq 4$ is much more difficult since one must identify a k-separation and an *a priori* specified connecting minor. We do not know how this can be efficiently accomplished for binary matroids, let alone for general matroids. Truemper (1985a) treats a particular 4-sum case where the connecting minors have eight elements.

Most of the material of Section 8.5 is based on Grötschel and Truemper (1989b). That reference examines the Δ-sum and Y-sum in much more detail. The Δ-sum of Section 8.5 is popular in graph theory (e.g., see Wagner (1937a)) and is used in Seymour (1980b) to effect the decomposition of the regular matroids. Lemma (8.5.6) is proved by a quite different method in Seymour (1980b).

We have omitted entirely some interesting decomposition results of Truemper (1985a),(1985b). These references contain a number of basic results about k-sums and three decomposition classification theorems plus applications. The theorems cover general matroids, binary matroids, and graphs, respectively. Each of them says that any given matroid is decomposable, or a number of elements can be removed in any order without loss of 3-connectivity, or the matroid belongs to a small class of matroids with few elements. The proofs are quite long and are the main reason that we have omitted these results. Significant applications of the theorems are rather short proofs of the profound excluded minor theorems for planar graphs, graphic matroids, and regular matroids due to Kuratowski (1930) and Tutte (1958), (1965). The latter theorems are proved in this book using different machinery in Chapters 7, 9, and 10.

Chapter 9

Matrix Total Unimodularity and Matroid Regularity

9.1 Overview

At this point, we have assembled the basic matroid results and tools for the remaining developments. We now begin the investigation into the two main technical subjects, which are matrix total unimodularity and matroid regularity.

Total unimodularity and its many variants are very important for combinatorial optimization. For a long time, matrix techniques evidently did not permit any profound insight into total unimodularity. That fact motivated the translation of the matrix property of total unimodularity into the matroid property of regularity. The translation opened the way for the application of powerful matroid techniques. After an effort spanning three decades, matroid regularity and thus total unimodularity were at last understood. The main results, in historical order, are due to Tutte and Seymour. The subsequent chapters contain these results as well as closely related material.

We proceed as follows. In Section 9.2, we define total unimodularity, sketch important applications, and translate total unimodularity into matroid language, thus getting matroid regularity. Section 9.3 contains the first major result for regularity, which is Tutte's characterization of the regular binary matroids. Recall that F_7 is the Fano matroid and F_7^* its dual, and that U_n^m is the uniform matroid on n elements where any subset with at most m elements is independent. The characterization says that a matroid is regular if and only if it has no U_4^2, F_7, or F_7^* minors. The original proof of that characterization is long and complicated. Here we rely on

188

a proof of amazing brevity and clarity by Gerards. A minor modification of the proof produces Reid's characterization of the ternary matroids, i.e., the matroids representable over GF(3). The characterization says that a matroid is ternary if and only if it has no U_5^2, U_5^3, F_7, or F_7^* minors. This result is presented in Section 9.4. The final section, 9.5, contains extensions and references.

The chapter requires knowledge of the material of Chapters 2 and 3.

9.2 Basic Results and Applications of Total Unimodularity

In this section, we define matrix total unimodularity and matroid regularity and establish elementary results about these two properties. We also point out representative applications of total unimodularity.

A real matrix is *totally unimodular* if every square submatrix has 0 or ±1 as determinant. Thus, the entries of a totally unimodular matrix must be 0 or ±1. For example, the zero matrices and the identity matrices are totally unimodular. Nontrivial examples are given by the next lemma.

(9.2.1) Lemma.   Let A be a real $\{0, \pm 1\}$ matrix where every column contains exactly one +1 and one −1. Then A is totally unimodular.

Proof. Let D be a square submatrix of A. We induct on the order of D. In the nontrivial case, D has order $k \geq 2$ and has no zero columns. If D has a column with exactly one ±1, we use cofactor expansion and induction to calculate the determinant as 0 or ±1. Otherwise, D has exactly one +1 and one −1 in each column, and thus the determinant is 0. ☐

Define F to be the support matrix of the matrix A of Lemma (9.2.1). View F to be binary. Since A has exactly one +1 and one −1 in each column, the matrix F has exactly two 1s in each column. Then according to the definitions of Section 3.2, F is the node/edge incidence matrix of some graph G. Specifically, each row of F defines a node of G, and each column g of F, say with 1s in rows i and j, defines an edge connecting nodes i and j of G. We now declare the matrix A to represent a certain directed version of G, as follows. Assume that column g of A has a +1 in row i and a −1 in row j. Then we replace the undirected edge of G connecting the nodes i and j by a directed arc from i to j. Let H be the resulting directed graph. We call A the *node/arc incidence matrix* of H. By Lemma (9.2.1), every node/arc incidence matrix is totally unimodular.

The next lemma summarizes elementary operations that maintain total unimodularity.

(9.2.2) Lemma.   *Total unimodularity is maintained under the taking of submatrices, transposition, pivots, and the adjoining of zero or unit vectors or of parallel* $\{0, \pm 1\}$ *vectors.*

Proof. The pivot is the only nontrivial operation. Let a pivot convert a totally unimodular matrix A to a matrix A'. Adjoin to A an identity matrix to get $[I \mid A]$. Now A is totally unimodular if and only if every basis matrix of $[I \mid A]$ has ± 1 as determinant. The latter property is maintained under the elementary row operations in $[I \mid A]$ that correspond to the pivot in A, and that, together with scaling by $\{\pm 1\}$ factors and a column exchange, convert $[I \mid A]$ to $[I \mid A']$. Thus, A' is totally unimodular.   ☐

According to the next lemma, counting may be used to check total unimodularity for the $\{0, \pm 1\}$ matrices whose bipartite graph $\mathrm{BG}(\cdot)$ is a chordless cycle, i.e., for the $k \times k$ real matrices, $k \geq 2$, of the form

(9.2.3)

$$
\begin{array}{|ccccc|}
\hline
\pm 1 & & & & \pm 1 \\
\pm 1 & \pm 1 & & & \\
& \pm 1 & \cdot & & \\
& & \cdot & \pm 1 & \\
& & & \pm 1 & \pm 1 \\
\hline
\end{array}
$$

Matrix whose bipartite graph
is a chordless cycle

(9.2.4) Lemma.   *The real matrix of (9.2.3) is totally unimodular if and only if its entries sum to* $0(\mathrm{mod}\ 4)$.

Proof. Let A be the matrix of (9.2.3). Scaling of a row or column of A by -1 does not affect total unimodularity and changes the sum of the entries of A by a multiple of 4. Thus, for the proof of the lemma, we scale A to the matrix A' given by

(9.2.5)

$$
\begin{array}{|ccccc|}
\hline
1 & & & & \alpha \\
-1 & 1 & & & \\
& -1 & \cdot & & \\
& & \cdot & 1 & \\
& & & -1 & 1 \\
\hline
\end{array}
$$

Particular matrix whose bipartite graph
is a chordless cycle

By cofactor expansion and counting, we confirm that $\det_{\mathbb{R}} A'$ is 2 (resp. 0) if and only if $\alpha = 1$ ($\alpha = -1$), which holds if and only if the entries of A' sum to 2 (resp. 0). The "only if" part of the proof is then immediate. The "if" part follows from Lemma (9.2.1).   ☐

A matrix B is *regular* if it is binary and if its 1s can be replaced by ± 1s so that a real matrix results that is totally unimodular. By the above discussion, every node/edge incidence matrix is regular. The signing of a regular matrix to achieve a real totally unimodular matrix is essentially unique, a fact proved next.

(9.2.6) Lemma. *Let A and A' be two totally unimodular matrices with the same support matrix. Then A' may be obtained from A by a scaling of the rows and columns by $\{\pm 1\}$ factors.*

Proof. We may assume A to be connected. Let T be a tree of $\mathrm{BG}(A)$, and let T' be the corresponding tree of $\mathrm{BG}(A')$. Because of scaling, we may suppose that A and A' agree on the entries corresponding to the edges of T and T'. Suppose A and A' differ, say on the (i, j) element. The corresponding edges e and e' of $\mathrm{BG}(A)$ and $\mathrm{BG}(A')$ form cycles C and C' with T and T', respectively. Select T and e, and thus T' and e', so that the cardinality of the cycles is minimum. Suppose C and C' have chords. By the minimality condition, the entries of A and A' corresponding to such chords must agree. But then C and C' do not have minimum cardinality, a contradiction. Thus, C and C' are chordless cycles of $\mathrm{BG}(A)$ and $\mathrm{BG}(A')$, and the submatrices of A and A' corresponding to C and C' are of the form (9.2.3). The sums of the entries of the two submatrices differ by 2, since they differ on just one entry. But then one of the sums is $0 (\mathrm{mod}\,4)$, and the other one is $2 (\mathrm{mod}\,4)$. By Lemma (9.2.4), one of A and A' is not totally unimodular, a contradiction.   □

Lemma (9.2.6) and its proof imply the following result.

(9.2.7) Corollary. *There is a polynomial algorithm that by signing converts any regular matrix B to a real matrix A that is totally unimodular. The signing can be carried out as follows. First some submatrix is signed. Then, iteratively, the method signs an arbitrarily selected additional row or column that is adjoined to the submatrix on hand.*

Proof. Use Lemma (9.2.6) and the arguments of its proof to establish the correct signs for the additional row and column.   □

A matroid M is *regular* if it has a regular representation matrix B. The essentially unique totally unimodular matrix A deduced from the regular B may be used to represent M as follows.

(9.2.8) Lemma. *Let M be a matroid represented by a real totally unimodular matrix A. Define A' to be a numerically identical copy of A, but view A' to be a matrix over an arbitrary field $\mathcal{F}$. Then M is represented by A' over $\mathcal{F}$.*

Proof. Since all square submatrices of A have real determinant 0 or ± 1, any such submatrix of A must be $\mathbb{R}$-nonsingular if and only if the corresponding submatrix of A' is $\mathcal{F}$-nonsingular. Thus, A over $\mathbb{R}$ and A' over $\mathcal{F}$ define the same matroid.   □

Lemma (9.2.8) leads to a short proof of the following theorem.

(9.2.9) Theorem.   *The following statements are equivalent for a matroid*
M.

(i) M *is regular.*
(ii) M *has a real representation matrix that is totally unimodular.*
(iii) M *is representable over every field.*
(iv) M *is representable over* GF(2) *and* GF(3).
(v) M *representable over* GF(3), *and every representation matrix over*
    GF(3), *when viewed as real, is totally unimodular.*

Proof. By definition, (i) $\iff$ (ii), and by Lemma (9.2.8), (ii) $\implies$ (iii).
Trivially, (iii) $\implies$ (iv) and (v) $\implies$ (ii). We show (iv) $\implies$ (v). By (iv),
some matrix C over GF(3) and the binary support matrix B of C represent
M. Declare A to be a copy of C, but view A to be over the reals. We have
shown (v) once we have proved A to be totally unimodular. Suppose A has
a submatrix with real determinant different from 0, ± 1. We may assume
that A itself is such a matrix, and that every proper submatrix of A is
totally unimodular. If A is a 2×2 matrix, then by a trivial case analysis
we must have up to scaling by $\{\pm 1\}$ factors in A and C,

$$(9.2.10) \qquad A = \begin{bmatrix} 1 & 1 \\ 1 & -1 \end{bmatrix}; \quad B = \begin{bmatrix} 1 & 1 \\ 1 & 1 \end{bmatrix}; \quad C = \begin{bmatrix} 1 & 1 \\ 1 & -1 \end{bmatrix}$$

A over ℝ, *B over* GF(2), *and C over* GF(3)

Then $\det_2 B = 0$ and $\det_3 C = 1$. We conclude that B and C represent
different matroids, a contradiction. Suppose the order of A is greater than
2. We produce the 2×2 case by pivots in A, B, and C, as follows. We
perform an ℝ-pivot on any nonzero entry of A and delete the pivot row
and pivot column. In B and C, we carry out the corresponding operations.
Let A', B', C' be so obtained from A, B, and C. It is easy to see that A'
is not totally unimodular, and that every proper submatrix of A' is totally
unimodular. Furthermore, B' is the binary support matrix of A', and A'
and C' have their 0s, +1s, and −1s in the same positions. By induction,
the contradictory 2×2 case applies. $\qquad \square$

The preceding results and arguments yield the following corollaries.

(9.2.11) Corollary.   *For every regular matroid M, the following holds.*

(a) *Every binary representation matrix of M is regular.*
(b) *Every minor of M is regular.*
(c) *The dual of M is regular.*

Proof. Lemma (9.2.2) and Theorem (9.2.9) imply (a), (b), and (c). $\square$

(9.2.12) Corollary.   *The graphic matroids as well as the cographic ones
are regular.*

Proof. By Corollary (9.2.11), we only need to consider the case of graphicness. So let G be a graph producing a graphic matroid M. Add a new node to G and connect it to all other nodes. Let G' be the resulting graph. The added edges constitute a spanning tree T' of G'. The representation matrix of the graphic matroid M' of G' is nothing but the node/edge incidence matrix of G, with rows indexed by T'. By Lemma (9.2.1), the latter matrix is regular. Thus, M' is regular. By Corollary (9.2.11), the minor $M'\backslash T'$, which is M, is regular as well. □

We know that the transpose of a graphic matrix need not be graphic. Examples are the representation matrices of $M(K_5)$ and $M(K_{3,3})$ given by (3.2.38) and (3.2.41). By 1-sum composition, we thus can create regular matrices B that are not graphic and not cographic. For example, we may choose as one block of B the matrix B' of (3.2.38), and as the second block of B the transpose of B'. There are less trivial ways of creating regular nongraphic and noncographic matrices. We present them in Chapter 11. At this time, we introduce two regular nongraphic and noncographic matrices, called B^{10} and B^{12}, that play a very special role in Chapters 10 and 11. The matrix B^{10} is

$$(9.2.13) \qquad B^{10} = \begin{bmatrix} 1 & 0 & 0 & 1 & 1 \\ 1 & 1 & 0 & 0 & 1 \\ 0 & 1 & 1 & 0 & 1 \\ 0 & 0 & 1 & 1 & 1 \\ 1 & 1 & 1 & 1 & 1 \end{bmatrix}$$

Matrix B^{10} of regular matroid R_{10}

To prove regularity, we declare B^{10} to be over $\mathbb{R}$ and check that all square submatrices have real determinant equal to 0 or ± 1. Brute-force checking turns out to be quite tedious. But in Chapter 12, we learn much about minimal $\{0, \pm 1\}$ matrices that are not totally unimodular. The conditions presented there almost immediately prove B^{10} to be totally unimodular. The regular matroid represented by B^{10} is called R_{10}. We prove nongraphicness and noncographicness of R_{10} as follows. Delete the last column from B^{10}. Up to indices, the matrix of (3.2.41) results. Hence, R_{10} has an $M(K_{3,3})$ minor. The matrix B^{10} is also its transpose. Thus, R_{10} also has an $M(K_{3,3})^*$ minor. We conclude that R_{10} is not graphic and not cographic.

We have met the matrix B^{12} already in (4.4.9). Below we include that matrix in (9.2.14) in partitioned form, for reasons explained shortly. Following that matrix, we display a signed version of B^{12} in (9.2.15). We claim that the matrix of (9.2.15) is totally unimodular, which implies that B^{12} is regular. Analogously to the B^{10} case, one may verify the claim using the results of Chapter 12.

(9.2.14)

$$B^{12} = \begin{array}{|cccc|cc|} \hline 1 & 0 & 1 & 1 & 0 & 0 \\ 0 & 1 & 1 & 1 & 0 & 0 \\ \hline 1 & 0 & 1 & 0 & 1 & 1 \\ 0 & 1 & 0 & 1 & 1 & 1 \\ 1 & 0 & 1 & 0 & 1 & 0 \\ 0 & 1 & 0 & 1 & 0 & 1 \\ \hline \end{array}$$

Matrix B^{12} of regular matroid R_{12}

(9.2.15)

$$\begin{array}{|cccc|cc|} \hline 1 & 0 & 1 & 1 & 0 & 0 \\ 0 & 1 & 1 & 1 & 0 & 0 \\ \hline 1 & 0 & 1 & 0 & 1 & 1 \\ 0 & -1 & 0 & -1 & 1 & 1 \\ 1 & 0 & 1 & 0 & 1 & 0 \\ 0 & -1 & 0 & -1 & 0 & 1 \\ \hline \end{array}$$

Totally unimodular version of B^{12}

An easier method for verifying total unimodularity of the matrix of (9.2.15) uses decomposition, as follows. By (8.3.10) and (8.3.11), the partition of the matrix B^{12} of (9.2.14) induces a 3-sum decomposition with component matrices B^1 and B^2 given by (9.2.16) below. A GF(2)-pivot on the 1 in the lower right corner of B^2 converts that matrix, up to indices, to the matrix of (3.2.41). The latter matrix represents up to indices $M(K_{3,3})$, and so does B^2. The matrix B^1 is the transpose of B^2. Thus, the matroid of B^1 is isomorphic to $M(K_{3,3})^*$.

(9.2.16)

$$B^1 = \begin{array}{|cccc|c|} \hline 1 & 0 & 1 & 1 & 0 \\ 0 & 1 & 1 & 1 & 0 \\ \hline 1 & 0 & 1 & 0 & 1 \\ 0 & 1 & 0 & 1 & 1 \\ \hline \end{array} \quad ; \quad B^2 = \begin{array}{|cc|cc|} \hline 1 & 1 & 0 & 0 \\ \hline 1 & 0 & 1 & 1 \\ 0 & 1 & 1 & 1 \\ 1 & 0 & 1 & 0 \\ 0 & 1 & 0 & 1 \\ \hline \end{array}$$

Component matrices B^1 and B^2 of B^{12}

These facts imply that R_{12} is not graphic and not cographic. They also establish that R_{12} is a 3-sum of a copy of the regular $M(K_{3,3})$ with one of the regular $M(K_{3,3})^*$. In Chapter 11, we see that any 3-sum of two regular matroids is also regular. This fact proves R_{12} to be regular. The simple signing algorithm implicit in the proof of Corollary (9.2.7) confirms that the signed version of B^{12} given by (9.2.15) is indeed totally unimodular.

Total unimodularity is an important property for combinatorial optimization. Numerous problems of that area of mathematics can be expressed as an *integer program* of the form

(9.2.17)
$$\begin{aligned} \min \quad & d^t \cdot x \\ \text{s. t.} \quad & A \cdot x \le b \\ & 0 \le x \le c \\ & x \text{ integer} \end{aligned}$$

where A is a given integral matrix, b, c, and d are given integral column vectors, and x is a column vector representing the solution. The dimensions of the arrays are such that the indicated multiplications and inequalities make sense. The abbreviation "s. t." stands for "subject to." In general, (9.2.17) is not easy. But if A is a totally unimodular matrix, then one can effectively drop the integrality requirement from (9.2.17) and solve the resulting *linear program*. In this book, we do not dwell on details and implications of this approach, which has been treated extensively elsewhere. In Section 9.5, we provide appropriate references.

A special class of the problems (9.2.17) involves as A the node/arc incidence matrices of directed graphs, or matrices derived from node/arc incidence matrices by pivots and deletion of rows and columns. Any problem of (9.2.17) in that class is called a *network flow problem*. That class has also been treated extensively, and numerous special algorithms and famous inequalities exist. That these algorithms work and that the inequalities are valid can almost always be traced back to the total unimodularity of node/arc incidence matrices. Again, we must resist an even cursory treatment and must point to Section 9.5 for references. However, in Section 10.6 of the next chapter, we do present an efficient algorithm for testing whether or not a given $\{0, \pm 1\}$ matrix A is the matrix of some network flow problem.

The importance of total unimodularity naturally leads us to ask a number of questions. How can we recognize a totally unimodular matrix? Is there a simple construction of the entire class of totally unimodular matrices? What are the characteristics of the matrices that are not totally unimodular, but all of whose proper submatrices are totally unimodular? Are there other well-behaved problem classes of type (9.2.17) where A is not totally unimodular, but where A is related to some totally unimodular matrix?

Theorem (9.2.9) links matrix total unimodularity and matroid regularity. Thus, one may pose the following related matroid questions. Is there a simple construction of the entire class of regular matroids? Which are the nonregular binary matroids all of whose proper minors are regular? Which are the nongraphic binary matroids all of whose proper minors are graphic? The latter two questions make sense since regularity as well as graphicness are maintained under minor-taking. Are there other classes of binary matroids that are not regular, but that are closely related to the regular matroids?

The above questions have complete answers, all of which are provided in the sequel. The answer to the question concerning the minimal nonregular binary matroids is given in the next section.

9.3 Characterization of Regular Matroids

By Theorem (9.2.9), every regular matroid is binary. We already have a characterization of the binary matroids: Theorem (3.5.2) says that a matroid is binary if and only if it has no U_4^2 minors. Thus, a regular matroid has no U_4^2 minors. By Corollary (9.2.11), regularity is maintained under minor-taking and dualizing. For a complete characterization of regularity, we thus must identify the minimal binary matroids M that are not regular. In this section, we prove the famous theorem of Tutte according to which there is just one such M up to isomorphism and dualizing. That matroid is the *Fano matroid* F_7 represented by the matrix B^7 of (9.3.1) below. The name is due to the fact that the matroid is the Fano plane, which is the projective geometry $PG(2,2)$. The seven elements of the matroid are the points of the geometry. We saw the Fano matroid earlier in Sections 3.3 and 4.4, under (3.3.22) and (4.4.13). In the latter case, *"con"* and *"del"* labels were assigned to its elements. We have no such labels here.

(9.3.1)
$$B^7 = \begin{bmatrix} 1 & 0 & 1 & 1 \\ 1 & 1 & 0 & 1 \\ 0 & 1 & 1 & 1 \end{bmatrix}$$

Matrix B^7 of Fano matroid F_7

The matrix B^7 cannot be signed to become totally unimodular, as follows. The last three columns of B^7 give a totally unimodular submatrix. If B^7 is regular, then by Corollary (9.2.7), we can sign the first column of B^7 to achieve a totally unimodular matrix. But for all such ways, a 2×2 or 3×3 matrix with real determinant equal to $+2$ of -2 is created. Thus, B^7 is not regular. Tutte's theorem is as follows. The amazingly simple proof is due to Gerards.

(9.3.2) Theorem. *A binary matroid is regular if and only if it has no F_7 or F_7^* minors.*

Proof. For proof of the nontrivial "if" part, let M be a nonregular binary matroid all of whose proper minors are regular. Evidently, any binary representation matrix B of M must be connected. $BG(B)$ cannot be a single cycle or a path, for the first case corresponds to a wheel matroid, and the second one to a minor of a wheel matroid; both matroids are regular. Thus, $BG(B)$ is connected and has a node of degree 3, and hence has a spanning tree with a degree 3 node. Such a tree has at least three tip nodes, and hence has at least two tip nodes that correspond to two rows or to two columns of B. Thus, deletion of two columns or of two rows, say indexed by p and q, reduces B to a connected matrix $\overline{B}$. Because of dualizing of M, we may assume the former case. Thus,

(9.3.3)

$$B = \begin{array}{c} \\ \end{array} \begin{array}{|cc|c|} \hline p & q & \\ g & h & \overline{B} \\ \hline \end{array}$$

Binary Matrix B of minimal nonregular matroid M

where the submatrices $\overline{B}$, $[g \mid \overline{B}]$, and $[h \mid \overline{B}]$ are connected. By Corollary (9.2.7), we may sign B so that a real matrix A of the form

(9.3.4)

$$A = \begin{array}{|cc|c|} \hline p & q & \\ a & b & \overline{A} \\ \hline \end{array}$$

Matrix A derived from B of (9.3.3) by signing

results where both $[a \mid \overline{A}]$ and $[b \mid \overline{A}]$ are totally unimodular. Since M is not regular, A is not totally unimodular. By the construction, any submatrix of A proving the latter fact must intersect both columns p and q. Let D be a minimal such submatrix. If D is not a 2×2 matrix, we may convert it to one by real pivots in $\overline{A}$. The corresponding binary pivots in B must lead to a matrix that is the support of the one deduced from A, for otherwise, at least one of the matrices $[a \mid \overline{A}]$, $[b \mid \overline{A}]$ is not totally unimodular. Furthermore, by Lemma (9.2.2), the pivots convert $[a \mid \overline{A}]$ and $[b \mid \overline{A}]$ to some other totally unimodular matrices, and do not affect connectedness of $\overline{A}$. Thus, we may assume the 2×2 D to be already present in A of (9.3.4). That is, the columns a and b of A contain a 2×2 matrix with determinant equal to $+2$ or -2, say in rows v and w.

Up to scaling, A can thus be further partitioned as shown in (9.3.5) below. The submatrix $\overline{A}$ has been subdivided into $\overline{\overline{A}}$, $\overline{c}$, and $\overline{e}$. The vectors a and b have become $\overline{a}$ and $\overline{b}$ plus the explicitly shown ± 1s. The latter entries make up D.

(9.3.5)

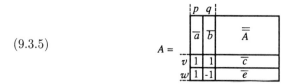

Partition of matrix A displaying
non-totally unimodular 2×2 submatrix

Examine the row subvectors $\bar{c}$ and $\bar{e}$ in rows v and w. Suppose both sub-vectors have ± 1 entries in some column $y \neq p, q$. Then those two entries plus the two ± 1s of column p or column q must form a non-totally unimodular 2×2 matrix, a contradiction that both $[a \mid \bar{A}]$ and $[b \mid \bar{A}]$ are totally unimodular. None of the subvectors $\bar{c}$ and $\bar{e}$ can be zero, since otherwise $\overline{A}$ is not connected. By scaling, we thus can assume $\bar{c}$ and $\bar{e}$ to be as indicated below.

(9.3.6)

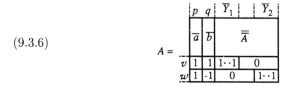

Further partitioning of matrix A

Since $\overline{A}$ is connected, there is a path from some $r \in \overline{Y}_1$ to some $s \in \overline{Y}_2$ in the bipartite graph of the submatrix $\overline{\overline{A}}$. By the path shortening technique of Chapter 5, we may assume the path to have exactly two edges. Thus, we can refine A of (9.3.6) to

(9.3.7)

$$A = \begin{array}{c|cccc|c} & p & q & r & s & \\ & \bar{\bar{a}} & \bar{\bar{b}} & & & \overline{\overline{A}} \\ \hline u & \alpha & \beta & \pm 1 & \pm 1 & \\ v & 1 & 1 & 1 & 0 & \bar{\bar{c}} \\ w & 1 & -1 & 0 & 1 & \bar{\bar{e}} \end{array}$$

Final partitioning of matrix A

By scaling in row u, we may presume the entry in row u and column r to be a $+1$. We now concentrate on the submatrix A' of A indexed by $\{u, v, w\}$ and $\{p, q, r, s\}$. We show that submatrix below. The entries α and β of the submatrix are yet to be determined.

(9.3.8)

$$A' = \begin{array}{c} \\ u \\ v \\ w \end{array} \begin{array}{|cccc|} p & q & r & s \\ \hline \alpha & \beta & 1 & \pm 1 \\ 1 & 1 & 1 & 0 \\ 1 & -1 & 0 & 1 \end{array}$$

Submatrix A' of A displaying F_7 minor

If $\alpha = \beta = 0$, then by Lemma (9.2.4), the 3×3 column submatrix of A' indexed by $\{p, r, s\}$ or the one indexed by $\{q, r, s\}$ is not totally unimodular. But then one of $[a \mid \overline{A}]$ and $[b \mid \overline{A}]$ is not totally unimodular, a contradiction. Suppose both α and β are nonzero. Due to column r and total unimodularity of $[a \mid \overline{A}]$ and $[b \mid \overline{A}]$, we have $\alpha = \beta = 1$. But due to column s and rows u and w, the entries α and β must have opposite sign, a contradiction. Thus, exactly one of α and β is zero. Then A' of (9.3.8) has up to index sets the matrix B^7 as support. We conclude that M or M^* is isomorphic to F_7.   □

During a first reading of the book, the next section may be skipped. There we characterize the ternary matroids.

9.4 Characterization of Ternary Matroids

The proof of Theorem (9.3.2) of the preceding section can be extended to establish the following characterization by R. Reid of the ternary matroids, i.e., the matroids representable over GF(3). Recall from Section 3.4 that representability over a given field is maintained under minor-taking and dualizing.

(9.4.1) Theorem. *A matroid is representable over* GF(3) *if and only if it has no* F_7, F_7^*, U_5^2, *or* U_5^3 *minors.*

Proof. We assume that the reader is quite familiar with the material on abstract representations of Section 3.4. It is easy to check that the excluded minors are not representable over GF(3). Indeed, it suffices to verify this for F_7 and U_5^2, since F_7^* and U_5^3 are the duals of these matroids. Thus, the "if" part holds. For proof of the converse, we argue as in the proof of Theorem (9.3.2), except that B is an abstract matrix representing M. The graph $BG(B)$ cannot be a cycle or path, since otherwise M is a GF(3)-representable wheel or whirl matroid or a minor of such a matroid. Thus, up to dualizing of M, B can be partitioned as in (9.3.3), where $\overline{B}$ is connected.

We need the following auxiliary result: Two matrices over GF(3) and with the same support represent the same matroid if and only if either one of the two matrices may be obtained from the other one by scaling of rows and columns by $\{\pm 1\}$ factors. The proof is identical to that of Lemma (9.2.6), except that the final sentence of the proof is replaced by the observation that a matrix of (9.2.3) has the GF(3)-determinant equal to 0 if and only if its entries sum (in $\mathbb{R}$) to 0(mod 4).

Because of the auxiliary result and its proof, the signing of an abstract matrix to achieve a representation matrix over GF(3) can be done column by column, provided of course the matroid is GF(3)-representable. This observation is the GF(3)-analogue of Corollary (9.2.7). Thus, for the case at hand, we may deduce from B, which is partitioned as in (9.3.3), a matrix A of (9.3.4) over GF(3) where $[a \mid \overline{A}]$ represents $M \backslash q$ and where $[b \mid \overline{A}]$ represents $M \backslash p$. Since M is not representable over GF(3), A does not represent M.

Due to abstract pivots in $\overline{B}$ of B and corresponding GF(3)-pivots in $\overline{A}$ of A, we may assume that B and A satisfy the following two additional conditions. First, A is the matrix of (9.3.5), except that the explicitly shown -1 in row w and column q is a $+1$ or -1. We denote that entry by γ. Second, the GF(3)-determinant of the 2×2 submatrix of rows v, w and columns p, q is not correct for M. That is, the GF(3)-determinant of that submatrix is zero (resp. nonzero), i.e., $\gamma = 1$ (resp. $\gamma = -1$), if and only if the related 2×2 submatrix of B has abstract determinant 1 (resp. 0). Let us include the just-described matrix A for easy reference.

(9.4.2)

$$A =$$

Partition of matrix A displaying 2×2 submatrix
with incorrect determinant

Due to the γ entry, the ensuing arguments are a bit more subtle than those proving Theorem (9.3.2). Assume that both row subvectors $\overline{c}$ and $\overline{e}$ in rows v and w of A have ± 1s in a column $y \neq p, q$. Extract from A of (9.4.2) the submatrix A' indexed by $\{v, w\}$ and $\{p, q, y\}$, i.e.,

(9.4.3)

$$A' = \begin{array}{c} \quad p \ \ q \ \ y \\ \begin{array}{c} v \\ w \end{array}\left[\begin{array}{ccc} 1 & 1 & \pm 1 \\ 1 & \gamma & \pm 1 \end{array}\right] \end{array}$$

Submatrix A' of A displaying U_5^2 minor

That submatrix corresponds to a minor M' of M. According to the previous discussion, p and q are parallel in M' if and only if $\gamma = -1$. Assume

$\gamma = -1$. By (9.4.3), one of $\{p, y\}$, $\{q, y\}$, say $\{p, y\}$, contains two parallel elements of M', and the other one is a base of M'. Since $\gamma = -1$, p and q are parallel in M'. But "is parallel to" is an equivalence relation, so y and q must also be parallel in M', a contradiction. Thus, $\gamma = +1$. Arguing by contradiction as in the previous case, both $\{p, y\}$ and $\{q, y\}$ must be bases of M'. Then M' is isomorphic to U_5^2, and we are done.

We return to A of (9.4.2), knowing now that it can be further partitioned as in (9.3.6), except that the explicitly shown -1 in (9.3.6) must be replaced by γ. By path-shortening GF(3)-pivots in $\overline{A}$ and subsequent deletion of rows and columns, we get the following analogue of the small case (9.3.8).

(9.4.4)

$$A' = \begin{array}{c} \\ u \\ v \\ w \end{array}\begin{array}{c} p\ q\ r\ s \\ \hline \alpha\ \beta\ \delta\ \epsilon \\ 1\ 1\ 1\ 0 \\ 1\ \gamma\ 0\ 1 \end{array}$$

Submatrix A' of A displaying U_5^2, U_5^3, or F_7 minor

Here $\delta, \epsilon = \pm 1$, while α and β are as-yet-undetermined $\{0, \pm 1\}$ entries. We examine the possible cases for α and β. Let M' be the minor of M corresponding to A'.

Case 1: $\alpha, \beta = \pm 1$. Suppose the 2×2 submatrix $\overline{A}'$ of A' indexed by u, v and p, q has incorrect GF(3)-determinant for M'. Then the rows u, v and the columns p, q, r are up to indices an instance of (9.4.3). Thus, M' has a U_5^2 minor. Hence, we may assume the determinant of $\overline{A}'$ to be correct for M'. Similarly, the GF(3)-determinant of the 2×2 matrix indexed by u, w and p, q may be assumed to be correct for M'. But then up to indices, the columns p, q constitute the transpose of the case (9.4.3), thus giving an U_5^3 minor.

Case 2: $\alpha = \beta = 0$. We GF(3)-pivot in A' on the 1 in row w and column s to obtain case 1.

Case 3: $\alpha = 0$, $\beta \neq 0$: This is the last case, since $\alpha \neq 0$, $\beta = 0$ is symmetric to it by scaling. We may assume $\beta = 1$ because of scaling of row u. If $\delta = -1$ (resp. $\epsilon \neq \gamma$), we GF(3)-pivot on the 1 in row v (resp. w) and column r (resp. s) of A' to get case 1. Hence, assume $\delta = 1$ and $\epsilon = \gamma$. If $\gamma = 1$, a GF(3)-pivot on the 1 in row v and column r of A', followed by a GF(3)-pivot on the 1 in row w and column s, produces case 1. If $\gamma = -1$, then a simple analysis of the GF(3)-determinants proves M' to be isomorphic to F_7.

The conclusions drawn via the preceding GF(3)-pivot arguments are valid for the following reasons. Let one of the above GF(3)-pivots transform A' of (9.4.4) to A''. Let B' be the abstract matrix that represents M' and that corresponds to A'. We claim that the related abstract pivot in B' produces a B'' that is linked to A'' as follows. First, the 2×2 submatrix in the lower left corner has determinant 0 in B'' if and only if the corresponding submatrix of A'' has a nonzero GF(3)-determinant. Thus, that submatrix of A'' has incorrect GF(3)-determinant for M' as does that of A'. Second, let deletion of the first or second column reduce A'' and B'' to A''' and B''', respectively. Then any square submatrix of B''' has determinant 0 if and only if the corresponding square submatrix of A''' has GF(3)-determinant 0. Thus, that submatrix of A''' has a correct GF(3)-determinant for M'. In particular, A''' and B''' have the same support. The preceding claims follow directly from the pivot rules for abstract determinants as described in Section 3.4. $\qquad\square$

Theorem (9.4.1) implies Theorem (9.3.2) by elementary arguments. By Theorem (9.2.9), a binary matroid M is regular if and only if it is representable over GF(3). By Theorem (9.4.1), a matroid is representable over GF(3) if and only if it has no U_5^2, U_5^3, F_7, F_7^* minors. But U_5^2 and U_5^3 are not binary, and thus Theorem (9.3.2) follows.

In the final section, we indicate extensions and cite references.

9.5 Extensions and References

The prior work on total unimodularity can be roughly divided into two categories. The first one contains papers investigating the structure of totally unimodular matrices, of minimal $\{0, \pm 1\}$ matrices that are not totally unimodular, and of closely related matrix classes. Relevant references are Cederbaum (1957), Ghouila-Houri (1962), Camion (1963a), (1963b), (1965), (1968), Heller (1963), Veinott and Dantzig (1968), Chandrasekaran (1969), Commoner (1973), Gondran (1973), Padberg (1975), (1976), (1988), Brown (1976), (1977), Tamir (1976), Truemper (1977), (1978), (1980b), (1982b), (1990), (1992), Kress and Tamir (1980), de Werra (1981), Cunningham (1982a), Chandrasekaran and Shirali (1984), Fonlupt and Raco (1984), Seymour (1985a), Yannakakis (1985), Crama, Hammer, and Ibaraki (1986), Conforti and Rao (1987), and Hempel, Herrmann, Hölzer, and Wetzel (1989).

The second category concerns applications of total unimodularity. Pertinent references are Dantzig and Fulkerson (1956), Hoffman and Kruskal (1956), Hoffman and Kuhn (1956), Motzkin (1956), Heller and Tompkins (1956), Heller (1957), Hoffman (1960), (1974), (1976), (1979), Ford and Fulkerson (1962), Heller and Hoffman (1962), Rebman (1974), Lawler (1976), Baum and Trotter (1978), Hoffman and Oppenheim (1978), Truemper and Chandrasekaran (1978), Truemper and Soun (1979), Soun and Truemper (1980), Maurras, Truemper, and Akgül (1981), Bixby (1982a), (1984a),

Schrijver (1983), Tamir (1987), Recski (1989), and Ahuja, Magnanti, and Orlin (1989). A comprehensive treatment is given in Schrijver (1986).

Lemma (9.2.1) is well known, while Lemmas (9.2.2) and (9.2.4) are implicit in most references of the first category. Lemma (9.2.6) is taken from Camion (1963b). It also follows from Brylawski and Lucas (1973). Lemma (9.2.8) and Theorem (9.2.9), though with a quite different proof, are from Tutte (1958), (1965), (1971).

References given above for the second category contain numerous results about problems of the form (9.2.17).

The matroids R_{10} and R_{12} are the two central matroids in the proof of the regular matroid decomposition theorem in Seymour (1980b). The matroid R_{10} had appeared earlier in Hoffman (1960). See also Bixby (1977). Contrary to sometimes-voiced claims, the matroids R_{10} and R_{12} do arise from well-known combinatorial problems. We present two examples. The first one involves the real constraint matrices of two-commodity flow problems on directed graphs. In Soun and Truemper (1980) it is shown that an infinite subclass of such problems has constraint matrices that give rise to regular matroids that are nongraphic and noncographic. Indeed, it can be shown that these matroids have R_{12} minors.

The second example is due to Chvátal (1986). It involves the graph G below.

(9.5.1)

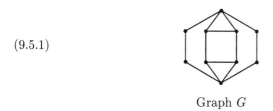

Graph G

Construct the following real matrix B from G. Each column of B corresponds to a node of G, and each row to a clique (= maximal complete subgraph) of G. Each row i of B is then the incidence vector of the nodes in clique i. The matrix B is the *clique/node incidence matrix* of G. The clique/node incidence matrices of graphs are very useful for the solution of the so-called independent vertex set problem. For the graph G at hand, the clique/node incidence matrix B can be proved to be totally unimodular. Indeed, it is not difficult to verify that the regular matroid represented by B is nongraphic and noncographic and has an R_{12} minor.

Theorem (9.3.2) is Tutte's famous characterization of the regular matroids (Tutte (1958), (1965), (1971)). The beautiful proof is due to Gerards (1989a). Variations of the theorem appear in Bixby (1976) and Truemper (1982b), (1991). Related is work on extremal matroids in Murty (1976).

Theorem (9.4.1) is due to R. Reid, who never published his proof. Other proofs of that result appear in Bixby (1979), Seymour (1979a),

Truemper (1982b), Kahn (1984), and Kahn and Seymour (1988). As remarked in Section 9.4, any proof of Theorem (9.4.1) essentially constitutes a proof of Theorem (9.3.2) as well. A characterization of the matroids representable over GF(3) in terms of circuit signatures is given in Roudneff and Wagowski (1989).

Representability over GF(3) and GF(q) is treated in Kung (1990b), and Kung and Oxley (1988). Ternary matroids without $M(K_4)$ minors are covered in Oxley (1987c). Partial results for representability over GF(4) are given in Oxley (1986), (1990a) and Kahn (1988). Early examples of nonrepresentable matroids are in MacLane (1936), Lazarson (1958), Ingleton (1959), (1971), and Vamos (1968). It has been conjectured (Rota (1970)) that for any finite field $\mathcal{F}$, the number of matroids that are not representable over $\mathcal{F}$, and that are minimal with respect to that property, is finite.

General questions concerning representability are discussed in Vamos (1978), Kahn (1982), and Ziegler (1990).

The complexity of representability tests using various oracles is examined in Robinson and Welsh (1980), Seymour (1981c), Seymour and Walton (1981), Jensen and Korte (1982), and Truemper (1982a). Virtually every such test requires exponential time when the matroid is given by an oracle that decides independence of sets. An exception is the test of representability over every field. A polynomial algorithm for that problem is given in Truemper (1982a). The algorithm relies on the regular matroid decomposition theorem of Seymour (1980b), which is covered in Chapter 11.

Chapter 10

Graphic Matroids

10.1 Overview

At this point, we are ready to investigate the first complicated class of binary matroids treated in this book: the class of graphic matroids. Recall the following definitions and results. K_n is the complete graph on n nodes, and $K_{m,n}$ is the complete bipartite graph with m nodes on one side and n nodes on the other side. For any graph G, the corresponding graphic matroid is regular and is denoted by $M(G)$. F_7 denotes the nonregular Fano matroid. Finally, the asterisk is used as dualizing operator.

In this chapter, we first identify certain minimal regular matroids that are not graphic, or that are not graphic and not cographic. Specifically, in Section 10.2, we characterize the planar regular matroids, i.e., the matroids produced by planar graphs. In Section 10.3, we investigate the behavior of nongraphic regular matroids with $M(K_{3,3})$ minors. We build upon these results in Section 10.4. There we prove two profound theorems of matroid theory: Tutte's theorem that $M(K_5)^*$ and $M(K_{3,3})^*$ are the minimal regular matroids that are not graphic, and Seymour's theorem that the two matroids R_{10} and R_{12} defined in Section 9.2 are the minimal regular, 3-connected matroids that are not graphic and not cographic. The latter theorem is one of the two central ingredients in the proof of Seymour's profound decomposition theorem for regular matroids. We take up the latter theorem and its proof in Chapter 11.

In Section 10.5, we introduce a simple but very useful decomposition scheme that will be used repeatedly in Chapters 11–13. Indeed, the scheme is the second ingredient in the proof of the just-mentioned decomposition theorem for regular matroids. In Section 10.5, we employ the scheme to deduce some graph decomposition theorems, among them Wagner's famous decomposition theorem for the graphs without K_5 minors.

In Section 10.6, we present an efficient algorithm for deciding graphicness of binary matroids and for deciding whether or not a real $\{0, \pm1\}$ matrix A is the coefficient matrix of a network flow problem. Finally, in Section 10.7, we indicate extensions and provide references.

The chapter requires knowledge of Chapters 2, 3, and 5–8. We also make use of the easy part of Theorem (9.3.2), according to which the Fano matroid is nonregular.

10.2   Characterization of Planar Matroids

In this section, we prove that a regular matroid is planar if and only if it has no $M(K_{3,3})$, $M(K_{3,3})^*$, $M(K_5)$, or $M(K_5)^*$ minors. The result constitutes one of two preparatory steps toward proofs of the theorems by Tutte and Seymour cited in the introduction to this chapter. The latter results are established in Section 10.4.

Definition of Graph with T Nodes

For the arguments of this section and the next one, we need a convenient way to encode and manipulate 1-element binary additions of any graphic matroid N. Let $\overline{B}$ be a binary representation matrix of N, say with row index set X and column index set Y. Thus, the matroid $M = N+z$ is represented by the following matrix B.

(10.2.1)

$$ B = \begin{array}{c} \\ X \end{array} \begin{array}{|c|c|} \multicolumn{1}{c}{Y} & \multicolumn{1}{c}{z} \\ \hline \overline{B} & b \\ \hline \end{array} $$

Matrix B for matroid $M = N+z$

Let G be any connected graph for N, i.e., $M(G)$ is N. The graph G need not be unique. The row index set X of B is a tree of G. Suppose we premultiply the matrix $[I \mid \overline{B} \mid b]$ with the node/edge incidence matrix of the tree X. By the results of Section 3.2, that multiplication turns the submatrix $[I \mid \overline{B}]$ into the node/edge incidence matrix, say F, of G. The column vector b becomes some vector, say d. Evidently, F and $[F \mid d]$ have the same GF(2)-rank. Accordingly, since every column of F has exactly two 1s or none, the vector d must have an even number of 1s. Since each row of F corresponds to a node of G, we can associate the 1s of d with a node subset T of G. Each node of T we call a T *node*. In drawings of G, we denote each T node by a square box. By the construction, M is completely represented by G and the set T.

Section 3.2 contains the following alternate way of representing M via G. Each 1 of the vector b of (10.2.1) corresponds to an edge of the tree X. To single out these edges, we temporarily paint them red. In general, the red edges form a red subgraph $\overline{X}$ of G without cycles. The following lemma links that subgraph to the set T.

(10.2.2) Lemma.   *A node of G is a T node if and only if the node has an odd number of red edges incident.*

Proof. In the matroid M, the element z forms a fundamental circuit C with X. Indeed, $C - \{z\}$ is nothing but the red subgraph $\overline{X}$. That subgraph indexes a column submatrix $\overline{F}$ of F. Thus, the columns of $[\overline{F} \mid d]$, which are indexed by C, are GF(2)-mindependent, i.e., the columns are GF(2)-dependent, but any proper subset of the columns is GF(2)-independent. This is so if and only if the 1s of d are in the rows of $\overline{F}$ with an odd number of 1s. Equivalently, a node of G is in T if and only if the node has an odd number of red edges incident.   □

Lemma (10.2.2) implies a convenient method for determining the T nodes. Suppose we have B of (10.2.1) and a graph G for the matroid N of $\overline{B}$. For each 1 in the vector b, say in row $x \in X$ of B, we temporarily paint the edge x of G red. Then we define the nodes of G with odd number of red edges incident to be the T nodes. Finally, we declare the red edges to be unpainted again.

We claim that M is graphic if $|T| = 2$. Indeed, the red subgraph $\overline{X}$ is then a red path, say from node u of G to node v. We thus may add an edge z connecting u and v to get a graph representing M.

Example Graphs with T Nodes

Below, we carry out the derivation of T for a few nongraphic matroids that are important for our purposes. We depict each instance using the following scheme.

(10.2.3)

$$B = \begin{array}{c} \\ X \end{array} \begin{array}{|c|c|} \hline \overline{B} & b \\ \hline \end{array} \begin{array}{c} Y \quad z \end{array}$$

Matrix B

$\longrightarrow$ Graph G with red subgraph $\overline{X}$ (indicated by bold edges)

$\longrightarrow$ Graph G with T nodes (indicated by squares enclosing nodes)

Representation of M by matrix B, by graph G with red edges, and by graph G with T nodes

Here are the example matroids M.

(10.2.4)

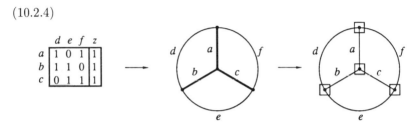

$$\begin{array}{c|ccc|c} & d & e & f & z \\ \hline a & 1 & 0 & 1 & 1 \\ b & 1 & 1 & 0 & 1 \\ c & 0 & 1 & 1 & 1 \end{array}$$

Fano Matroid F_7 (defined by (9.3.1))

(10.2.5)

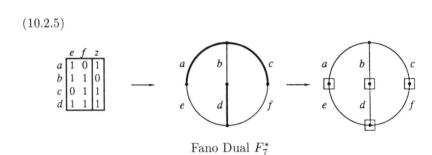

$$\begin{array}{c|ccc} & e & f & z \\ \hline a & 1 & 0 & 1 \\ b & 1 & 1 & 0 \\ c & 0 & 1 & 1 \\ d & 1 & 1 & 1 \end{array}$$

Fano Dual F_7^*

(10.2.6)

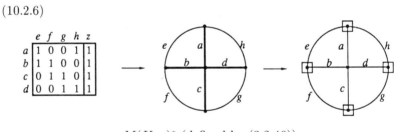

$$\begin{array}{c|cccc|c} & e & f & g & h & z \\ \hline a & 1 & 0 & 0 & 1 & 1 \\ b & 1 & 1 & 0 & 0 & 1 \\ c & 0 & 1 & 1 & 0 & 1 \\ d & 0 & 0 & 1 & 1 & 1 \end{array}$$

$M(K_{3,3})^*$ (defined by (3.2.46))

(10.2.7)

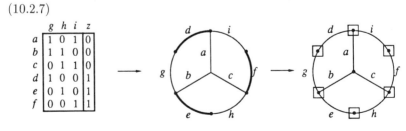

$$\begin{array}{c|ccc|c} & g & h & i & z \\ \hline a & 1 & 0 & 1 & 0 \\ b & 1 & 1 & 0 & 0 \\ c & 0 & 1 & 1 & 0 \\ d & 1 & 0 & 0 & 1 \\ e & 0 & 1 & 0 & 1 \\ f & 0 & 0 & 1 & 1 \end{array}$$

$M(K_5)^*$ (defined by (3.2.44))

(10.2.8)

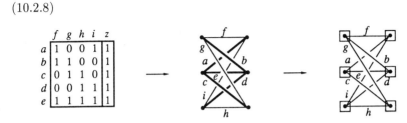

	f	g	h	i	z
a	1	0	0	1	1
b	1	1	0	0	1
c	0	1	1	0	1
d	0	0	1	1	1
e	1	1	1	1	1

R_{10} (defined by (9.2.13))

(10.2.9)

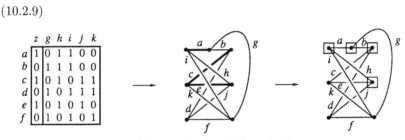

	z	g	h	i	j	k
a	1	0	1	1	0	0
b	0	1	1	1	0	0
c	1	0	1	0	1	1
d	0	1	0	1	1	1
e	1	0	1	0	1	0
f	0	1	0	1	0	1

R_{12} (defined by (9.2.14); note that
column z is the first column)

In each of the example cases, we have established the set T via a particular tree X of G. Since T was originally defined via the matrix $[F \mid d]$, any other tree of G would have produced the same set T. For this reason, we may always select a particularly suitable tree X when discussing some matroid operation and its impact on G and T. We have three such operations in mind: deletion of an element $y \neq z$ of M that is not a coloop of G, contraction of an element $x \neq z$ of M that is not a loop of G, and the switching operation of Section 3.2. We take up these operations next.

Deletion, Contraction, and Switching in Graphs with T Nodes

Suppose we delete from M an element $y \neq z$ that is not a coloop of G. Thus, G has a tree X that is also a tree of $G \backslash y$. Then $G \backslash y$ with the original T node labels represents $M \backslash y$.

Suppose in M we contract an element $x \neq z$ that is not a loop. We may assume that the tree X of G contains x. Then $X - \{x\}$ is a tree of G/x and a base of M/x. We claim that the red edges of $X - \{x\}$ plus z

constitute the fundamental circuit that z forms with $X - \{x\}$ in M/x. For a proof, we delete row $x \in X$ from the matrix B to obtain a representation matrix for M/x. Inspection of the column z of the latter matrix verifies the claim. Thus, we may use the red edges of $X - \{x\}$ to represent the element z of M/x by a node subset T' of G/x analogously to the use of the red edges of X to define the node subset T of G. We deduce T' directly from T by the following rules. Let i be any node of G different from the two endpoints u and v of x. Then i is in T' if and only if i is in T. Define w to be the node of G/x created from u and v of G when x is contracted, i.e., $w = (u \cup v) - \{x\}$. Then w is in T' if and only if exactly one of u and v is in T. Validity of these rules follows directly from the just-mentioned fact about the red edges of $X - \{x\}$, and from a simple parity argument involving the red edges of G incident at nodes u or v.

Recall the switching operation of Section 3.2. On hand must be a 2-separation of G, say involving subgraphs G_1 and G_2. The graph G_1 is removed, turned over, and reattached to G_2. The resulting graph G' is 2-isomorphic to G. Thus, G' also represents $M\backslash z$. We want to deduce the set T' for G' from the set T of G. That is, G' and T' are to represent M analogously to G and T. Evidently, any tree X of G is one of G', and the red edges of X may be used to deduce T' for G'. Thus, any node different from the two nodes joining G_1 and G_2 is in T' if and only if it is in T. The rules for the latter two nodes are also quite simple. For the general situation, we leave their derivation to the reader. Instead, we just examine the special case where the graph G has two series edges e and f with a common endpoint $w \in T$ with degree 2. Let u be the second endpoint of edge e, and let v be that of edge f. Assume $u \neq v$. The switching operation resequences e and f. Thus, u becomes $u' = (u - \{e\}) \cup \{f\}$ and v becomes $v' = (v - \{f\}) \cup \{e\}$. For the derivation of T' of G' from T of G, we may suppose that the tree X of G includes both e and f. Since $w \in T$, exactly one of the edges e and f is red. Correspondingly, the parity of the number of red edges at u (resp. v) in G is different from the parity of the number of red edges at u' (resp. v') in G'. Thus, u' (resp. v') is in T' if and only if u (resp. v) is not in T. An example case is depicted below. As before, nodes of T are indicated by squares.

(10.2.10)

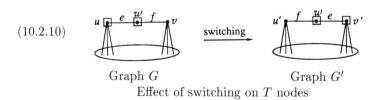

Graph G Graph G'

Effect of switching on T nodes

The set T' produced by any switching may have cardinality different from that of T. Thus, we are justified in assuming for convenience that $|T|$ is minimal under switchings. In that case, M is graphic if and only if $|T|$ is 0 or 2. The 0 case corresponds to z being a loop of M. It cannot occur when M is connected.

Characterization of Planar Matroids

We now have sufficient machinery to prove the main result of this section, which characterizes planarity of regular matroids in terms of excluded minors.

(10.2.11) Theorem. *A regular matroid M is planar if and only if M has no $M(K_{3,3})$, $M(K_{3,3})^*$, $M(K_5)$, or $M(K_5)^*$ minors.*

Proof. The "only if" part holds since planarity is maintained under minor-taking and since $M(K_{3,3})$, $M(K_{3,3})^*$, $M(K_5)$, and $M(K_5)^*$ are not planar. For the proof of the nontrivial "if" part, let M be a regular matroid all of whose proper minors are planar. Thus, M is minimally nonplanar with respect to the taking of minors. We must show that M is isomorphic to $M(K_{3,3})$, $M(K_{3,3})^*$, $M(K_5)$, or $M(K_5)^*$.

If M is graphic or cographic, then the desired conclusion is provided by Theorem (7.4.1), which characterizes nonplanar graphs by exclusion of $K_{3,3}$ and K_5 minors. Thus, we may assume from now on that M is not graphic and not cographic.

We claim that M is 3-connected. If that is not the case, then M has a 1- or 2-separation. By Lemma (8.2.2) or (8.2.6), M is a 1- or 2-sum. In either case, the components of the sum are proper minors of M and thus planar. But by Lemma (8.2.2) or (8.2.7), the latter conclusion implies M to be planar as well, a contradiction.

By the census of Section 3.3, every 3-connected regular matroid on at most eight elements is planar. Thus, M has at least nine elements.

We apply the binary matroid version of the wheel Theorem (7.3.3) to the 3-connected, nongraphic, and noncographic M on at least nine elements. Accordingly, M must have an element z so that at least one of the minors M/z and $M\backslash z$ is 3-connected. If the M/z case applies, we replace M by its dual. Since all assumptions made so far for M are invariant under dualizing, this change does not affect the proof. Thus, we may assume that $M\backslash z$ is 3-connected. By the minimality of M, the 3-connected minor $M\backslash z$ is planar. Let G be the corresponding planar graph. We extend G to a representation of M by selecting an appropriate subset T of nodes of G for the element z. Recall that the cardinality of T is necessarily even. Since M is not graphic, $|T| \geq 4$.

Suppose G is a wheel. It is easily verified that one can delete spokes from G and contract rim edges so that the wheel with four spokes and with four T nodes results. Then we either have, up to indices, the $M(K_{3,3})^*$ case of (10.2.6) and are done, or we can delete one spoke and contract one

rim edge to obtain an instance of the F_7 case of (10.2.4), which contradicts
the regularity of M.

We are left with the case where G is not a wheel. By the wheel
Theorem (7.3.3), G has an edge e so that G/e or $G\backslash e$ is 3-connected.
Assume the latter case. The deletion of the edge e from G leaves the number
of T nodes unchanged. Note that $G\backslash e$ plus these T nodes represents $M\backslash e$.
Since $G\backslash e$ is 3-connected and $|T| \geq 4$, the minor $M\backslash e$ must be 3-connected
and nongraphic, a contradiction of the minimality of M. Thus, the case
of a 3-connected G/e must be at hand. We claim that the contraction of
the edge e in G must reduce the number of T nodes to 2. If this is not
the case, then arguments analogous to those for $M\backslash e$ prove M/e to be
3-connected and nongraphic, a contradiction. Since G/e has exactly two
T nodes, the graph G must have exactly four T nodes, two of which must
be the endpoints of e, say u and v. Define i and j to be the other two T
nodes of G.

By the 3-connectedness of G and Menger's theorem, there is a path P
from i to u and a second path Q from j to u such that these paths have
only node u in common and do not involve node v. Imagine G drawn in
the plane. Then deletion of node u would create a new face. The boundary
would be a cycle, say C. Clearly, v has become a node of C. If in G the T
node i does not lie on C, then we contract the edge of the path P incident
at i. After suitable repetition of this process, the T node has become a
node of C. Then we declare that T node to be i again. Similarly, we make
the T node j a node of C. At that time, the edges incident at node u
and those of the cycle C constitute a subdivision of a wheel with at least
three spokes. The rim of that wheel subdivision contains the T nodes v, i,
and j. The latter nodes induce a partition of the rim into three paths, say
P_1 from i to v, P_2 from j to v, and P_3 from i to j. Simple case checking
confirms that the arguments made earlier for the wheel case of G apply
here fully unless every edge incident at node u has its second endpoint in
just one of the paths P_1 or P_2, say P_1. Below, we show a typical instance
of that exceptional case, together with an additional path P_4. That path is
nothing but a portion of the previously defined path Q that in G connected
nodes j and u. In the general case, the path P_4 connects an interior node
of the path $P_2 \cup P_3$ with an interior node k of the path P_1.

(10.2.12)

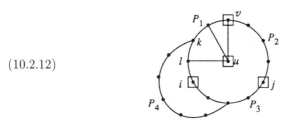

Wheel subdivision plus path P_4

The drawing does not show any edges of the previously defined path P. But there are still enough edges left from that path so that node u can be reached from node i while avoiding all nodes of P_4. This fact and the planarity of G imply that an interior node l of the subpath of P_1 from i to k must be connected with node u, as depicted in (10.2.12).

Evidently, we can contract enough edges of the paths P_1–P_4 and delete some edges incident at node u so that node k becomes the center node of a wheel graph with four spokes and with four T nodes on its rim. Thus, up to indices, we have an instance of the $M(K_{3,3})^*$ case of (10.2.6). $\square$

Related to Theorem (10.2.11) is the following pioneering combinatorial characterization of planar graphs due to Whitney. It constitutes the first matroid result about graph planarity.

(10.2.13) Corollary.   *A graph G is planar if and only if $M(G)^*$ is graphic.*

Proof. The "only if" part is clear. For proof of the "if" part, let $M = M(G)$. Since $M^* = M(G)^*$ is graphic, M^* has no $M(K_5)^*$ or $M(K_{3,3})^*$ minors. Thus, M has no $M(K_5)$ or $M(K_{3,3})$ minors. Since M is graphic, M has no $M(K_5)^*$ or $M(K_{3,3})^*$ minors. Thus, by Theorem (10.2.11), M and G are planar. $\square$

One may derive Corollary (10.2.13) directly from Theorem (7.4.1), without use of Theorem (10.2.11). Indeed, Theorem (7.4.1) is the graph version of Corollary (10.2.13). Whitney's contribution is the deduction of this result from Kuratowski's original planarity characterization, which involved subdivisions of K_5 and $K_{3,3}$.

In the preceding chapters, we took great care when we dualized planar graphs. Each time, we embedded a given planar graph in the plane, then dualized that plane graph. Corollary (10.2.13) frees us from this adherence to planar embeddings. We now may take the following viewpoint. Given a planar graph G, let H be any graph for $M(G)^*$. Thus, the matroids $M(G)$ and $M(H)$ are duals of each other. We declare H to be a *dual graph* of G. For our purposes, any H satisfying $M(H)^* = M(G)$ will do. By this definition, any two such graphs are 2-isomorphic. Thus, by Theorem (3.2.36), H is unique if it, or equivalently G, is 3-connected, the case typically of interest to us. In that situation, we are justified to call H *the* dual of G. Suppose H is 2-connected. By Theorem (3.2.36), any other graph for $M(G)^*$ is related to H by switchings. We leave it to the reader to explore this issue further. Whitney first recognized these relationships among embeddings, 2-isomorphism, and switchings.

We move on to the next section, where we investigate regular matroids with $M(K_{3,3})$ minors.

10.3   Regular Matroids with $M(K_{3,3})$ Minors

The title of this section may seem strange. Why would one be interested in $M(K_{3,3})$ minors of regular matroids? Early in this section, we give a partial answer in the form of two lemmas. More satisfactory as answer are the arguments of the next section, which prove $M(K_{3,3})$ to play a central role in the analysis of regular matroids.

Following the two lemmas and some other preparatory material, we introduce the main theorem of this section. It says that any 3-connected, regular, nongraphic and noncographic matroid with an $M(K_{3,3})$ minor has a minor isomorphic to one of the nongraphic and noncographic matroids R_{10} and R_{12}. This profound result is due to Seymour. In the next section we combine it with Theorem (10.2.11) to obtain two important characterizations of nongraphic regular matroids by Seymour and Tutte.

To start, we recall the splitter definition of Section 7.2. Let $\mathcal{M}$ be a class of binary matroids that is closed under isomorphism and under the taking of minors. Let N be a 3-connected matroid of $\mathcal{M}$ on at least six elements. Then N is a *splitter* of $\mathcal{M}$ if every connected matroid $M \in \mathcal{M}$ with a proper N minor is 2-separable. Intuitively and informally speaking, the presence of an N minor forces M to split. By Theorem (7.2.11), the graph K_5 is a splitter of the graphs without $K_{3,3}$ minors. That result has the following matroid extension.

(10.3.1) Lemma. $M(K_5)$ *is a splitter of the regular matroids without* $M(K_{3,3})$ *minors.*

Proof. By Theorem (7.2.1), we only need to show that every 3-connected regular 1-element extension of $M(K_5)$ has an $M(K_{3,3})$ minor. This is accomplished by a straightforward case analysis. To assist the reader, we sketch one way of checking.

By (3.2.38), the matrix

$$(10.3.2) \qquad\qquad B = \begin{bmatrix} 1 & 0 & 0 & 1 & 1 & 0 \\ 1 & 1 & 0 & 0 & 0 & 1 \\ 0 & 1 & 1 & 0 & 1 & 0 \\ 0 & 0 & 1 & 1 & 0 & 1 \end{bmatrix}$$

Matrix B for $M(K_5)$

represents $M(K_5)$. Suppose we adjoin a column vector b that represents an added element z. If b has at most two 1s, then z is a coloop or a parallel element. If b has three or four 1s, then $[B \mid b]$ contains a submatrix representing the Fano matroid F_7, which is nonregular. Thus, no 3-connected regular 1-element addition is possible.

Now assume we adjoin a row vector c that represents a 1-element regular and 3-connected expansion by an element e. We may view B as the node/edge incidence matrix of a graph H that is isomorphic to K_4. It is convenient to encode each 1 of the row vector c by a red edge of H. Since $M(K_5)\&e$ is 3-connected, H must have at least two red edges. We analyze the possible configurations of red edges.

Suppose that H contains exactly two red edges, and that these edges share an endpoint. Then $M(K_5)\&e$ is easily confirmed to be graphic, with an $M(K_{3,3})$ minor. Next, suppose H contains exactly four red edges that form a cycle C. Let y and z be the edges of H that are not in C. Then $(M(K_5)\&e)\backslash\{y,z\}$ is isomorphic to $M(K_{3,3})$.

For the remaining configurations of red edges, one proves the presence of an F_7^* minor. Specifically, if the red edges form a triangle, then the columns of $[B/c]$ corresponding to that triangle establish the presence of an F_7^* minor. The other cases are slightly more difficult to prove. As an example case, let us examine the 1-element expansion M of $M(K_5)$ by the element e given by

(10.3.3)

$$
\begin{array}{c|cccccc}
 & f & g & h & i & j & k \\
\hline
a & 1 & 0 & 0 & 1 & 1 & 0 \\
b & 1 & 1 & 0 & 0 & 0 & 1 \\
c & 0 & 1 & 1 & 0 & 1 & 0 \\
d & 0 & 0 & 1 & 1 & 0 & 1 \\
e & 1 & 0 & 1 & 0 & 0 & 0 \\
\end{array}
$$

Matrix of 1-element expansion M of $M(K_5)$

It turns out that we can prove nonregularity without using the last column k. So let us consider that column deleted. One readily verifies that the remaining matrix is represented by

(10.3.4)

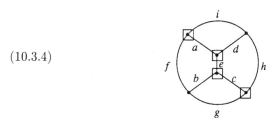

Graph plus T nodes for $M\backslash k$

where the T nodes correspond to the element j. From the graph of (10.3.4), we delete the edges a and c, and contract the edge b. The resulting graph with adjusted T set is given by (10.3.5) below. By (10.2.5), that graph represents an F_7^* minor. Thus, the matroid given by the matrix of (10.3.3) is nonregular.

(10.3.5)

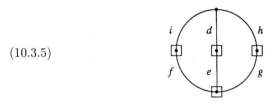

Graph plus T nodes for $M/\{b\}\backslash\{a,c,k\}$

The remaining open cases are handled the same way.   □

For later reference, we include another lemma about $M(K_{3,3})$. We omit the elementary case analysis via graphs plus T sets.

(10.3.6) Lemma.   *Every 3-connected binary 1-element expansion of $M(K_{3,3})$ is nonregular.*

The proof of the next result about $M(K_{3,3})$ requires a bit of preparation. Recall from Section 3.3 the following definition. Let M be a binary matroid and z be an element of M. Then $M \textcircled{c} z$ is the matroid obtained from M/z by deletion of all elements but one from each parallel class. The minor $M \textcircled{d} z$ is derived from $M \backslash z$ by contraction of all elements but one in each series class. For convenient reference, we restate Lemma (3.3.31), which links $M \textcircled{c} z$ and $M \textcircled{d} z$ to 3-connectedness.

(10.3.7) Lemma.   *Let M be a 3-connected binary matroid on a set E. Take z to be any element of E. Then $M \textcircled{c} z$ or $M \textcircled{d} z$ is 3-connected.*

Define a *line* of a graph to be a path of maximal length where all internal nodes have degree 2. A *corner node* of a graph is a node of degree at least 3. Let G be a subdivision of a 3-connected graph with at least four corner nodes. Evidently, each line of G is a series class and vice versa. We emphasize the latter fact, since generally a series class of a graph need not be a line. This is due to the definition of Section 2.2, where two edges are declared to be in series if they form a cocycle. *A priori,* the same subtle point needs to be considered when one specializes Lemma (10.3.7) to graphs. In that case, the notation $G \textcircled{c} z$ and $G \textcircled{d} z$ is interpreted analogously to that of $M \textcircled{c} z$ and $M \textcircled{d} z$.

We now show that the just-mentioned complications concerning series edges do not arise when we apply Lemma (10.3.7) to graphs. Let G be 3-connected, and let z be an edge of G, say with endpoints u and v. Recall that u and v are edge subsets. By Lemma (10.3.7), one of $G \textcircled{c} z$, $G \textcircled{d} z$ is 3-connected. Let $G \textcircled{c} z$ be that graph. The contraction of z may introduce parallel edges only at the new vertex $(u \cup v) - \{z\}$. Thus, $G \textcircled{c} z$ is readily determined. Now let $G \textcircled{d} z$ be 3-connected. We claim that $G \backslash z$ contains series classes with more than one edge only if u or v has degree 3, and that such series classes correspond to paths with two edges. Indeed, if u has degree 3, then $u - \{z\}$ is one such class. Similarly, v may produce a second series class. No other series class is possible, since 2-connected series expansions of a 3-connected graph can only produce a subdivision of that graph. Thus, $G \oplus z$ is readily determined.

The above conclusion is valid only if $G \oplus z$ is 3-connected. Indeed, if $G \oplus z$ is not 3-connected, then $G \backslash z$ may have a series class that is not a line. Fortunately, we never deal with the latter case since we make use of $G \oplus z$ only when that graph is 3-connected.

Here is a second preparatory lemma about $G \copyright z$ and $G \oplus z$.

(10.3.8) Lemma. *Let H be a subdivision of a 3-connected graph. Assume H has at least four corner nodes. Let G be any graph derived from H by the addition of nonloop edges. No such added edge is to connect two nodes of a line of H. Also, G is not allowed to have parallel or series edges. Then (a)–(c) below hold.*

(a) *G is 3-connected.*
(b) *$G \oplus z$ is 3-connected for every arc of G that is not in H.*
(c) *$G \copyright z$ is 3-connected for every arc of H both of whose endpoints have degree 2 in H.*

Proof. Clearly, G of part (a) is 2-connected. Suppose G has a 2-separation. We know that H is a subdivision of a 3-connected graph. Thus, any 2-separation of H has on one side a subset of one line of H. Assume that the 2-separation of G induces one in H. Then G has series edges, or G has an edge that is not in H and that connects two nodes of one line of H. Both cases contradict the assumptions. If the 2-separation of G does not induce one in H, then G must have parallel edges, again a contradiction. Thus, G is 3-connected.

Under the given assumptions, $G \oplus z$ and $G \copyright z$ of parts (b) and (c) satisfy the construction rules imposed on G. By (a), these minors are 3-connected. □

The next result concerns graphs with $K_{3,3}$ minors.

(10.3.9) Theorem. *Let G be a 3-connected graph with a $K_{3,3}$ minor.*

(a) *If G contains a triangle formed by edges e, f, and g, then G has one of the graphs of (10.3.10) below as a minor. The bold edges denote e, f, and g.*
(b) *If G has a node u of degree 3, then G has as subgraph a subdivision of $K_{3,3}$ that has u as corner node.*

(10.3.10)

Two extensions of $K_{3,3}$ containing a triangle

Proof. First we show part (a). Due to minor-taking, we may assume that every proper minor of G is not 3-connected, or does not have a $K_{3,3}$ minor, or does not contain the triangle $\{e, f, g\}$. We say that G is *minimal* to denote this fact. Let u, v, and w be the nodes of the triangle. Denote by H any subgraph of G that is a subdivision of $K_{3,3}$.

Suppose u is not a node of some H, say of H_1. Thus, u has an arc $z \neq e, f, g$ incident that is not in H_1. By Lemma (10.3.7), one of $G \copyright z$, $G \oplus z$ is 3-connected. Assume $G \copyright z$ to be 3-connected. In G/z we can delete parallel edges so that the triangle $\{e, f, g\}$ is retained. Now suppose $G \oplus z$ is 3-connected. If u has degree 3, then in $G \backslash z$ two edges of e, f, g are in series, and $G \oplus z$ has two edges of e, f, g in parallel, a contradiction. Thus, u has degree of at least 4, and $G \oplus z$ contains the triangle $\{e, f, g\}$. Clearly, both $G \copyright z$ and $G \oplus z$ have $K_{3,3}$ minors. But these facts contradict the minimality of G.

We conclude that u, v, and w are nodes of every H. If all three nodes occur on one line of some H, then there exists another H that avoids one of the three nodes. Therefore, at most two of the nodes occur on any one line of any H. Accordingly, we can always delete and contract edges in G such that e, f, g are retained and such that their endpoints become corner nodes of some $K_{3,3}$ minor. That process produces one of the graphs of (10.3.10).

For part (b), we once more define H to be any subgraph of G that is a subdivision of $K_{3,3}$. Suppose the given degree 3 node u of G is not a corner node of some H. Then by a ΔY exchange (see Section 4.3), the 3-star u can be replaced by a triangle $\{e, f, g\}$. It is easily seen that upon deletion of edges parallel to e, f, or g, we have a 3-connected graph with a $K_{3,3}$ minor. Apply part (a) to the latter graph. Thus, that graph has as minor one of the graphs of (10.3.10). Now replace $\{e, f, g\}$ by the 3-star u again. The resulting graph is a minor of G and is readily verified to have a $K_{3,3}$ minor $\overline{G}$ with u as a corner node. In straightforward fashion, we extend $\overline{G}$ to a subgraph of G that is a subdivision of $K_{3,3}$ and that has u as a corner node. $\qquad\square$

We are ready to state and prove the main result of this section.

(10.3.11) Theorem. *Let M be a 3-connected regular matroid with an $M(K_{3,3})$ minor. Assume that M is not graphic and not cographic, but that each proper minor of M is graphic or cographic. Then M is isomorphic to R_{10} or R_{12}.*

Proof. Let M be a smallest regular matroid that satisfies the assumptions of the theorem but not its conclusion. By Lemma (10.3.6), M does not have a 3-connected 1-element expansion of any $M(K_{3,3})$ minor. Take N to be any $M(K_{3,3})$ minor of M. Apply Theorem (7.3.4) to M and N. That theorem establishes the existence of a certain nested sequence of minors. In particular, the result implies that M has an element z for which the minor $M \backslash z$ is a series extension of a 3-connected matroid with an $M(K_{3,3})$ minor. The proof of Theorem (10.3.11) consists of a thorough analysis of $M \backslash z$ and of the role of the element z. We begin with two simple claims about $M \backslash z$.

First, we claim that each series class of $M \backslash z$ has at most two elements. Suppose otherwise. Thus, the connected $M \backslash z$ has a representation matrix with three parallel rows. Adjoin a column z to that matrix to get a representation matrix for M. Regardless of the entries of column z, the matrix for M has two parallel rows. Thus, M is not 3-connected, a contradiction.

Second, we claim that $M \backslash z$ is graphic. This is so since each proper minor of M is graphic or cographic, and since $M \backslash z$ has an $M(K_{3,3})$ minor, which is not cographic.

By the first claim, a graph G for $M \backslash z$ is obtained from a 3-connected graph by subdividing each edge at most once. Hence, each line of G has one or two edges. We represent M by G plus a node subset T that handles the extra element z. Recall that $|T|$ is even. Choose G so that $|T|$ is minimal. Since M is not graphic, we have $|T| \geq 4$.

The strategy in the remainder of the proof is as follows. We attempt to reduce G and T to G' and T', where G' has fewer edges than G or $|T'| < |T|$. The graph G' is to have an $M(K_{3,3})$ minor, and $|T'| \geq 4$ is to hold. Indeed, G' and T' must represent a nongraphic matroid. At times, we simply say that we *reduce* G to denote this process. Evidently, any such reduction contradicts the minimality of M or T, and thus is not possible. The reduction attempts reveal enough structural information about M to prove that matroid to be isomorphic to R_{10} or R_{12}, contrary to the initial assumption. We present the details.

Claim 1. G is 3-connected.

Proof. Suppose G is not 3-connected. We know that G is a subdivision of a 3-connected graph. Indeed, each line of G with at least two edges has exactly two edges. Let G have m such lines. If a midpoint of one of these lines is not in T, then in M the elements corresponding to the two edges of that line are in series. Thus, M is not 3-connected, a contradiction. Hence, the midpoints of the m lines are in T.

If $m \geq 4$, we contract in one of the m lines an edge. By switchings, the set of T nodes may now be reduced, say to T'. But the midpoints of the remaining $m - 1$ lines must remain in T', so $|T'| \geq 4$. Thus, G has been reduced. Similarly, one may handle the cases $m = 1$ and 2, and also $m = 3$ when the three lines do not form a cycle.

Consider the remaining case, where $m = 3$ and where the three lines form a cycle. If any node other than those of the three lines is in T, again we can reduce G. Thus, all such nodes are not in T. Temporarily contract one edge of each line, getting a graph $\overline{G}$. The three lines have become a triangle. According to Theorem (10.3.9), the graph $\overline{G}$ has a minor isomorphic to one of the two graphs of (10.3.10). The bold lines of those graphs are those of the triangle. We can further reduce each one of the two graphs to

(10.3.12)

Minor of the two graphs of (10.3.10)

where again the bold lines indicate the triangle. The graph of (10.3.12) is still a minor of G if we replace each triangle edge by the corresponding line with appropriate designation of T nodes. By the minimality of $|T|$, this substitution must result in

(10.3.13)

Minor of G with three lines

which is not graphic. Contract the edges labeled e and f in (10.3.13). The resulting G' and T' has $|T'| = 2$ and represents a matroid with an $M(K_{3,3})$ minor. Thus, the matroid of (10.3.13) is not cographic. But that matroid is a proper minor of M since the graph of (10.3.12) is a proper minor of the two graphs of (10.3.10). Thus, we have a contradiction of the minimality of M.   Q. E. D. Claim 1

Denote by H any subgraph of G that is a subdivision of $K_{3,3}$. Recall that $G\copyright e$ is obtained from G by contraction of edge e and deletion of parallel edges. Similarly, $G\circledd e$ is produced by deletion of edge e and contraction of each line with two edges to just one edge. Indeed, in $G\backslash e$, at most two lines of length 2 may exist. In each such line, we may choose the edge to be contracted as is convenient. This aspect is important when we want to preserve T nodes. As was done before, we use squares to denote nodes in T. We assign a question mark to a node if that node may or may not be in T. We say that a node is *cubic* if it is a 3-star.

Claim 2. Every node v of G that is not part of some H is cubic, and that node and two of its neighbors are in T. Furthermore, $|T| = 4$ in that case.

Proof. Let e be an edge incident at v, and u be its other endpoint. If $G \textcircled{d} e$ is 3-connected, one of the situations below must prevail; otherwise a reduction is possible, as is readily checked.

(10.3.14)

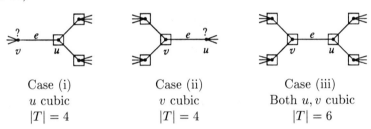

Case (i)	Case (ii)	Case (iii)
u cubic	v cubic	Both u, v cubic
$|T| = 4$	$|T| = 4$	$|T| = 6$

If $G \textcircled{c} e$ is 3-connected, then $u, v \in T$ and $|T| = 4$, since otherwise $G \textcircled{c} e$ with adjusted T set is a smaller case. We call the last situation case (iv). Thus, we have a total of four possible cases for the edge e.

Suppose $v \notin T$. Apply the above arguments to every edge e incident at v. In each instance, case (i) of (10.3.14) must apply, since cases (ii), (iii), and (iv) demand v to be in T. Thus, $|T| = 4$, and all neighbors of v and their neighbors (except v) must be in T. In addition, the neighbors of v must be cubic. A simple case analysis shows that these conditions imply G to be 2-separable. Thus, $v \in T$.

Suppose v is not cubic. Again by arguments for each edge e incident at v, we have $|T| = 4$, and v and all neighbors of v must be in T. By assumption, there are at least four neighbors, so this is not possible.

So far we know v to be cubic and to be in T. If $|T| > 4$, then case (iii) of (10.3.14) must hold for every edge e incident at v, and $|T| = 6$. Once more, a case analysis shows that G is not 3-connected. Thus, $|T| = 4$.

Suppose v and its three neighbors are in T. Then no other node of G is in T. Take any node w different from these four nodes. There exist three internally node-disjoint paths in G from v to w. Then clearly G and T can be reduced to a graph isomorphic to that of (10.2.5). The latter graph represents the nonregular matroid F_7^*, a contradiction.   Q. E. D. Claim 2

Claim 3. There exists an H such that

(i) G has no node beyond those of H, and

(ii) no edge of G that is not an edge of H connects two nodes of a line of H.

Proof. Suppose some H violates (ii). We then can find another one violating (i). Thus, G has a node u not in some H. By Claim 2, u is cubic, u and precisely two of its neighbors are in T, and $|T| = 4$. By Theorem (10.3.9), there is another H with u as corner node. Pick a minimal such H, say H_1. Clearly, H_1 satisfies (ii) of Claim 3. We now prove that (i) holds as well. If not, then G has, again by Claim 2, a cubic node v not in H_1 such that v and precisely two of its neighbors are in T. Thus, G has the following subgraph, where dashed lines represent internally node-disjoint paths.

(10.3.15)

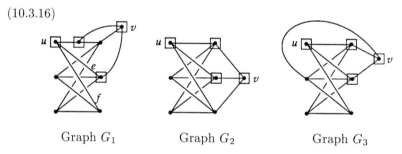

Graph H plus node v

By Menger's Theorem, there is a path from w to a non-T node of the dashed graph. Thus, we can reduce G to produce one of the following graphs.

(10.3.16)

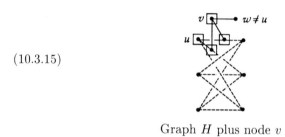

Graph G_1 Graph G_2 Graph G_3

From G_1, delete the edge e and contract the edge f. This produces the graph

(10.3.17)

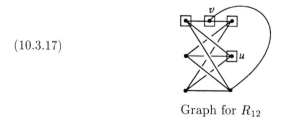

Graph for R_{12}

which by (10.2.9) represents R_{12}. It is easily checked that both G_2 and G_3 can be reduced to the graph

(10.3.18)

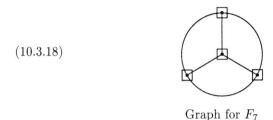

Graph for F_7

which by (10.2.4) represents the nonregular F_7.   Q. E. D. Claim 3

Claim 4. $|T| \leq 6$. Take any H that satisfies (i) and (ii) of Claim 3. Suppose G has an edge e that is not in H. Then at least one endpoint v of e is cubic, and v and the neighbors linked to v by edges of H are in T.

Proof. Suppose an edge e exists that is not in H. By part (b) of Lemma (10.3.8), $G\textcircled{d}e$ is 3-connected. If $|T| \geq 8$, or if the remaining conclusions of Claim 4 do not hold, then $G\textcircled{d}e$ provides a smaller case. If G has no edges beyond those of H, then G is isomorphic to $K_{3,3}$, and $|T| \leq 6$ holds trivially.   Q. E. D. Claim 4

Claim 5.   There is an H satisfying (i) and (ii) of Claim 3 such that H contains all edges of G but at most one.

Proof. Let H be the graph of Claim 3. Simple case checking confirms that H must contain all edges of G except for possibly two edges. Assume there are two such edges. By Claim 4, one of these arcs has a cubic endpoint, say u, such that u and the neighbors linked to u by arcs of H are in T.

    Now select a minimal H, say H_2, that has u as corner node. By the proof of Claim 3, H_2 may be assumed to satisfy (i) and (ii) of that claim. Suppose again there are two edges in G, say e and f, that are not in H_2. Let v and w be the cubic endpoints of e and f that have the properties described in Claim 4. Clearly $|T| = 6$. If v and w are on one line of H_2, we can contract an intermediate arc of that line, and by part (c) of Lemma (10.3.8) have a smaller case. Otherwise, one endpoint of e or f, say of e, is not in T, and $G\textcircled{d}e$ produces a smaller case. Thus, H_2 is the desired graph.   Q. E. D. Claim 5

    Let H be the graph of Claim 5. If every edge of G is in H, then G and H are isomorphic to $K_{3,3}$. If $|T| = 6$, we have the graph of (10.2.8), and M is isomorphic to R_{10}. If $|T| = 4$, then G can be reduced to the nonregular instance of (10.3.18).

    Finally, assume just one edge of G, say e, is not in H. By Claim 4, at least one endpoint of e, say u, is cubic, and u and the neighbors linked to

u by arcs of H are in T. If the second endpoint v of e is cubic or $|T| \geq 6$, then we can reduce G to a smaller case. If $v \in T$, then G can be reduced to the nonregular case of (10.3.18). Thus, G must be

(10.3.19)

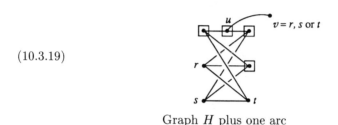

Graph H plus one arc

If $v = r$ or s, then the nonregular case of (10.3.18) can be produced. If $v = t$, we have by (10.2.9) an instance of R_{12}.   □

We are prepared for the next section, where we prove two profound excluded minor theorems of Seymour and Tutte.

10.4 Characterization of Graphic Matroids

We prove profound characterizations of two classes of regular matroids in terms of excluded minors. The first characterization is due to Tutte. It says that a regular matroid is graphic if and only if it has no $M(K_{3,3})^*$ or $M(K_5)^*$ minors. The second characterization is due to Seymour. According to that result, a 3-connected regular matroid is graphic or cographic if and only if it has no R_{10} or R_{12} minors. The latter matroids are defined by (10.2.8) and (10.2.9). In this section, we prove these characterizations and deduce some related material. Given the results of the preceding two sections, it is advantageous for us to start with the second characterization.

(10.4.1) Theorem.   *A 3-connected regular matroid is graphic or co-graphic if and only if it has no R_{10} or R_{12} minors.*

Proof. The graphs, T sets, and representation matrices of (10.2.8) and (10.2.9) for R_{10} and R_{12} prove these matroids to be nongraphic and isomorphic to their respective duals. Thus, R_{10} and R_{12} are also not cographic. These observations establish the easy "if" part.

For proof of the converse, let M be a 3-connected regular matroid that is not graphic and not cographic. Thus, M is not planar, and by Theorem (10.2.11) has a minor isomorphic to $M(K_5)$, $M(K_{3,3})$, $M(K_5)^*$, or $M(K_{3,3})^*$. By Lemma (10.3.1), $M(K_5)$ is a splitter for the regular matroids without $M(K_{3,3})$ minors. By these results, M has a minor isomorphic to $M(K_{3,3})$ or $M(K_{3,3})^*$, or M is isomorphic to $M(K_5)$ or $M(K_5)^*$. The latter case is a contradiction. Thus, M or M^* has an $M(K_{3,3})$ minor. By Theorem (10.3.11), M or M^* has an R_{10} or R_{12} minor. Since R_{10} and R_{12} are isomorphic to their duals, M itself has an R_{10} or R_{12} minor.   □

We turn to the characterization of regular matroids that are graphic.

(10.4.2) Theorem. *A regular matroid is graphic (resp. cographic) if and only if it has no $M(K_5)^*$ or $M(K_{3,3})^*$ minors (resp. $M(K_5)$ or $M(K_{3,3})$ minors).*

Proof. Lemma (3.2.48) implies the easy "only if" part. For proof of the converse, let M be a nongraphic regular matroid all of whose proper minors are graphic. If M is 1- or 2-separable, then arguments analogous to those of the proof of Theorem (10.2.11) establish M to be graphic, a contradiction. Thus, M is 3-connected.

Suppose M is not cographic. By Theorem (10.4.1), M has an R_{10} or R_{12} minor. The drawings of (10.2.8) and (10.2.9) clearly establish that both R_{10} and R_{12} have proper $M(K_{3,3})$ minors. Now R_{10} and R_{12} are isomorphic to their duals. Thus, M has an $M(K_{3,3})^*$ minor.

Now consider M to be cographic, i.e., consider M^* to be graphic. If M^* is planar, then M is planar, and hence graphic, a contradiction. If M^* is nonplanar, then by Theorem (7.4.1), M^* has an $M(K_5)$ or $M(K_{3,3})$ minor. Thus, M has an $M(K_5)^*$ or $M(K_{3,3})^*$ minor, as desired.

The parenthetic claim of the theorem follows by duality.   □

We complete this section with two corollaries. The first one effectively restates Corollary (10.2.13).

(10.4.3) Corollary. *A matroid is planar if and only if it is graphic and cographic.*

Proof. Apply Corollary (10.2.13), or compare the excluded minors of Theorems (10.2.11) and (10.4.2) to obtain the conclusion.   □

For the second corollary, we need the following auxiliary result.

(10.4.4) Lemma. *Every 1-element reduction of R_{10} or R_{12} produces a matroid with an $M(K_{3,3})$ or $M(K_{3,3})^*$ minor.*

Proof. For R_{10}, the proof is easy. One first shows that R_{10} is highly symmetric as follows. Consider a binary matrix D with five rows. Each column of D has exactly three 1s, with each possible case occurring. Thus, D has $\binom{5}{3} = 10$ columns. Index the columns of D by a 10-element set E. We claim that E and the subsets of E indexing GF(2)-independent columns of D define a matroid isomorphic to R_{10}. For a proof, we perform row operations in D until a 5×10 matrix $[I \mid B]$ results. Simple checking confirms that the matrix B is either up to indices the matrix of (10.2.8) for R^{10}, or is the matrix of (10.4.5) below. A pivot on the 1 in the $(1,2)$ position of the latter matrix converts it to the former one.

(10.4.5)

$$
\begin{array}{|ccccc|}
1 & 1 & 0 & 0 & 1 \\
1 & 1 & 1 & 0 & 0 \\
0 & 1 & 1 & 1 & 0 \\
0 & 0 & 1 & 1 & 1 \\
1 & 0 & 0 & 1 & 1 \\
\end{array}
$$

Alternate matrix for R_{10}

By (10.2.8), the minor $R_{10} \backslash z$ of R_{10} is isomorphic to $M(K_{3,3})$. By the just-proved symmetry and duality, every 1-element deletion (resp. contraction) in R_{10} produces a matroid isomorphic to $M(K_{3,3})$ (resp. $M(K_{3,3})^*$).

The proof for R_{12} is about as easy. That matroid is also isomorphic to its dual. Thus, we can confine ourselves to the contraction case. With the aid of (10.2.9), one can prove presence of an $M(K_{3,3})$ or $M(K_{3,3})^*$ minor after each 1-element contraction.   □

The proof of Lemma (10.4.4) contains the following result about R_{10}.

(10.4.6) Lemma.   *Up to indices, the matrices of (10.2.8) and (10.4.5) are the only binary representation matrices for R_{10}.*

We use Theorem (10.4.1) as the main ingredient in the proof of the following corollary.

(10.4.7) Corollary.   *Every regular 1-element extension of a connected planar matroid is graphic or cographic.*

Proof. Let M be a 1-element regular extension of a connected planar matroid N. The proof is by induction on the number of elements of N. The result clearly holds for small M, say with at most eight elements. Planarity and the conclusions of the corollary are preserved under dualizing. Thus, we may assume that for some x, $N = M/x$.

If M is 1-separable, then x is a loop or coloop of M. Since N is planar, M is planar as well. Hence, we may suppose M to be connected.

Assume M to be 2-separable. We may select a representation matrix B of M as given by (8.1.1) with GF(2)-rank $D = 1$ and $x \in X_1$. Below, (10.4.8) displays that matrix. Since $x \in X_1$, B displays N. If $|X_1 \cup Y_1| = 2$, then M is a series extension of N, and thus is planar. If $|X_1 \cup Y_1| \geq 3$, then $((X_1 \cup Y_1) - \{x\}, X_2 \cup Y_2)$ is a 2-separation of N. Let N_1 and N_2 be the components of the related 2-sum decomposition of N.

(10.4.8)

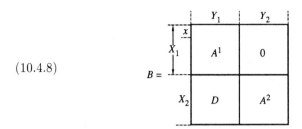

Matrix B for M with 2-separation

By Lemma (8.2.6), N_1 and N_2 are connected. Then for some 1-element expansion of N_1 by x, say N_1', the matroid M is a 2-sum with N_1' and N_2 as components. Since N_1 is connected, by induction N_1' is graphic or cographic. We know that N_2 is planar. By Lemma (8.2.7), M is graphic or cographic.

As final case, consider M to be 3-connected. If M is not graphic or cographic, then by Theorem (10.4.1), M has an R_{10} or R_{12} minor. But by Lemma (10.4.4), every 1-element reduction of either minor, and thus of M, is not planar, a contradiction.   □

Corollary (10.4.7) is false when the given planar matroid is not assumed to be connected. For example, for any element z of K_5, the 1-sum of $M(K_5/z)$ and $M(K_5\backslash z)^*$ is planar. But there is a regular 1-element extension that has $M(K_5)$ and $M(K_5)^*$ as a minor. Thus, that extension is nongraphic and noncographic.

In the next section, we introduce a major switch of topic. We examine a simple yet powerful idea that produces interesting graph decomposition theorems. In Chapters 11–13, we rely on the matroid generalization of this idea to prove profound matroid decomposition theorems.

10.5 Decomposition Theorems for Graphs

In subsequent chapters, we make repeated use of a recursive construction of matroid decomposition theorems. In this section, we use a graph example to motivate and explain that construction. In the process, we deduce by rather elementary checking a famous decomposition theorem of Wagner about the class of graphs without K_5 minors.

The basic idea of the construction is as follows. Given is a class $\mathcal{G}$ of connected graphs. The class is closed under isomorphism and under the taking of minors. On hand are also two subclasses $\mathcal{L}$ and $\mathcal{H}$ of $\mathcal{G}$. The two subclasses are so selected that each graph $G \in \mathcal{G}$ is in $\mathcal{L}$ or has, for some $H \in \mathcal{H}$, an H minor.

One iteration of the construction is as follows. We select a graph H of $\mathcal{H}$ and use one of the decomposition theorems of Chapter 6 or 7 to establish a result of the following type: If a graph $G \in \mathcal{G}$ has an H minor, then G has a certain decomposition caused by H, or G is a member of a certain collection of graphs, say $\{L_1, L_2, \ldots, L_m\}$, or G has a minor that is a member of a certain second collection of graphs, say $\{H_1, H_2, \ldots, H_n\}$. We now derive from $\mathcal{L}$ the set $\mathcal{L}' = \mathcal{L} \cup \{L_1, L_2, \ldots, L_m\}$, and from $\mathcal{H}$ the set $\mathcal{H}' = (\mathcal{H} - H) \cup \{H_1, H_2, \ldots, H_n\}$. By the derivation of $\mathcal{L}'$ and $\mathcal{H}'$, we have established the following theorem: Each graph $G \in \mathcal{G}$ can be decomposed, or belongs to $\mathcal{L}'$, or has an H' minor for some $H' \in \mathcal{H}'$.

At this point, we have completed one iteration of the construction. If $\mathcal{H}'$ is empty, we stop; most likely the cited theorem is an interesting decomposition result for the graphs of $\mathcal{G}$. If $\mathcal{H}'$ is nonempty, we have two choices: We may stop, or we may carry out another iteration by declaring $\mathcal{L}'$ to be $\mathcal{L}$ and $\mathcal{H}'$ to be $\mathcal{H}$. Evidently, the recursive construction is nothing but a concatenation of decomposition results, each of which is deduced from suitably selected theorems of Chapters 6 and 7.

An example will help to clarify the construction. We want a decomposition theorem for the graphs without K_5 minors. That class of graphs turns out to be important for a number of combinatorial problems. Details are included in Section 10.7. We ignore the applications for the time being, and concentrate on the construction of a decomposition theorem for that class. In agreement with the preceding outline, we define $\mathcal{G}$ to be the set of connected graphs without K_5 minors. The subclass $\mathcal{L}$ is the collection of connected planar graphs, and $\mathcal{H}$ is the set $\{K_{3,3}\}$. By Theorem (7.4.1) and the exclusion of K_5 minors from the graphs of $\mathcal{G}$, each graph $G \in \mathcal{G}$ is planar or has a $K_{3,3}$ minor. Thus, $\mathcal{L}$ and $\mathcal{H}$ do satisfy the condition stated earlier, i.e., each graph $G \in \mathcal{G}$ is in $\mathcal{L}$ or has an H minor for the single graph $H = K_{3,3}$ of $\mathcal{H}$.

We are ready for the first iteration of the construction. We will rely on the splitter theorem for graphs of Section 7.2, listed there as Corollary (7.2.10). We repeat that result below.

(10.5.1) Theorem.   *Let $\mathcal{G}$ be a class of connected graphs that is closed under isomorphism and under the taking of minors. Let H be a 3-connected graph of $\mathcal{G}$ with at least six edges.*

(a) *If H is not a wheel, then H is a splitter of $\mathcal{G}$ if and only if $\mathcal{G}$ does not contain any graph derived from H by one of the following two extension steps:*

(1) *Connect two nonadjacent nodes of N by a new edge.*

(2) *Partition a vertex of degree at least 4 into two vertices, each of degree at least 2, and connect these two vertices by a new edge.*

(b) If H is a wheel, then H is a splitter of $\mathcal{G}$ if and only if $\mathcal{G}$ does not contain any of the extensions of H described under (a) and does not contain the next larger wheel.

The application of the splitter theorem involves several graphs, which we define next. The graphs are $K_{3,n}$, $n \geq 3$, as well as certain variants of $K_{3,n}$, called $K_{3,n}^1$, $K_{3,n}^2$, and $K_{3,n}^3$. The graph $K_{3,n}^3$ is given by (10.5.2) below. The other graphs are obtained from $K_{3,n}^3$ by deletion of some edges. In the notation of (10.5.2), $K_{3,n}^2$ is the graph $K_{3,n}^3 \backslash c$, and $K_{3,n}^1$ is the graph $K_{3,n}^3 \backslash \{b, c\}$. At times, we refer to $K_{3,n}$ as $K_{3,n}^0$. We collect the graphs just defined in a set $\mathcal{K} = \{K_{3,n}^i \mid 0 \leq i \leq 3, n \geq 3\}$.

(10.5.2)

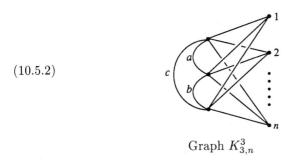

Graph $K_{3,n}^3$

We need one additional graph V defined by

(10.5.3)

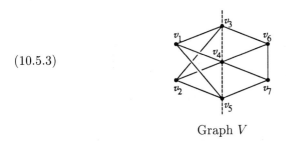

Graph V

For the moment, the reader should ignore the dashed line in the drawing of V. It indicates a 3-separation that we will utilize later. Note that V has no K_5 minors. Indeed, V has eleven edges and seven vertices, and if V has a K_5 minor, then such a minor must be produced by a single contraction or deletion. But any such reduction results in a graph with at least six vertices.

The splitter Theorem (10.5.1) involves 3-connected 1-edge extensions of graphs. The next lemma supplies information about such extensions for the graphs of $\mathcal{K}$.

(10.5.4) Lemma.   *For any graph G of $\mathcal{K}$, any 3-connected 1-edge extension of G is isomorphic to another graph of $\mathcal{K}$, or has a K_5 minor, or has a V minor.*

Proof. The proof involves a simple checking of cases for the graphs $K_{3,n}^i$, $0 \leq i \leq 3$, with n fixed, plus induction. As examples, we cover the cases of $K_{3,3}$ and $K_{3,3}^1$. Since every vertex of $K_{3,3}$ has degree 3, any 3-connected 1-edge extension must be an addition. Indeed, just one addition case exists, the graph $K_{3,3}^1$. We depict that graph below.

(10.5.5)

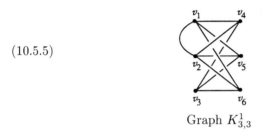

Graph $K_{3,3}^1$

The 3-connected 1-edge extensions of $K_{3,3}^1$ are as follows. We start with the expansion cases. By symmetry, all cases are isomorphic to the following one, where we split the vertex v_1. The resulting graph is isomorphic to V, as is evident from the next drawing.

(10.5.6)

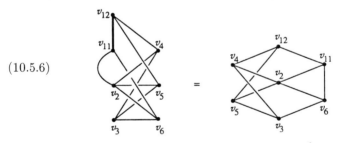

3-connected 1-edge expansion of $K_{3,3}^1$

We turn to the addition cases. All such instances are isomorphic to two graphs, one of which is $K_{3,3}^2$, while the second one becomes K_5 upon contraction of one edge. We leave the verification of this claim to the reader.   □

With these preparations, we carry out the first iteration of the construction process. We must select the single graph $K_{3,3}$ of $\mathcal{H}$ as H. From the splitter Theorem (10.5.1) and Lemma (10.5.4), we deduce the following decomposition result.

(10.5.7) Theorem.   *Every connected graph without K_5 minors is 1-or 2-separable, or is planar, or is isomorphic to one of the graphs $K_{3,n}^i$, $0 \leq i \leq 3$, $n \geq 3$, or has a V minor.*

Proof. Let G be any graph without K_5 minors. In the nontrivial case, G is nonplanar. By Theorem (7.4.1), we know that G has a $K_{3,3}$ minor, and thus a $K_{3,n}^i$ minor with an edge set of maximum cardinality. If G itself is that minor, we are done. Otherwise, we apply the splitter Theorem (10.5.1) to the class of graphs consisting of all minors of G and their isomorphic versions. The graph $K_{3,n}^i$ plays the role of the splitter. Thus, G is 2-separable, or is 3-connected and has a 3-connected 1-edge extension of a $K_{3,n}^i$ minor. In the first case, we are done. In the second case, we know by Lemma (10.5.4) and the assumed maximality of the edge set of $K_{3,n}^i$ that G has a V minor. $\square$

We have reached the end of the first iteration. In the notation of the general construction process, the current set $\mathcal{L}'$ contains the planar graphs, and all graphs $K_{3,n}^i$ plus their isomorphic versions. The set $\mathcal{H}'$ contains just one graph, V.

We begin the second iteration. The current $\mathcal{H}$, i.e., the set $\mathcal{H}'$ of iteration 1, contains just V. Thus, that graph is selected as H. We intend to invoke the induced separation result for graphs of Section 6.3, listed there as Corollary (6.3.26). We include that result below.

(10.5.8) Theorem.   *Let $\mathcal{G}$ be a class of connected graphs that is closed under isomorphism and under the taking of minors. Let a 3-connected graph $H \in \mathcal{G}$ have a 3-separation (F_1, F_2) with $|F_1|, |F_2| \geq 4$. Assume that H/F_2 has no loops and $H \backslash F_2$ has no coloops. Furthermore, assume that for every 3-connected 1-edge extension of H in $\mathcal{G}$, say by edge z, the pair $(F_1, F_2 \cup \{z\})$ is a 3-separation of that extension. Then for any 3-connected graph $G \in \mathcal{G}$ with an H minor, the following holds. Any 3-separation of any such minor that corresponds to (F_1, F_2) of H under one of the isomorphisms induces a 3-separation of G.*

We need a 3-separation (F_1, F_2) of V of (10.5.3) for the application of Theorem (10.5.8). Thus, we define F_1 to be the set of edges of V incident at node v_1 or v_2, and declare F_2 to be the set of the remaining edges. The 3-separation (F_1, F_2) of V is informally indicated in (10.5.3) by the dashed line.

We will encounter one additional graph G_8 given by (10.5.9) below. For the moment, the unusual indexing of the node labels of G_8 should be ignored. It will make sense shortly. The graph G_8 has twelve edges, and every vertex has degree 3. We claim that G_8 does not have K_5 minors; otherwise, the contraction of two suitably selected edges could produce a graph where at least five vertices have degree of at least 4. But that is not possible.

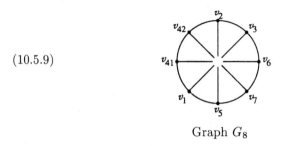

Graph G_8

(10.5.9)

The decomposition result for the second iteration is as follows.

(10.5.10) Theorem. *Let G be a 3-connected graph without K_5 minors, but with a V minor. Then the 3-separation of that minor defined from (F_1, F_2) of V induces a 3-separation of G, or G has a G_8 minor.*

Proof. We apply Theorem (10.5.8) with the class of connected graphs without K_5 minors as $\mathcal{G}$, and with the graph V of (10.5.3) as H. We readily verify that V/F_2 has no loops and that $V\backslash F_2$ has no coloops. Thus, by Theorem (10.5.8), the claimed induced 3-separation exists, or V can be extended by one edge z to a 3-connected graph for which $(F_1, F_2 \cup \{z\})$ is not a 3-separation. We consider all such 3-connected extensions of V by an edge z.

We start with the expansion case. Since $V \& z$ is to be 3-connected, we must split the single degree 4 vertex v_4 of V and insert z. There are three ways to do this. The corresponding graphs V_1, V_2, V_3 are given below.

(10.5.11)

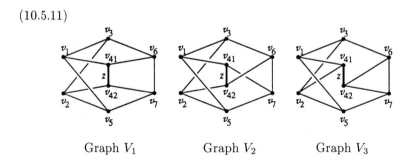

Graph V_1 Graph V_2 Graph V_3

Evidently, $(F_1, F_2 \cup \{z\})$ is not a 3-separation for V_1 or V_2, but this is so for V_3. Thus, V_1 and V_2 are the graphs of interest to us. The graphs V_1 and V_2 are isomorphic. Indeed, one isomorphism from the vertices of V_1 to those of V_2 is an identity except that it takes v_3, v_5, v_6, v_7 of V_1 to v_5, v_3, v_7, v_6 of V_2, respectively. Furthermore, a comparison of V_1 with G_8 of (10.5.9) proves these two graphs to be identical.

We turn to the addition case. Since $V+z$ is to be 3-connected, the added edge z cannot be parallel to another edge. Also, $(F_1, F_2 \cup \{z\})$ is not to be a 3-separation of $V+z$. Thus, one endpoint of z must be v_1 or v_2. The second endpoint must be v_1, v_2, v_6, or v_7. Suppose v_1 is one endpoint and v_2 the second one. Contract in $V+z$ the edges (v_3, v_6) and (v_5, v_7). A K_5 minor results, which is a contradiction. Suppose v_1 is one endpoint and v_6 is the second one. Contract in $V+z$ the edges (v_2, v_3) and (v_5, v_7). Again a K_5 minor results. The remaining cases are isomorphic to the latter one.

We conclude that in each case G has a G_8 minor as desired.   □

We combine Theorems (10.5.7) and (10.5.10) to the following result.

(10.5.12) Theorem. *Every connected graph without K_5 minors is 1- or 2-separable, or has a 3-separation with at least four edges on each side, or is planar, or is isomorphic to a graph $K_{3,n}^i$, $0 \leq i \leq 3$, $n \geq 3$, or has a G_8 minor.*

Proof. By Theorem (10.5.7), we may assume G to have a V minor. Then Theorem (10.5.10) establishes presence of the 3-separation or of a G_8 minor.   □

We have reached the end of the second iteration. In terms of the general description of the construction, the current set $\mathcal{L}'$ contains the planar graphs, and all graphs $K_{3,n}^i$ plus their isomorphic versions. The set $\mathcal{H}'$ contains just one graph, G_8.

The third iteration involves an application of the splitter Theorem (10.5.1). The result so produced is as follows.

(10.5.13) Theorem. *G_8 is a splitter for the graphs without K_5 minors.*

Proof. We only need to show that every 3-connected 1-edge extension of G_8 has K_5 minors. Since every vertex of G_8 has degree 3, a 3-connected 1-edge expansion is not possible. Up to isomorphism, just two addition cases are possible. Both cases have K_5 minors. We leave the easy verification of this claim to the reader.   □

We combine Theorems (10.5.12) and (10.5.13) with the results for 1-, 2-, and 3-sums of Chapter 8 to obtain Wagner's famous decomposition theorem for the graphs without K_5 minors.

(10.5.14) Theorem. *Every connected graph without K_5 minors is a 1-, 2-, or 3-sum, or is planar, or is isomorphic to $K_{3,3}$ or G_8.*

Proof. Assume G to be a connected graph without K_5 minors. Theorems (10.5.12) and (10.5.13) imply that G is 1- or 2-separable, or has a 3-separation with at least four edges on each side, or is planar, or is isomorphic to G_8 or to a graph $K_{3,n}^i$, $0 \leq i \leq 3$, $n \geq 3$. If G is isomorphic to a graph $K_{3,n}^i$ different from $K_{3,3}$, then by (10.5.2), G has a 3-separation with at least four edges on each side. We apply some results of Chapter 8. If G is 1- or 2-separable, then according to Section 8.2, G is a 1- or 2-sum. If G has a 3-separation with at least four edges on each side, then by Lemma (8.3.12) and the discussion following that lemma, G is a 3-sum. Thus, we may conclude that G is a 1-, 2-, or 3-sum, or is planar, or is isomorphic to $K_{3,3}$ or G_8, as claimed in the theorem.   □

Note that at the end of the third iteration, the set $\mathcal{H}'$ is empty. Thus, the construction process stops.

Recall the Δ-sum decomposition of Section 8.5. In a connected graph G, such a decomposition is carried out as follows. Given is a 3-separation of G with at least four edges on each side. Let H_1 and H_2 be the corresponding subgraphs of G. Thus, both H_1 and H_2 are connected subgraphs of G. Identification of three connecting nodes of H_1 with three connecting nodes of H_2 produces G. For $i = 1, 2$, we enlarge H_i by attaching a triangle to the three connecting nodes. Let G_i be the resulting graph. Then G is a Δ-sum of G_1 and G_2, denoted by $G = G_1 \oplus_\Delta G_2$. The components G_1 and G_2 of a 3-connected Δ-sum $G_1 \oplus_\Delta G_2$ are 3-connected, except possibly for edges parallel to the edges of the connecting triangle in G_1 or G_2.

Also recall the 2-sum decomposition of Section 8.2. The graphs H_1 and H_2 have two connecting nodes each. For $i = 1, 2$, we enlarge H_i by joining the connecting nodes by an edge. Let G_i be the resulting graph. Then G is a 2-sum of G_1 and G_2, denoted by $G = G_1 \oplus_2 G_2$. The components G_1 and G_2 are 2-connected if G is 2-connected.

By inverting the above operations, we obtain the Δ-sum and 2-sum compositions. At times, one may desire to construct a graph recursively by these operations. Initially, one combines two graphs G_1 and G_2 in a 2- or Δ-sum. Then one composes the resulting graph in a 2- or Δ-sum with a graph G_3. Continuing in this fashion, one recursively enlarges the graph on hand by G_4, G_5, etc. We call the G_i, $i \geq 1$, the *building blocks* of this process.

We may use Theorem (10.5.14) to establish such a construction process for the graphs without K_5 minors. The details are specified in the next theorem.

(10.5.15) Theorem. *Any 2-connected graph without K_5 minors is planar, or isomorphic to $K_{3,3}$ or G_8, or may be constructed recursively by 2-sums and Δ-sums. The building blocks of that construction are as follows.*

2-sums: *planar graphs, and graphs isomorphic to $K_{3,3}$ or G_8.*
Δ-sums: *planar graphs.*

The proof of Theorem (10.5.15) utilizes the following two lemmas.

(10.5.16) Lemma. *Let G be a 3-connected nonplanar graph without K_5 minors and not isomorphic to $K_{3,3}$ or G_8. Assume G to have a triangle C.*

Then G has a 3-separation (E_1, E_2) where $|E_1|, |E_2| \geq 4$ and where one of E_1, E_2 contains C.

Proof. We use induction. If G has ten edges, the smallest case, then direct checking proves the lemma. Otherwise, G is by Theorem (10.5.14) a 3-sum, and thus has a 3-separation (E_1, E_2) where $|E_1|, |E_2| \geq 4$. If one of E_1, E_2 contains C, we are done. Thus, we may assume that just one edge of C, say c, is in E_2. If $|E_2| \geq 5$, we shift the edge c from E_2 to E_1 and have the case where one side of a 3-separation contains C. Thus, we assume that $|E_2| = 4$. It is easy to see that G must be of the form

(10.5.17)

3-separation (E_1, E_2) of G with $|E_2| = 4$ and $c \in E_2$

If G has an edge that forms a triangle with the explicitly shown edges e and g, or with f and g, then we exchange c and such an edge between E_1 and E_2, and again have the desired 3-separation. Otherwise, the minor G/g has no parallel edges. Note that $\{e, f, c\}$ is a triangle of G/g. Indeed, G/g is isomorphic to one of the components of the Δ-sum induced by (E_1, E_2), and thus is 3-connected. We consider two cases, depending on whether G/g is planar.

If G/g is planar, draw it in the plane. If the triangle $\{e, f, c\}$ lies on one face, then it is easily seen that G itself is planar, a contradiction. Thus, $\{e, f, c\}$ partitions the plane into two regions, both of which contain at least one vertex of G/g. Then we readily confirm that G/g, and hence G, has a 3-separation of the form claimed in the lemma.

For the second case, we assume G/g to be nonplanar. Since G/g has a triangle while $K_{3,3}$ and G_8 do not, G/g cannot be isomorphic to either one of the latter graphs. We apply induction and see that G/g as well as G have the desired 3-separation as well. $\Box$

(10.5.18) Lemma.   *Let G be a 3-connected nonplanar graph without K_5 minors and not isomorphic to $K_{3,3}$ or G_8. Assume G to have either a designated triangle C or a designated edge e. Then G is a Δ-sum $G_1 \oplus_\Delta G_2$, where G_1 contains C or e, whichever applies, and where G_2 is planar.*

Proof. We prove the case for the triangle C and leave the easier situation with the edge e to the reader. We use induction. The smallest case, which has ten edges, is handled by direct checking. For larger G, we apply Lemma (10.5.16). Thus, G is a Δ-sum $G_1 \oplus_\Delta G_2$ where C is part of the component

G_1. If the second component G_2 is planar, we are done. Otherwise, we may assume G_2 to be 3-connected. We define C' to be the triangle of G_2 involved in the Δ-sum. Due to the presence of the triangle C', the graph G_2 cannot be isomorphic to $K_{3,3}$ or G_8. By induction, G_2 has a Δ-sum decomposition $G_{21} \oplus_\Delta G_{22}$ where C' is in G_{21}, and where G_{22} is planar. Evidently, $G_1 \oplus_\Delta G_{21}$ and G_{22} are the components of a Δ-sum decomposition of G of the desired form. □

Proof of Theorem (10.5.15). Let G be any 2-connected graph without K_5 minors and not isomorphic to $K_{3,3}$ or G_8. If G is 3-connected, the result follows from Lemma (10.5.18). Otherwise, G is a 2-sum. Choose the 2-sum decomposition, say $G_1 \oplus_2 G_2$, so that G_2 has a minimal number of edges. Evidently, any 2-separation of G_2 contradicts the minimality assumption, so G_2 is 3-connected. If G_2 is planar or isomorphic to $K_{3,3}$ or G_8, we are done. Otherwise, let e be the edge of G_2 that is identified with an edge of G_1 in the 2-sum composition creating G. By Lemma (10.5.18), G_2 is a Δ-sum $G_{21} \oplus_\Delta G_{22}$, where G_{21} contains e and where G_{22} is planar. Clearly, G is a Δ-sum where one component is $G_1 \oplus_2 G_{21}$, and where the second component is the planar G_{22}, as demanded in the theorem. □

The recursive construction scheme described at the beginning of this section produces a number of additional decomposition theorems. We include two example theorems that may be obtained that way. The first theorem refers to the graph G_9 of (10.5.19) below, and to $K_5 \backslash y$, which is K_5 minus an arbitrarily selected edge y.

(10.5.19)

Graph G_9

(10.5.20) Theorem. *Every 3-connected graph G with at least six edges and without $K_5 \backslash y$ minors is, for some $k \geq 3$, isomorphic to the wheel W_k, or is isomorphic to G_9 or $K_{3,3}$.*

Proof. One first shows that each 3-connected 1-edge extension of any wheel graph with at least four spokes has a $K_{3,3}$ or G_9 minor. Then one suitable application of the splitter Theorem (10.5.1) proves the result. □

The next theorem is much more complicated. We omit the proof, since it involves rather tedious calculations. The theorem refers to a number of graphs, which are listed subsequently under (10.5.22).

(10.5.21) Theorem. *Every connected graph without G_{12} minors is a 1-, 2-, or 3-sum, or is planar, or is isomorphic to K_5, $K_{3,3}$, G_8, G_{13}, G_{14}^1, G_{14}^2, G_{15}^1 G_{15}^2, G_{15}^3, or G_{15}^4.*

Here are the graphs mentioned in Theorem (10.5.21).

(10.5.22)

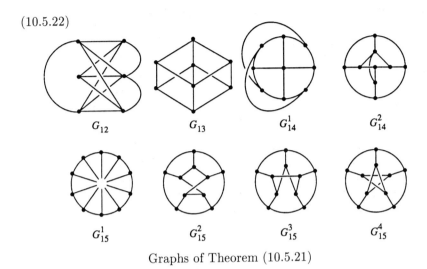

Graphs of Theorem (10.5.21)

The graph G_{15}^4 is the well-known Petersen graph.

From Theorems (10.5.20) and (10.5.21), one may deduce the following 2- and Δ-sum construction results.

(10.5.23) Theorem. *Any 2-connected graph on at least two edges and without $K_5 \backslash y$ minors may be constructed recursively by 2-sums. Each building block is a cycle with two or three edges, or is isomorphic to W_k, $k \geq 3$, G_9, or $K_{3,3}$.*

Proof. Let G be a 2-connected graph on at least two edges and without $K_5 \backslash y$ minors. If G is not isomorphic to one of the listed graphs, then G has by Theorem (10.5.20) a 2-sum decomposition $G_1 \oplus_2 G_2$. Choose the decomposition so that G_2 has a minimum number of edges. Then G_2 must be 3-connected, and by Theorem (10.5.20) must be one of the prescribed building blocks. ☐

(10.5.24) Theorem. *Any 2-connected graph without G_{12} minors is planar, or isomorphic to K_5, $K_{3,3}$, G_8, G_{13}, G_{14}^1, G_{14}^2, G_{15}^1, G_{15}^2, G_{15}^3, or G_{15}^4, or may be constructed recursively by 2-sums and Δ-sums. The building blocks are as follows.*

> *2-sums: planar graphs, and graphs isomorphic to K_5, $K_{3,3}$, G_8, G_{13}, G_{14}^1, G_{14}^2, G_{15}^1, G_{15}^2, G_{15}^3, or G_{15}^4.*
> *Δ-sums: planar graphs and graphs isomorphic to K_5.*

Proof. We use appropriately modified Lemmas (10.5.16) and (10.5.18), and the proof of Theorem (10.5.15). Below, we indicate the necessary adjustments. First, the graphs $K_{3,3}$ and G_8 must throughout be replaced by K_5, $K_{3,3}$, G_8, $G_{13}, \ldots, G_{15}^4$. We note that no graph of that new list has a triangle except for K_5. Second, the proof of the modified Lemma (10.5.16) must be adjusted as follows. In the case of a nonplanar graph G/g, that graph cannot be isomorphic to K_5, $K_{3,3}$, G_8, $G_{13}, \ldots, G_{15}^4$, except for K_5. For the exceptional case, one directly shows G to have a 3-separation of the type demanded by the modified Lemma (10.5.16). Third, in the modified Lemma (10.5.18) and its proof, one now permits the graph G_2 to be isomorphic to K_5. The analogous change applies to the graph G_{22} of the proof of Theorem (10.5.15). □

The decomposition tools provided in this section plus those cited in the references should enable the reader to construct additional decomposition theorems as they are needed. We sketch representative applications for the preceding decomposition theorems in Section 10.7.

Once more we switch topics, and turn to the problem of deciding graphicness of a binary matroid.

10.6   Testing Graphicness of Binary Matroids

Chapters 3, 5, 7, and 8 implicitly contain a quite efficient algorithm for testing graphicness of binary matroids, the topic of this section. Thus, this section is mainly a synthesis of material gleaned from those chapters. We also cover the related problem of deciding whether or not a real $\{0, \pm 1\}$ matrix is the coefficient matrix of a network flow problem.

We start with the graphicness test. Let B be a binary representation matrix of the matroid M to be tested. Small instances are easily decided, so assume that M has at least six elements. If B is not connected, then it clearly suffices that we test each connected component of B for graphicness. Indeed, for some $m \geq 2$, let G_1, $G_2, \ldots, G_m$ be the graphs corresponding to the connected components of B. From each G_i, we select some vertex v_i. Then we combine G_1, $G_2, \ldots, G_m$ to a connected graph G for M and B by identifying the vertices $v_1, v_2, \ldots, v_m$ to one vertex.

Suppose that B is connected, and that we know of a 2-separation of B. By Lemma (8.2.6), M is a 2-sum of two matroids M_1 and M_2. Let B^1 and B^2 be the respective submatrices of B representing the latter matroids. By Lemma (8.2.7), M is graphic if and only if B^1 and B^2 are graphic. Thus, we may analyze B^1 and B^2 instead of B. Indeed, let G_1 and G_2 be graphs for B^1 and B^2, respectively. Then the 2-sum composition of G_1 and G_2 displayed in (8.2.8) produces a graph G for B. In general, G is not unique. But by Theorem (3.2.36), we know that any other graph for B can be obtained from G by a sequence of switchings.

The case remains where B is 3-connected. By Corollary (5.2.15), M has an $M(W_3)$ minor. Furthermore, by Theorem (7.3.4), M has a 3-connected 1-element expansion of an $M(W_3)$ minor, or has a sequence of nested 3-connected minors M_0, M_1, ..., $M_t = M$, where M_0 is an $M(W_3)$ minor, and where each M_{i+1} is obtained from M_i by some series expansions, possibly none, followed by a 1-element addition.

By the census of Section 3.3, every 3-connected 1-element expansion of an $M(W_3)$ minor must be an F_7^* minor, which is not regular, and hence not graphic. Thus, evidence of such a minor proves M to be nongraphic. On the other hand, graphicness of each matroid of the sequence M_0, M_1, ..., M_t can be efficiently decided as follows. Clearly, the $M(W_3)$ minor M_0 is graphic, and a graph for it is readily found. Suppose for given $0 \leq i < t$, we know M_i to be graphic. By Theorem (3.2.36), the 3-connectedness of M_i implies that just one graph exists for M, say G_i. We assume that we have that 3-connected graph on hand. Now M_{i+1} is obtained from M_i by some series expansions and a 1-element addition. The expansions steps can also be done in G_i, so they preserve graphicness. By Lemma (3.2.49), there is a polynomial, indeed very simple, subroutine for deciding whether the addition step preserves graphicness as well. In the affirmative case, the subroutine also produces the graph for M_{i+1}.

So far, we have assumed that we can locate 2-separations and $M(W_3)$ minors, as well as a 3-connected 1-element expansion of a given $M(W_3)$ minor or the sequence M_0, M_1, ..., M_t. But these tasks can be efficiently accomplished by the path shortening technique of Chapter 5 and by the efficient method sketched in the proof of Theorem (7.3.6). These two methods plus the simple graphicness testing subroutine of Lemma (3.2.49) constitute an efficient way of deciding whether a given binary matroid is graphic.

We turn to the second testing problem covered in this section. We are to decide whether or not a real $\{0, \pm 1\}$ matrix is the coefficient matrix of a network flow problem. Recall from Section 9.2 that a node/arc incidence matrix of a directed graph is a real $\{0, \pm 1\}$ matrix where each column has only 0s or exactly one $+1$ and one -1. Furthermore, recall that a real matrix is defined to be the coefficient matrix of a network flow problem if it is the node/arc incidence matrix $\tilde{A}$ of a directed graph, or is derived from such a matrix by pivots and deletion of rows and columns. The support matrix $\tilde{B}$ of $\tilde{A}$ is the node/edge incidence matrix of the undirected version of that graph. View $\tilde{B}$ to be over GF(2).

Lemmas (9.2.1), (9.2.2), (9.2.6), (9.2.8), and Corollary (9.2.7) permit the following conclusions about $\tilde{A}$ and $\tilde{B}$. The matrix $\tilde{A}$ is totally unimodular, and $\tilde{B}$ is regular. Any matrix A deduced from $\tilde{A}$ by real pivots and deletion of rows and columns is totally unimodular. The support matrix B of A can be deduced from $\tilde{B}$ by the corresponding GF(2)-pivots. Thus,

B is graphic. Finally, an elementary signing process exists by which one may determine from B the signs of the entries of A, up to a scaling of some rows and columns by -1.

Because of these results, we may test a given real $\{0, \pm 1\}$ matrix A for the network flow property as follows. View the support matrix B of A to be over GF(2). Test B for graphicness with the efficient method described earlier in this section. If B is not graphic, then A cannot have the network flow property, and we stop. Otherwise, sign B to convert it to a totally unimodular matrix A'. By a simple variation of that signing process, which effectively is the proof procedure of Lemma (9.2.6), determine whether or not A' can be converted to A by scaling. Then A is a network flow problem if and only if the answer is affirmative. This procedure can be improved by use of the undirected graph on hand once B has been determined to be graphic. We leave the details to the reader.

A variation of the preceding problem is as follows. We are given a real matrix $\tilde{A}$. We are to settle whether by row scaling, or by row and column scaling, that matrix can be converted to a $\{0, \pm 1\}$ matrix with the network flow property. The question can be reduced to the above situation as follows. If $\mathrm{BG}(\tilde{A})$ is not connected, we apply the test given below to the submatrices of $\tilde{A}$ that correspond to the connected components of $\mathrm{BG}(\tilde{A})$. Hence, assume $\tilde{A}$ to be connected. First, we determine whether by scaling under the assumed restrictions (i.e., scaling of rows only, or scaling of rows and columns), the given $\tilde{A}$ can be converted to a $\{0, \pm 1\}$ matrix. This is readily accomplished by a scaling of the entries of $\tilde{A}$ corresponding to an arbitrarily selected tree of $\mathrm{BG}(\tilde{A})$. We leave it to the reader to fill in the simple details. If a $\{0, \pm 1\}$ matrix cannot be produced, then the network flow property cannot be attained by scaling under the assumed restrictions. If a $\{0, \pm 1\}$ matrix is obtained, we test that matrix for the network flow property. The answer for the original question is affirmative if and only if the scaled matrix has the network flow property. The conclusion is valid since the scaled matrix is unique up to scaling of some rows and columns by -1, as is readily confirmed via the proof of Lemma (9.2.6).

We summarize the above discussion in the following theorem.

(10.6.1) Theorem.   *There are polynomial algorithms for each one of the problems* (a), (b), *and* (c) *below.*

(a) *Given is any binary matrix B. Let M be the binary matroid represented by B. It must be decided whether M is graphic. In the affirmative case, an undirected graph G must be produced so that $M(G) = M$.*

(b) *Given is any $\{0, \pm 1\}$ real matrix A. It must be decided whether A has the network flow property. In the affirmative case, a directed graph G must be produced so that the node/arc incidence matrix of G can by row operations be transformed to the matrix $[I \mid A]$.*

(c) *Given is any real matrix A. It must decided whether A can by row scaling, or by row and column scaling, be transformed to a $\{0, \pm 1\}$ matrix. In the affirmative case, the $\{0, \pm 1\}$ matrix must be produced, and for that matrix, problem (b) must be solved.*

In the final section, we sketch applications and extensions and cite relevant references.

10.7   Applications, Extensions, and References

We link the material of this chapter to prior work, and point out applications and extensions.

The representation of a binary 1-element addition of a graphic matroid in Section 10.2 by a graph plus a node subset T is taken from Seymour (1980b). It is one of the many innovative concepts and ideas of that reference. The planarity characterization of Theorem (10.2.11) is properly implied by Tutte's characterization of the graphic matroids (Tutte (1958), (1959), (1965)). The same applies to Corollary (10.2.13), which originally was proved in Whitney (1932), (1933b).

Lemma (10.3.1) is taken from Seymour (1980b). Theorem (10.3.9) has been generalized to nongraphic 3-connected matroids with triangles in Asano, Nishizeki, and Seymour (1984). Seymour (1980b) contains the difficult Theorem (10.3.11). The proof given here is a considerably shortened version of the one of that reference. The key difference lies in the repeated application of Theorem (10.3.9).

The profound characterizations of Section 10.4 in historical order are due to Tutte and Seymour. Theorem (10.4.2) is due to Tutte (1958), (1959), (1965), and Theorem (10.4.1) is due to Seymour (1980b). The latter theorem is one of two main ingredients in the proof of the decomposition theorem of regular matroids, which is covered in the next chapter. Reasonably short proofs of Theorem (10.4.2) are given in Seymour (1980a), Wagner (1985a), and Gerards (1990).

The material of Section 10.5 is based on Truemper (1988). That reference contains details about induced graph decompositions. The famous Theorem (10.5.14) of Wagner was originally proved by quite different arguments (Wagner (1937a), (1970)). Short proofs are given in Halin (1964), (1967), (1981), Ore (1967), and Young (1971). When it was first proved, Theorem (10.5.14) established the equivalence of Hadwiger's conjecture about graph coloring and the four-color conjecture for planar graphs (now a theorem). The details are as follows. A graph G is *colorable with n colors* if the vertices can be colored with n colors so that any two vertices with same color are not connected by an edge. The *chromatic number* of a graph is the least number of colors that permit a coloring of the graph. Hadwiger's conjecture (Hadwiger (1943)) says that any graph with chromatic number n has a K_n minor.

The conjecture is readily seen to be correct for $n = 1, 2$, and 3. For $n = 4$, it was proved in Dirac (1952). Indeed, by Theorem (4.2.6), a 2-connected graph is a series-parallel graph if and only if it has no K_4 minor. It is an elementary exercise to show that any graph whose 2-connected components are series-parallel graphs is colorable with three colors. Thus, a graph with chromatic number equal to 4 must have a K_4 minor.

The case $n = 5$ is more complicated. The first step toward a proof was accomplished in Wagner (1937a) with Theorem (10.5.14). That result allows one to prove that the graphs without K_5 minors are colorable with four colors if this is so for all planar graphs, as follows. Suppose the graph, say G, is 3-connected. By Theorem (10.5.14), G is planar, or is isomorphic to $K_{3,3}$ or G_8, or has a 3-separation with at least four edges on each side. The planar case is handled by the assumption. The graphs $K_{3,3}$ and G_8 are colorable with three colors. In the 3-separation case, G is a Δ-sum, say $G_1 \oplus_\Delta G_2$. By Lemma (8.5.6) or by direct checking, G_1 and G_2 are minors of G. By induction, they have a coloring with at most four colors. In G_1, the colors of the nodes of the connecting triangle must be distinct. The same holds for G_2. By a suitable renaming of the colors of G_2, the graph G can thus be colored with four colors. The case of a 2-separable or 1-separable graph G is even simpler.

The second step in the proof of the case $n = 5$ involves showing that all planar graphs can be colored with four colors. The conjecture of that result was open for about one hundred years. It finally was proved in Appel and Haken (1977), and Appel, Haken, and Koch (1977). An exposition of the proof is included in Saaty and Kainen (1977). Thus, Hadwiger's conjecture is correct for $n \leq 5$. For $n \geq 6$, the conjecture is still open.

Theorem (10.5.15) is part of Wagner (1937a). Theorem (10.5.20) is proved in Wagner (1960). Related decomposition results are described in Halin (1981). Theorem (10.5.21) is taken from Truemper (1988). These theorems are useful for the solution of combinatorial problems via decomposition. An interesting instance is the *max cut problem*. Given is a graph G with nonnegative edge weights. One must find a disjoint union C of cocycles such that the sum of the weights of the edges of C is maximum. This problem is solved in Barahona (1983) for graphs without K_5 minors using Theorem (10.5.15). The same approach applies to graphs without G_{12} minors when Theorem (10.5.24) is substituted for Theorem (10.5.15).

There are numerous other ways in which graphs may be decomposed. The number of results is so large that we cannot even sketch the many ideas, theorems, and applications. Thus, we cite some representative references, but omit details. For example, the *ear decomposition* reduces a given

2-connected graph to a cycle by removing one path at a time while maintaining 2-connectedness. This decomposition is simple, but with its aid significant results have been proved (e.g., in Lovász and Plummer (1975), Kelmans (1987), Lovász (1983), and Frank (1990a)).

For some classes of graphs or for particular graph problems, special decompositions have been developed. Examples are decompositions for perfect graphs (e.g., in Burlet and Uhry (1982), Bixby (1984b), Burlet and Fonlupt (1984), Whitesides (1984), Chvátal (1985), (1987), Chvátal and Hoang (1985), Hoang (1985), Cornuéjols and Cunningham (1985), Hsu (1986), (1987a), (1987b), (1988), and Chvátal, Lenhart, and Sbihi (1990)) and circle graphs (e.g., in Gabor, Supowit, and Hsu (1989)), and for optimization problems (e.g., in Uhry (1979), Boulala and Uhry (1979), Cunningham (1982c), Edmonds, Lovász, and Pulleyblank (1982), Ratliff and Rosenthal (1983), Cornuéjols, Naddef, and Pulleyblank (1983), (1985), Bern, Lawler, and Wong (1987), Lovász (1987), Mahjoub (1988), Barahona and Mahjoub (1989a), (1989b), (1990a), (1990b)), Coullard and Pulleyblank (1989), and Fonlupt and Naddef (1991)). Diverse graph decomposition ideas are contained in Lovász and Plummer (1986). An axiomatic treatment of graph decompositions, indeed of decompositions of general combinatorial structures, is given in Cunningham and Edmonds (1980).

The 3-connected components of a graph can be found in linear time by an algorithm of Hopcroft and Tarjan (1973). A decomposition of minimally 3-connected graphs is given in Coullard, Gardner, and Wagner (1991).

The first polynomial test for graphicness of binary matroids was given by Tutte (1960). The graphicness test of Section 10.6 is a simplified version of a scheme of Truemper (1990). Other relevant references have already been cited in Section 3.6. The problem of deciding presence of the network flow property was treated completely and for the first time in Iri (1968). That material appears also in Bixby and Cunningham (1980). An efficient method for deciding graphicness of a matroid not known *a priori* to be binary was first proposed in Seymour (1981c). The matroid is assumed to be specified by a black box for deciding the independence of subsets of the groundset. Related material is included in Bixby (1982a), and Truemper (1982a). The easier case where all circuits are explicitly given is covered in Inukai and Weinberg (1979). The recognition problem of generalized networks, which constitute an extension of directed graphs, is treated in Chandru, Coullard, and Wagner (1985).

Last but by no means least, we should mention a truly astounding proof by Robertson and Seymour of the following daring conjecture due to Wagner: If a given graph property is maintained under minor-taking, then the number of nonisomorphic minor-minimal graphs that do not have the property is finite. The proof of that result involves a powerful graph decomposition concept called tree decomposition that we cannot treat here. The length of the proof is extraordinary. Together with numerous applications, it is being published in a long series of papers (Robertson and Seymour (1983), (1984), (1985), (1986a), (1986b), (1986c), (1988), (1990a), (1990b), (1990c), (1991a)–(1991k)). Related to this work on graphs are the matroid results of Kahn and Kung (1982), and Kung (1986a), (1986b), (1987), (1988), (1990a), (1990b). A survey of applications is made in Fellows (1989). There is some evidence supporting the following conjecture for the matroids that are representable over a given finite field: For any property that is maintained under minor-taking, the number of nonisomorphic minor-minimal matroids not having the property is finite. Truemper (1986) contains an encouraging related result about induced decompositions.

Chapter 11

Regular Matroids

11.1 Overview

Building upon the material on graphic matroids of Chapter 10, we analyze in this chapter the class of regular matroids. Already, we know that class quite well, or so at least it seems. By Theorem (9.3.2), a binary matroid is regular if and only if it has no F_7 or F_7^* minors. By Corollary (9.2.12), every graphic matroid is regular. There are also regular matroids that are not graphic and not cographic. If such a matroid is 3-connected, then by Theorem (10.4.1), it has an R_{10} or R_{12} minor. These results are interesting; indeed, the first and third one are profound. But they do not tell us how to construct regular matroids, or how to test binary matroids efficiently for regularity. That gap in our knowledge is filled by the extraordinary decomposition theorem of regular matroids due to Seymour. The entire chapter is devoted to that theorem and to some of its ramifications and applications.

We proceed as follows. In Section 11.2, we prove that 1-, 2-, and 3-sum compositions produce regular matroids when the components are regular. As a lemma for the 3-sum case, we also establish that the ΔY exchange defined in Section 4.4 maintains regularity.

Section 11.3 contains the main result of this chapter, the regular matroid decomposition theorem. It essentially says that every regular matroid can be produced by 1-, 2-, and 3-sums where the building blocks are graphic matroids, cographic matroids, and copies of R_{10}. Furthermore, only regular matroids can be generated by this process.

In Section 11.4, we develop efficient tests for regularity of binary matroids and for total unimodularity of real matrices. We obtain the regularity test by combining the method for finding 1-, 2-, and 3-sums of Section 8.4 with the graphicness test of Section 10.6. That scheme is then rather easily extended to an efficient test for total unimodularity.

The uses and implications of the regular matroid decomposition theorem are far-ranging. In Section 11.5, we describe representative applications. In the final section, 11.6, we list extensions and references.

The chapter requires knowledge of Chapters 2–10.

11.2   1-, 2-, and 3-Sum Compositions Preserve Regularity

In this section, we show that any 1-, 2-, or 3-sum with regular components is regular. This result is the comparatively easy part of the regular matroid decomposition theorem. We also prove that the ΔY matroids of Section 4.4 are regular.

The reader may wonder why we confine ourselves to 1-, 2-, and 3-sums, and do not treat general k-sum compositions. It turns out that the 1-, 2-, 3-sum cases have simple proofs and actually are the only k-sums needed for the regular matroid decomposition theorem. On the other hand, for k-sums with $k \geq 4$, the situation becomes much more complicated. In Section 11.6 we sketch what is known about that case.

Recall from Section 9.2 that a real matrix is totally unimodular if all of its determinants are $0, \pm 1$. Furthermore, a binary matrix is regular if it can be signed to become a totally unimodular real matrix. By Lemma (9.2.6) and Corollary (9.2.7), the signing is unique up to scaling by $\{\pm 1\}$ factors. Furthermore, the signing can be accomplished by signing one arbitrarily selected row or column at a time.

We need some elementary facts about the real $\{0, \pm 1\}$ matrices that are not totally unimodular, but each of whose proper submatrices has that property. We call such a matrix a *minimal violation matrix of total unimodularity*, for short *minimal violation matrix*. First, a minimal violation matrix is obviously square, and its real determinant is different from $0, \pm 1$. This fact implies that a 2×2 minimal violation matrix contains four ± 1s. Next, let a minimal violation matrix have order $k \geq 3$. Suppose we perform a real pivot in that matrix, then delete the pivot row and column. A simple cofactor argument proves that the resulting matrix is also a minimal violation matrix.

We are prepared for the proofs of this section. The first lemma deals with the case of 1- and 2-sums.

(11.2.1) Lemma.   *Any 1- or 2-sum of two regular matroids is also regular.*

Proof. Let M_1 and M_2 be the given regular components of a 1- or 2-sum M . We rely on the matrices of (8.2.1), (8.2.3), and (8.2.4), repeated below for convenient reference.

(11.2.2)

$$B = \begin{array}{c|c|c} & Y_1 & Y_2 \\ \hline X_1 & A^1 & 0 \\ \hline X_2 & 0 & A^2 \end{array}$$

Matrix B for 1-sum M

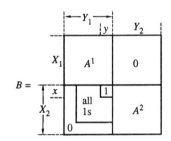

(11.2.3)

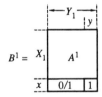   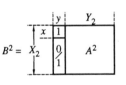

Matrices B, B^1, and B^2 for 2-sum M
with components M_1 and M_2

In the 1-sum case, M_1 and M_2 are represented by A^1 and A^2 of (11.2.2). Since M_1 and M_2 are regular, we can convert A^1 and A^2 by signing to totally unimodular real matrices, say $\tilde{A}^1$ and $\tilde{A}^2$. Derive a matrix $\tilde{B}$ from B of (11.2.2) by replacing A^1 and A^2 by $\tilde{A}^1$ and $\tilde{A}^2$. Evidently, $\tilde{B}$ is a totally unimodular signed version of B. Thus, M is regular.

The 2-sum case is slightly more complicated. Define D to be the submatrix of B of (11.2.3) indexed by X_2 and Y_1. We sign B^1 and B^2 of (11.2.3) so that totally unimodular matrices $\tilde{B}^1$ and $\tilde{B}^2$ result. The signing converts the submatrices A^1 and A^2 of B^1 and B^2 to, say, $\tilde{A}^1$ and $\tilde{A}^2$. Next we compute a signed version $\tilde{D}$ of D by the formula

(11.2.4) $\tilde{D} = (\text{column } y \text{ of } \tilde{B}^2) \cdot (\text{row } x \text{ of } \tilde{B}^1)$

Then $\tilde{A}^1$, $\tilde{A}^2$, and $\tilde{D}$ define a signed version $\tilde{B}$ of the matrix B for M. By the construction, the submatrices $[\tilde{A}^1/\tilde{D}]$ and $[\tilde{D} \mid \tilde{A}^2]$ of $\tilde{B}$ are totally unimodular.

We complete the proof by showing that the entire matrix $\tilde{B}$ is totally unimodular. Suppose it is not. Then $\tilde{B}$ contains a minimal violation matrix V that intersects $\tilde{A}^1$, $\tilde{A}^2$, and $\tilde{D}$. Thus, V also intersects the 0 submatrix of $\tilde{B}$ indexed by X_1 and Y_2, and accordingly must have order of at least 3. We perform a real pivot in $\tilde{B}$ on a ± 1 that is in both $\tilde{A}^1$ and V. The resulting real matrix $\tilde{B}'$ contains a smaller minimal violation matrix. The pivot changes $\tilde{A}^1$ and $\tilde{D}$, say to $\tilde{A}^{1'}$ and $\tilde{D}'$, but leaves $\tilde{A}^2$ unchanged. Since $[\tilde{A}^1/\tilde{D}]$ is totally unimodular, and since, by Lemma (9.2.2), pivots do not destroy total unimodularity, the matrix $[\tilde{A}^{1'}/\tilde{D}']$ is totally unimodular. If we perform the corresponding GF(2)-pivot in B, we thus get an unsigned version of $\tilde{B}'$. Furthermore, each column of $\tilde{D}'$ is a scaled version of a column of $\tilde{D}$. Thus, $[\tilde{D}' \mid \tilde{A}^2]$ is totally unimodular. A suitable repetition of the preceding reduction process eventually produces the contradictory case of a minimal violation matrix contained in a totally unimodular matrix.   □

We turn to the 3-sum case. By (8.3.10) and (8.3.11), we may assume M, M_1, and M_2 to be represented by the matrices B, B^1, and B^2, respectively, as follows.

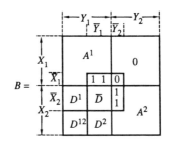

(11.2.5)

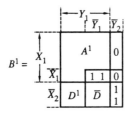

Matrices B, B^1, and B^2 for 3-sum M
with components M_1 and M_2

We may convert the component M_2 by a ΔY exchange of Section 4.4 to a matroid $M_{2\Delta}$ that is represented by the following matrix $B^{2\Delta}$, taken from (8.5.3).

(11.2.6)

$$B^{2\Delta} = \begin{array}{c} \\ X_2 \end{array}\begin{array}{c} \overset{\displaystyle Z_2 \qquad\quad Y_2}{\begin{array}{|c c|c|}\hline \multicolumn{2}{|c|}{\overline{D}} & 1 \\ & & 1 \\ \hline d & & A^2 \\ D^2 & & \\\hline\end{array}} \end{array}$$

Matrix $B^{2\Delta}$ for $M_{2\Delta}$

We first introduce a lemma that links regularity of M_2 to regularity of $M_{2\Delta}$.

(11.2.7) Lemma. M_2 of (11.2.5) is regular if and only if $M_{2\Delta}$ of (11.2.6) has that property.

Proof. For the "only if" part, let $\tilde{B}^2$ be a totally unimodular version of B^2 so that the two columns of $\tilde{B}^2$ indexed by $\overline{Y}_1$ do not contain any -1. This is possible, since we may begin the signing process with these two columns. Denote by $\tilde{\overline{D}}$ and $\tilde{D}^2$ the submatrices of $\tilde{B}^2$ that correspond to the submatrices $\overline{D}$ and D^2 of B. Declare $\tilde{d}$ to be the real difference of the two $\{0,1\}$ columns of $[\overline{D}/\tilde{D}^2]$. Thus, $\tilde{d}$ is a $\{0,\pm1\}$ vector that, together with the submatrices of $\tilde{B}^2$, defines a signed version $\tilde{B}_{2\Delta}$ of $B^{2\Delta}$. We are done once we prove $\tilde{B}_{2\Delta}$ to be totally unimodular. If this is not the case, then $\tilde{B}_{2\Delta}$ contains a minimal violation matrix V. By the construction, the latter matrix must intersect $\tilde{d}$. The two columns of $[\overline{D}/\tilde{D}^2]$ and the vector $\tilde{d}$ are $\mathbb{R}$-dependent, so V intersects $[\overline{D}/\tilde{D}^2]$ in at most one column. But in $\tilde{B}^2$, we can produce a scaled version of V as submatrix by a real pivot on one of the explicitly shown 1s in the first row, as may be readily checked. The latter fact contradicts the total unimodularity of $\tilde{B}^2$.

We prove the "if" part using duality. First we observe that M_2^* may be derived from $M_{2\Delta}^*$ by replacing a triad by a triangle. We just established that such a change maintains regularity. Thus, M_2^* is regular if $M_{2\Delta}^*$ is regular. $\quad\Box$

Lemma (11.2.7) implies the following result.

(11.2.8) Corollary. ΔY exchanges maintain regularity.

We are ready for the 3-sum case.

(11.2.9) Lemma. Any 3-sum of two regular matroids is regular.

Proof. We start with B, B^1, and B^2 of (11.2.5) for the assumed 3-sum M with regular components M_1 and M_2. By Lemma (11.2.7), the matrix $B^{2\Delta}$ of (11.2.6) is regular. Let d be the column vector displayed in $B^{2\Delta}$. Then up to indices, $[d \mid D \mid A^2]$ is the regular matrix $B^{2\Delta}$ plus possibly parallel columns. Thus, $[d \mid D \mid A^2]$ may be signed to become a totally unimodular matrix $[\tilde{d} \mid \tilde{D} \mid \tilde{A}^2]$. By the regularity of B^1 and duality, $[A^1/D]$ is also regular, and thus can be signed, starting with the submatrix D, to a totally unimodular matrix $[\tilde{A}^1/\tilde{D}]$. Clearly, the matrices $\tilde{A}^1$, $\tilde{A}^2$, and $\tilde{D}$ define a signed version $\tilde{B}$ of B.

We are done once we show $\tilde{B}$ to be totally unimodular. We accomplish this by essentially the same arguments as for the 2-sum case. Thus, if $\tilde{B}$ is not totally unimodular, then it has a minimal violation matrix V that intersects $\tilde{A}^1$, $\tilde{A}^2$, and $\tilde{D}$. The order of V must be at least 3. In $\tilde{B}$, we pivot on a ± 1 that is in both $\tilde{A}^1$ and V. The pivot changes $[\tilde{A}^1/\tilde{D}]$ to another totally unimodular matrix, say $[\tilde{A}^{1'}/\tilde{D}']$. It is easy to verify that the columns of $\tilde{D}'$ are nothing but scaled versions of the columns of $[\tilde{d} \mid \tilde{D}]$. Thus, the matrix $[\tilde{D}' \mid \tilde{A}^2]$ is up to scaling and parallel columns a submatrix of the previously defined $[\tilde{d} \mid \tilde{D} \mid \tilde{A}^2]$, and hence is totally unimodular. By a suitable repetition of the above reduction process, we eventually get the contradiction that a minimal violation matrix is contained in a totally unimodular matrix. $\qquad\square$

For future reference, we combine Lemmas (11.2.1) and (11.2.9) to the following theorem.

(11.2.10) Theorem. *Any 1-, 2-, or 3-sum of two regular matroids is regular.*

We return to a class of binary matroids introduced in Section 4.4, the class of ΔY matroids. Each of these matroids is constructed from the matroid represented by $B = [\,1\,]$ by repeated SP (= series-parallel) and ΔY exchanges. We do not repeat the details of these operations here. Thus, the reader may want to review Section 4.4 before proceeding. There the following result is claimed but not proved. We supply the proof next.

(11.2.11) Theorem. *The ΔY matroids are regular.*

Proof. We use induction on the number of SP extensions and ΔY exchanges used in the construction of a binary ΔY matroid M. The binary matroid represented by $B = [\,1\,]$ is regular. By Lemma (8.2.6), any SP extension of M can be viewed as a 2-sum composition where M is one of the components, and where the second component is a regular matroid with three elements. By Theorem (11.2.10), a 2-sum is regular if its components are regular. Thus, SP extensions maintain regularity. By Corollary (11.2.8), the same conclusion holds for ΔY exchanges. $\qquad\square$

Section 8.5 contains two variations of the 3-sum, called Δ-sum and Y-sum. The latter sum is the dual of the Δ-sum. A Δ-sum decomposition is obtained from a 3-sum $M_1 \oplus_3 M_2$ by replacing M_2 by $M_{2\Delta}$. For the Δ-

sum composition, we invert this process. Theorem (11.2.10) and Corollary (11.2.8) thus imply the following corollary for Δ-sums and Y-sums.

(11.2.12) Corollary. *Any Δ-sum or Y-sum M of two regular matroids is regular.*   □

In Section 4.4, it is mentioned that the class of ΔY matroids includes 3-connected matroids that are nongraphic and noncographic. By Lemma (4.4.10), every connected minor of a ΔY matroid is also a ΔY matroid. By (10.2.8), the regular matroid R_{10} has no triangles or triads, and thus is not a ΔY matroid. Indeed, R_{10} is 4-connected. We conclude that no binary matroid with an R_{10} minor is a ΔY matroid. The situation is different for R_{12}. In (10.2.9), the latter matroid is represented by a graph plus a node subset T. With the aid of that representation, one easily proves R_{12} to be a ΔY matroid, a fact already claimed in Section 4.4. Specifically, a triangle of that graph is a triangle of R_{12}, and a 3-star that is not a T node is a triad. Because of these facts, a ΔY sequence reducing R_{12} to the matroid represented by $B = [\,1\,]$ can be computed by graph operations. First, one reduces R_{12} while retaining the T set until a graphic or cographic matroid is attained. Second, one switches representations by selecting a suitable graph and finds the remaining reductions. We leave the details to the reader.

Theorems (11.2.10) and (11.2.11) enable us to construct many regular matroids from the matroids we already know to be regular, which are the graphic matroids, their duals, and R_{10} and R_{12}. In the next section, we see that we can construct all regular matroids that way. In fact, we require only a subset of the above initial matroids and a subset of the above construction steps to produce all regular matroids.

11.3 Regular Matroid Decomposition Theorem

In this section, we prove Seymour's profound decomposition theorem for regular matroids. It essentially says that every regular matroid can be obtained by 1-, 2-, and 3-sums where the building blocks are graphic matroids, cographic matroids, and matroids isomorphic to R_{10}. The theorem thus provides an elegant and useful construction for the entire class of regular matroids.

For the proof of the decomposition theorem, we rely on two ingredients. The first one we already know. It is Theorem (10.4.1). That result says that any 3-connected regular nongraphic and noncographic matroid has an R_{10} or R_{12} minor. The second ingredient consists of decomposition results of Sections 6.3 and 7.2, specifically Corollary (6.3.24) and the splitter Theorem (7.2.1). We could cast the results of this section in terms of the recursive construction of decomposition theorems introduced for graphs in Section 10.5. We will not do so. But we should mention that finding decomposition theorems such as the one of this section is facilitated by that recursive construction scheme. We discuss this aspect further in Chapter 13.

First, we analyze the influence of R_{10} and R_{12} minors. So assume R_{10} to be a minor of a regular matroid. For the analysis, we need the splitter Theorem (7.2.1), which we repeat next for convenient reference.

(11.3.1) Theorem (Splitter Theorem).   *Let $\mathcal{M}$ be a class of binary matroids that is closed under isomorphism and under the taking of minors. Let N be a 3-connected matroid of $\mathcal{M}$ on at least six elements.*

(a) *If N is not a wheel, then N is splitter of $\mathcal{M}$ if and only if $\mathcal{M}$ does not contain a 3-connected 1-element extension of N.*

(b) *If N is a wheel, then N is a splitter of $\mathcal{M}$ if and only if $\mathcal{M}$ does not contain a 3-connected 1-element extension of N and does not contain the next larger wheel.*

The matroid N is called a splitter of the class $\mathcal{M}$ of matroids. Here is the result for R_{10}.

(11.3.2) Theorem. R_{10} *is a splitter of the class $\mathcal{M}$ of regular matroids.*

Proof. By Theorem (11.3.1), we only need to show that every 3-connected 1-element extension of R_{10} is nonregular. Since R_{10} is isomorphic to its dual, it suffices that we consider 1-element additions. The case checking is conveniently accomplished when we represent R_{10} by a graph plus a T set as in (10.2.8), i.e., by

(11.3.3)

Graph plus T set representing R_{10}

Up to isomorphism, just three distinct 3-connected 1-element additions are possible.

In the first case, we join two nonadjacent nodes of the graph of (11.3.3)

by an edge shown in bold below.

(11.3.4)

1-element addition of R_{10}, case 1

We contract edge e. The two T endpoints of e become a non-T node. Evidently, the resulting graph contains a subdivision of

(11.3.5)

Graph plus T set representing F_7

which by (10.2.4) represents the Fano matroid F_7. Thus, that extension of R_{10} is nonregular.

The remaining two cases involve 1-element additions that are non-graphic even when one deletes from R_{10} the element represented by the set T. For both cases, we depict the additional element by a subset T' of the node set with $|T'| = 4$. The two possible ways are easily reduced to an instance of (11.3.5), and thus are also nonregular. □

Now we assume R_{12} to be a minor of a regular matroid. This time we rely on Corollary (6.3.24). We need a bit of preparation before we can restate that result. Let N be a binary matroid with a k-separation $(X_1 \cup Y_1, X_2 \cup Y_2)$ given by

(11.3.6)

$$B^N = \begin{array}{c|c|c|} & Y_1 & Y_2 \\ \hline X_1 & A^1 & 0 \\ \hline X_2 & D & A^2 \\ \hline \end{array}$$

Matrix B^N for N with k-separation

Consider the following three ways of extending N as depicted by the representation matrices below.

(a) The 1-element expansions represented by

(11.3.7)

	Y_1	Y_2
X_1	A^1	0
x	e	f
X_2	D	A^2

Matrix of 1-element expansion of N

In row x, e is not spanned by the rows of D, and f is nonzero.

(b) The 1-element additions given by

(11.3.8)

	Y_1	y	Y_2
X_1	A^1	g	0
X_2	D	h	A^2

Matrix of 1-element addition of N

In column y, g is nonzero, and h is not spanned by the columns of D.

(c) The 2-element extensions with representation matrix

(11.3.9)

	Y_1	y	Y_2
X_1	A^1	g	0
x	e	α	f
X_2	D	h	A^2

Matrix of 2-element extension of N

Either (c.1) or (c.2) below holds for e, f, g, h, and α.

    (c.1) e is not spanned by the rows of D; $f = 0$; $g = 0$; $h \neq 0$; $\alpha = 1$; e is not parallel to a row of A^1. If column $z \in Y_1$ of A^1 is nonzero, then e is not a unit vector with 1 in column z.

    (c.2) $g \neq 0$; h is spanned by the columns of D; e is spanned by the rows of D; $f = 0$ implies $e \neq 0$; $[e \mid \alpha]$ is not spanned by the rows of $[D \mid h]$. If $\overline{D}$, the matrix obtained from D by deletion of a column $z \in Y_1$, has the same GF(2)-rank as D, then $[g/h]$ is not parallel to column z of $[A^1/D]$. If the rows of D do not span a row $z \in X_1$ of A^1, then $[g/h]$ is not a unit vector with 1 in row z.

We restate Corollary (6.3.24).

(11.3.10) Theorem.   *Let M be a class of binary matroids that is closed under isomorphism and under the taking of minors. Suppose that N given by B^N of (11.3.6) is in M, but that the 1- and 2-element extensions of N given by (11.3.7), (11.3.8), (11.3.9), and the accompanying conditions are not in M. Assume that a matroid $M \in M$ has an N minor. Then any k-separation of any such minor that corresponds to $(X_1 \cup Y_1, X_2 \cup Y_2)$ of N under one of the isomorphisms induces a k-separation of M.*

We intend to use Theorem (11.3.10) with R_{12} as N and with the class of regular matroids as M. For the detailed arguments, we need the binary representation matrix B^{12} of (9.2.14) for R_{12}. We include that matrix below. According to that matrix, the pair $(X_1 \cup Y_1, X_2 \cup Y_2)$ constitutes a 3-separation of R_{12}.

(11.3.11)

$$B^{12} = \begin{array}{c|c} & \begin{array}{cc} Y_1 & Y_2 \end{array} \\ \hline X_1 & \begin{array}{cccc|cc} 1 & 0 & 1 & 1 & 0 & 0 \\ 0 & 1 & 1 & 1 & 0 & 0 \end{array} \\ \hline X_2 & \begin{array}{cccc|cc} 1 & 0 & 1 & 0 & 1 & 1 \\ 0 & 1 & 0 & 1 & 1 & 1 \\ 1 & 0 & 1 & 0 & 1 & 0 \\ 0 & 1 & 0 & 1 & 0 & 1 \end{array} \end{array}$$

Matrix B^{12} for R_{12}

Theorem (11.3.10) leads to the following decomposition result for the regular matroids with R_{12} minor.

(11.3.12) Theorem.   *Let M be a regular matroid with an R_{12} minor. Then any 3-separation of that minor corresponding to the 3-separation $(X_1 \cup Y_1, X_2 \cup Y_2)$ of R_{12} (see (11.3.11)) under one of the isomorphisms induces a 3-separation of M.*

Proof. We verify the sufficient conditions of Theorem (11.3.10). As a preparatory step, we calculate all 3-connected regular 1-element additions of R_{12}. By the symmetry of B^{12} of (11.3.11), and thus by duality, this result effectively gives us all 3-connected 1-element expansions as well. The addition cases are collected as columns in the following matrix C.

(11.3.13)

$$C = \begin{array}{c|c} & \begin{array}{c} Y_1 \end{array} \\ \hline X_1 & \begin{array}{cccc} 0 & 0 & 0 & 0 \\ 0 & 0 & 0 & 0 \end{array} \\ \hline X_2 & \begin{array}{cccc} 1 & 1 & 0 & 1 \\ 1 & 0 & 1 & 1 \\ 0 & 1 & 0 & 1 \\ 0 & 0 & 1 & 1 \end{array} \end{array}$$

Matrix C of 3-connected regular additions to R_{12}

Verification of the claim about C involves simple but somewhat tedious case checking. We omit the details, but should mention that the representation of R_{12} by a graph plus T set as in (10.2.9) greatly simplifies the task. The added element can be represented by a subset T' of the node set.

We verify the conditions of Theorem (11.3.10). The cases depicted in (11.3.13) rule out (11.3.7) and (11.3.8). They also narrow down the case analysis for (c.1) and (c.2) of (11.3.9), as follows.

According to (c.1), e is not spanned by the rows of D, $f = 0$, $g = 0$, $h \neq 0$, $\alpha = 1$, and e is not parallel to a row of A^1. If column $z \in Y_1$ of A^1 is nonzero, then e is not a unit vector with 1 in column z. We apply these conditions to B^{12}. By (11.3.13) and symmetry of B^{12}, we see that e must be the vector [0 0 1 1]. Furthermore, h must be a unit vector, or must be parallel to a column of A^2 or D, or must be the subvector of any column of C of (11.3.13) indexed by X_2. All such choices lead to nonregular matroids, as desired.

Suppose (c.2) applies. We ignore the conditions on e, f, and α. The remaining conditions are as follows. The vector g is nonzero, and h is spanned by the columns of D. If $\overline{D}$, the matrix obtained from D by deletion of a column $z \in Y_1$, has the same GF(2)-rank as D, then $[g/h]$ is not parallel to column z of $[A^1/D]$. If the rows of D do not span a row $z \in X_1$ of A^1, then $[g/h]$ is not a unit vector with 1 in row z. We apply these conditions to B^{12}. Thus, we determine that the matrix of (11.3.9) minus row x, which is B^{12} of (11.3.11) plus the column vector $[g/h]$, represents a 3-connected 1-element addition of R_{12}. By (11.3.13), any such addition with $g \neq 0$ is nonregular.

Thus, all conditions of Theorem (11.3.10) are satisfied. The conclusion of Theorem (11.3.10) then proves the result. $\Box$

At long last, we have completed all preparations for the regular matroid decomposition theorem.

(11.3.14) Theorem (Regular Matroid Decomposition Theorem). *Every regular matroid M can be decomposed into graphic and cographic matroids and matroids isomorphic to R_{10} by repeated 1-, 2-, and 3-sum decompositions.*

Specifically, if M is 3-connected and not graphic and not cographic, then M is isomorphic to R_{10} or has an R_{12} minor. In the latter case, any 3-separation of that minor corresponding to the 3-separation $(X_1 \cup Y_1, X_2 \cup Y_2)$ of R_{12} (see (11.3.11)) under one of the isomorphisms, induces a 3-separation of M.

Conversely, every binary matroid produced from graphic matroids, cographic matroids, and matroids isomorphic to R_{10} by repeated 1-, 2-, and 3-sum compositions is regular.

Proof. Let a regular matroid M be given. Assume M to be nongraphic and noncographic. If M is 1-separable, it is a 1-sum. If M is 2-separable, then it is by Lemma (8.2.6) a 2-sum. Hence, assume M to be 3-connected. By Theorem (10.4.1), M has an R_{10} or R_{12} minor. In the first case, M must by Theorem (11.3.2) be isomorphic to R_{10}. In the second case, M has by Theorem (11.3.12) an induced 3-separation, as claimed. By Lemma (8.3.12), M is a 3-sum. Finally, Theorem (11.2.10) establishes the converse part.   □

In Theorem (11.3.14), one may want to rely on Δ-sums or Y-sums instead of 3-sums. The next corollary supports that substitution.

(11.3.15) Corollary. *The claims of Theorem (11.3.14) remain valid when instead of 3-sums one specifies Δ-sums or Y-sums.*

Proof. By Section 8.5, any 3-sum decomposition may be converted to a Δ-sum or Y-sum decomposition by one ΔY exchange involving one of the components. By Lemma (11.2.8), such an exchange maintains regularity. Indeed, if M is 3-connected, then by Lemma (8.5.6), the components of any Δ-sum or Y-sum decomposition are isomorphic to minors of M. At any rate, the above arguments prove that we may substitute Δ-sums or Y-sums for 3-sums in Theorem (11.3.14).   □

Recall that the graph decomposition theorems (10.5.14) and (10.5.21) imply a construction of certain 2-connected graphs via 2- and Δ-sums. Each time, one of the components is a member of a class of building blocks, and the second component is a graph obtained by prior construction steps. We establish the analogous result for the connected regular matroids using Theorem (11.3.14). We treat Δ-sums as well as Y-sums and 3-sums. The terminology is adapted from that of Section 10.5.

(11.3.16) Theorem. *Any connected regular matroid is graphic, co-graphic, or isomorphic to R_{10}, or may be constructed recursively by 2-sums and Δ-sums (or Y-sums, or 3-sums) using as building blocks graphic matroids, cographic matroids, or matroids isomorphic to R_{10}.*

The proof of Theorem (11.3.16) is similar to that of Theorem (10.5.15). We confine ourselves to compositions involving 2-sums and Δ-sums. The Y-sum case is handled by duality, and the 3-sum case is proved by a simple adaptation of the proof for Δ-sums. We begin with two lemmas.

(11.3.17) Lemma. *Let M be a 3-connected, regular, nongraphic, and noncographic matroid that is not isomorphic to R_{10}. Assume M to have a triangle C. Then M has a 3-separation (E_1, E_2) where $|E_1|, |E_2| \geq 6$ and where one of E_1, E_2 contains C.*

Proof. By Theorem (11.3.14), M has a 3-separation (E_1, E_2) induced by the 3-separation of an R_{12} minor. The latter 3-separation corresponds to the one of (11.3.11) for R_{12}, and thus has six elements on each side. Hence, $|E_1|, |E_2| \geq 6$. If C is contained in E_1 or E_2, we are done. Otherwise, we shift one element of C from one side of the 3-separation (E_1, E_2) to the other one, to get a 3-separation (E_1', E_2') where C is contained in E_1' or E_2'. It is easily checked that the 3-separation of R_{12} depicted in (11.3.11) becomes an exact 4-separation when any one element is shifted from one side to the other. Thus, the element of C that we have shifted from E_1 or E_2 to the other set cannot be an element of the R_{12} minor inducing the 3-separation (E_1, E_2). We conclude that $|E_1'|, |E_2'| \geq 6$, as desired. $\square$

(11.3.18) Lemma. *Let M be a 3-connected, regular, nongraphic, and noncographic matroid that is not isomorphic to R_{10}. Assume M to have either a designated triangle C or a designated element y. Then M is a Δ-sum $M_1 \oplus_\Delta M_2$ where M_1 contains C or y, whichever applies, and where M_2 is graphic or cographic.*

Proof. We prove the case for the triangle C and leave the easier situation with the element y to the reader. We use induction. The smallest case involves an M isomorphic to R_{12}. By Lemma (11.3.17), the matroid R_{12} is a Δ-sum where one component contains C, and where the second component has nine elements and thus is graphic or cographic. Thus, we are done. For larger M, we again apply Lemma (11.3.17). Thus, M is a Δ-sum $M_1 \oplus_\Delta M_2$ where C is part of the component M_1, and where both M_1 and M_2 have at least nine elements. If the second component M_2 is graphic or cographic, we are done. Otherwise, we may assume M_2 to be 3-connected. We define C' to be the triangle of M_2 involved in the Δ-sum. Because of the presence of the triangle C', or by the splitter result for R_{10}, M_2 cannot be isomorphic to the 4-connected R_{10}. By induction, M_2 has a Δ-sum decomposition $M_{21} \oplus_\Delta M_{22}$ where M_{21} contains C', and where M_{22} is graphic or cographic. Via a representation matrix for M displaying the 3-separations involved in the Δ-sums $M_1 \oplus_\Delta M_2$ and $M_{21} \oplus_\Delta M_{22}$, we readily verify that M is a Δ-sum with $M_1 \oplus_\Delta M_{21}$ and M_{22} as components. That Δ-sum has the desired properties. $\square$

Proof of Theorem (11.3.16). Let M be any connected, regular, nongraphic, and noncographic matroid that is not isomorphic to R_{10}. If G is 3-connected, the result follows from Lemma (11.3.18). Otherwise, G is a 2-sum. Choose the 2-sum decomposition, say $M_1 \oplus_2 M_2$, so that M_2 has a minimal number of elements. Evidently, any 2-separation of M_2 contradicts the minimality assumption, so M_2 is 3-connected. If M_2 is graphic, cographic, or isomorphic to R_{10}, we are done. Otherwise, let y be the element of M_2 that together with an element of M_1 defines the 2-sum. By Lemma (11.3.18), M_2 is a Δ-sum $M_{21} \oplus_\Delta M_{22}$ where M_{21} contains the element y, and where M_{22} is graphic or cographic. Via a representation matrix for M, we confirm that M is a Δ-sum where one component is $M_1 \oplus_2 M_{21}$ and where the second component is the graphic or cographic M_{22} as demanded in the theorem. $\square$

An easy generalization of Theorems (11.3.14) and (11.3.16) is possible by the following splitter result.

(11.3.19) Lemma. F_7 *(resp. F_7^*) is a splitter of the binary matroids without F_7^* (resp. F_7) minors.*

Proof. According to the census of small 3-connected binary matroids in Section 3.3, there are just two 3-connected nonregular matroids on eight elements, with representation matrices given by (3.3.24) and (3.3.25). Clearly, both matroids have F_7 and F_7^* minors. Indeed, the matroids are selfdual. Thus, every 3-connected 1-element extension of F_7 (resp. F_7^*) has both F_7 and F_7^* minors. The result then follows from the splitter Theorem (11.3.1). $\qquad\qquad\Box$

We leave it to the reader to rewrite Theorems (11.3.14) and (11.3.16) so that they become results for the matroids without F_7 minors, or for the matroids without F_7^* minors.

One may concatenate matroid decomposition theorems of this section and graph decomposition theorems of Section 10.5. Later in this chapter, in Section 11.5, we meet one such case. At any rate, such concatenation is easy, and the reader may want to try his/her hand at producing potentially useful theorems.

In the next section, we use Theorem (11.3.14) to assemble efficient algorithms to decide regularity of binary matroids and total unimodularity of real matrices.

11.4 Testing Matroid Regularity and Matrix Total Unimodularity

Prior to the introduction of the regular matroid decomposition Theorem (11.3.14), no efficient algorithm was known for testing a binary matroid for regularity, or for deciding total unimodularity of real matrices. We know from Chapter 9 that these two tests are intimately linked. In fact, the results of Section 9.2 imply that an efficient test for one of the two problems can be easily converted to one for the other problem.

In this section, we construct the desired tests, relying, of course, on the regular matroid decomposition Theorem (11.3.14). In addition, we invoke the algorithm of Section 8.4 for finding 1-, 2-, and 3-sums, as well as the graphicness test of Section 10.6.

We begin with the regularity test for binary matroids. Let M be the binary matroid for which regularity is to be decided. We first apply the polynomial algorithm of Theorem (8.4.1) to determine whether or not M is a 1-, 2-, or 3-sum. In the affirmative case, we decompose M into two components. Then we apply the algorithm to each component, etc., until we eventually have a collection of binary matroids, say M_1, M_2, ..., M_n, none of which is a 1-, 2-, or 3-sum. It is not difficult to prove that n, the number of such matroids, is bounded by a function that is linear in the number of elements of M.

The regular matroid decomposition Theorem (11.3.14) says that M is regular if and only if each one of the matroids M_1, M_2, ..., M_n is graphic, or cographic, or isomorphic to R_{10}. We settle graphicness, and if necessary, cographicness, of each one of the matroids with the polynomial algorithm of Theorem (10.6.1). Deciding whether or not a matroid is isomorphic to R_{10} is trivial. If one of the matroids is found to be nongraphic, noncographic, and not isomorphic to R_{10}, then M is not regular. Otherwise, M has been proved to be regular.

The preceding algorithm is clearly polynomial. It can be made very efficient by an appropriate implementation. The algorithm is readily adapted to test total unimodularity of real matrices as follows. Given is a real matrix A. If A is not a $\{0, \pm 1\}$ matrix, we declare A to be not totally unimodular. So assume A to be a $\{0, \pm 1\}$ matrix. Let B be the support matrix of A. We view B to be binary. With the algorithm just described, we test the matroid M represented by B for regularity. If M is not regular, we know A to be not totally unimodular. So assume M, and hence B, to be regular. With the signing algorithm of Corollary (9.2.7), we deduce from B a totally unimodular matrix A'. By Lemma (9.2.6), A is totally unimodular if and only if A' can be obtained from A by scaling of some rows and columns by -1. Implicit in the proof of the lemma is an algorithm that finds the appropriate scaling factors, or determines that A' cannot be scaled to become A. Accordingly, we declare A to be totally unimodular or not.

We just have encountered one very important application of Theorem (11.3.14). We introduce several others in the next section.

11.5 Applications of Regular Matroid Decomposition Theorem

The uses of the regular matroid decomposition Theorem (11.3.14) range from the obvious to the unexpected. In this section, we cover representative instances.

Construction of Totally Unimodular Matrices

We begin with the most obvious case, the construction of the real totally unimodular matrices. These matrices represent precisely the regular matroids. To find a construction for them, we only need to translate the

3-sum version of the construction Theorem (11.3.16) into matrix language. First, we identify the matrix building blocks. We know that (10.2.8) and (10.4.5) provide the two possible GF(2)-representation matrices for R_{10}. With the scheme of Corollary (9.2.7), we sign these two matrices to obtain the following totally unimodular representation matrices.

$$(11.5.1) \qquad B^{10.1} = \begin{bmatrix} 1 & 0 & 0 & 1 & 1 \\ 1 & 1 & 0 & 0 & 1 \\ 0 & 1 & 1 & 0 & 1 \\ 0 & 0 & 1 & 1 & 1 \\ 1 & 1 & 1 & 1 & 1 \end{bmatrix} ; \qquad B^{10.2} = \begin{bmatrix} \text{-}1 & 1 & 0 & 0 & 1 \\ 1 & \text{-}1 & 1 & 0 & 0 \\ 0 & 1 & \text{-}1 & 1 & 0 \\ 0 & 0 & 1 & \text{-}1 & 1 \\ 1 & 0 & 0 & 1 & \text{-}1 \end{bmatrix}$$

The two totally unimodular matrices representing R_{10}

By Lemma (9.2.6), $B^{10.1}$ and $B^{10.2}$ are, up to scaling of rows and columns with $\{\pm 1\}$ factors, the unique totally unimodular representation matrices for R_{10}.

By the proof of Corollary (9.2.12), all graphic matroids are minors of the graphic matroids that are represented by the binary node/edge matrices. The latter matroids are also represented by the real totally unimodular node/arc incidence matrices, where each nonzero column has exactly one $+1$ and one -1. The minor-taking translates to IR-pivots in node/arc incidence matrices and to the deletion of rows and columns. Call any matrix so produced a *network matrix*. The effect of the IR-pivots can be readily established by graph operations. We leave it to the reader to work out the details.

At this point, we have the matrix building blocks that correspond to the matroid building blocks of Theorem (11.3.16). We now must understand the 2-sum and 3-sum decomposition/composition. The task is made complicated by the fact that we must interpret these operations in GF(2)-representation matrices without the use of GF(2)-pivots. Only that way can we deduce decomposition rules for totally unimodular matrices that do not involve IR-pivots. So let us assume B to be an arbitrary GF(2)-representation matrix of a regular 2- or 3-sum M with components M_1 and M_2. The underlying 2- or 3-separation is $(X_1 \cup Y_1, X_2 \cup Y_2)$. We assume M, and hence B, to be connected.

We start with the 2-sum case. The 2-separation manifests itself in B as depicted below, up to a switching of the roles of $X_1 \cup Y_1$ and $X_2 \cup Y_2$. The submatrix D of B has GF(2)-rank 1.

$$(11.5.2) \qquad B = \begin{array}{c|c|c} & Y_1 & Y_2 \\ \hline X_1 & A^1 & 0 \\ \hline X_2 & D & A^2 \end{array}$$

Matrix B with 2-separation

Correspondingly, the component matroids M_1 and M_2 are, by (8.2.3) and (8.2.4), represented by B^1 and B^2 below once appropriate indices are assigned to the row vector a of B^1 and the column vector b of B^2. These two vectors are nonzero vectors of D. Since GF(2)-rank $D = 1$, the two vectors are unique up to indices. Indeed, because of permutations of rows and columns in D, we may assume $D = b \cdot a$.

(11.5.3)

$$B^1 = \begin{array}{c|c} & Y_1 \\ \hline X_1 & A^1 \\ \hline & a \end{array} \quad ; \qquad B^2 = \begin{array}{c|cc} & & Y_2 \\ \hline X_2 & b & A^2 \end{array}$$

Matrices B^1 and B^2 of 2-sum

We sign B, B^1, and B^2 to get totally unimodular matrices. To simplify the notation, we now assume the matrices B, B^1, B^2 to be already such signed totally unimodular versions. We have the following 2-sum matrix composition rule.

(11.5.4) Matrix 2-Sum Rule. *Given are B^1 and B^2 of (11.5.3). Then we derive B of (11.5.2) from B^1 and B^2 by letting $D = b \cdot a$ (in $\mathbb{R}$).*

The 3-sum situation is more complicated, but yields to the same approach. At the outset, we assume all matrices to be binary. The underlying 3-separation manifests itself in one of two ways, up to a switching of the role of $X_1 \cup Y_1$ and $X_2 \cup Y_2$. Below, we indicate by B and $\tilde{B}$ the two cases, where GF(2)-rank $D = 2$ and GF(2)-rank $R = $ GF(2)-rank $S = 1$.

(11.5.5)

$$B = \begin{array}{c|cc} & Y_1 & Y_2 \\ \hline X_1 & A^1 & 0 \\ \hline X_2 & D & A^2 \end{array} \quad ; \qquad \tilde{B} = \begin{array}{c|cc} & Y_1 & Y_2 \\ \hline X_1 & A^1 & S \\ \hline X_2 & R & A^2 \end{array}$$

Matrices B and $\tilde{B}$ with 3-separation

We claim that, correspondingly, the component matroids M_1 and M_2 are represented below by B^1 and B^2, or by $\tilde{B}^1$ and $\tilde{B}^2$, once the vectors a, b, c, d, e, f, g, h of these matrices are appropriately defined and missing indices are added. Specifically, the vectors a and b of B^1 (resp. c and d of B^2) are two arbitrarily selected GF(2)-independent row (resp. column) vectors of D. Let $\overline{D}$ be the 2×2 submatrix of D created by the intersection of the row vectors a and b of D with the column vectors c and d of D. By Lemma (2.3.14), $\overline{D}$ is nonsingular. Because of (8.3.5)–(8.3.7) and row and column permutations, we may assume $D = [c \mid d] \cdot (\overline{D})^{-1} \cdot [a/b]$ (in GF(2)). We define the vector e of $\tilde{B}^1$ (resp. g of $\tilde{B}^2$) to be a nonzero vector of R, and f of $\tilde{B}^1$ (resp. h of $\tilde{B}^2$) to be a nonzero vector of S. Since GF(2)-rank $R =$ GF(2)-rank $S = 1$, the vectors e, g, f, h are unique up to indices. Furthermore, because of row and column permutations in R and S, we may assume $R = g \cdot e$ and $S = f \cdot h$.

(11.5.6)

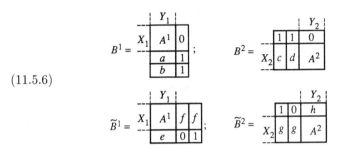

Matrices B^1, B^2 as well as $\tilde{B}^1$, $\tilde{B}^2$ of 3-sum

The case of B, B^1, and B^2 is our customary way of displaying 3-sums. The second one, with $\tilde{B}$, $\tilde{B}^1$, and $\tilde{B}^2$, we have not presented before. We show it to be correct by GF(2)-pivots that transform it to the first case. Specifically, suppose the submatrix S of $\tilde{B}$ has a 1 in the lower left-hand corner. Correspondingly, the last element of the leftmost vector f in $\tilde{B}^1$ and the first element of h in $\tilde{B}^2$ are 1s. Perform GF(2)-pivots on these 1s in the respective matrices. Each such GF(2)-pivot exchanges a row index against a column index. After the pivots, one readily confirms that the 3-separation $(X_1 \cup Y_1, X_2 \cup Y_2)$ of the new $\tilde{B}$ corresponds to the first case, and that the new $\tilde{B}^1$ and new $\tilde{B}^2$ are the component matrices for that case.

Consider the case of B, B^1, and B^2 of (11.5.5) and (11.5.6). We may sign B to obtain a totally unimodular matrix. In fact, by (11.5.5) and (11.5.6), the signing may be selected so that in the corresponding signing of B^1 and B^2 the explicitly shown 1s are not negated. The same signing convention may be followed for $\tilde{B}$, $\tilde{B}^1$, and $\tilde{B}^2$. As a matter of convenience, let us now assume that B, B^1, B^2, $\tilde{B}$, $\tilde{B}^1$, and $\tilde{B}^2$ are the signed totally unimodular matrices.

The above discussion validates the following matrix 3-sum construction.

(11.5.7) Matrix 3-Sum Rule. *Given are B^1 and B^2, or $\tilde{B}^1$ and $\tilde{B}^2$, of (11.5.6). In the first case, we calculate $D = [c \mid d] \cdot (\overline{D})^{-1} \cdot [a/b]$ (in $\mathbb{R}$) to determine B of (11.5.5). In the second case, we compute $R = g \cdot e$ and $S = f \cdot h$ (in $\mathbb{R}$) to find $\tilde{B}$ of (11.5.5).*

The next result summarizes the above discussion.

(11.5.8) Lemma. *The matrix 2- and 3-sum composition rules (11.5.4) and (11.5.7) correspond precisely to the regular matroid 2- and 3-sum compositions.*

The desired construction theorem for totally unimodular matrices can now be stated and proved.

(11.5.9) Theorem.   *Any connected totally unimodular matrix is up to row and column indices and scaling by* $\{\pm 1\}$ *factors a network matrix, or is the transpose of such a matrix, or is the matrix* $B^{10.1}$ *or* $B^{10.2}$ *of* (11.5.1), *or may be constructed recursively by matrix 2-sums and 3-sums. The rules are given by* (11.5.4) *and* (11.5.7). *The building blocks are network matrices, their transposes, and the matrices* $B^{10.1}$ *and* $B^{10.2}$ *of* (11.5.1).

Proof. The conclusion follows directly from Theorem (11.3.16) and Lemma (11.5.8). $\square$

Construction of $\{0, 1\}$ Totally Unimodular Matrices

Closely related to the construction of totally unimodular matrices is that of $\{0, 1\}$ totally unimodular matrices. So suppose B is a regular matrix that requires no signing to achieve total unimodularity. The matrix 2-sum rule obviously can be adopted without modification. The two 3-sum cases with B and $\tilde{B}$ of (11.5.5) are more troublesome. Even though B or $\tilde{B}$ requires no signing, the matrices B^1, B^2, $\tilde{B}^1$, or $\tilde{B}^2$ of (11.5.6) may not be totally unimodular without signing of some entries. By Corollary (9.2.7), it is easy to see that such signing can be confined to the explicitly shown 1s of the matrices of (11.5.6). Suppose such signing is needed for B^1 of (11.5.6). We may assume that the 1 in the lower right corner of B^1 must become a -1. From now on, let B^1 denote that real totally unimodular matrix, with the just-defined -1. We modify the last column of B^1 and add one additional row and column to get the following real matrix $\hat{B}^1$.

(11.5.10)

$$\hat{B}^1 = \begin{array}{c|c|c|c} & Y_1 & & \\ \hline X_1 & A^1 & 0 & 0 \\ \hline a & 1 & 0 \\ \hline b & 0 & 1 \\ \hline 0 & 1 & 1 \end{array}$$

Matrix $\hat{B}^1$ derived from B^1

We claim that the $\hat{B}^1$ is totally unimodular. For a proof, we perform an $\mathbb{R}$-pivot on the 1 of $\hat{B}^1$ in the lower right hand corner. We obtain B^1 as submatrix. Indeed, we see by the pivot that $\hat{B}^1$ represents a 3-sum. One component is the matroid of B^1. The second component is isomorphic to $M(W_4)$, the graphic matroid of the wheel with four spokes. Similarly, if needed, we modify B^2 to

(11.5.11)

$$\hat{B}^2 = \begin{array}{c|cccc} & \multicolumn{4}{c}{Y_2} \\ \hline & 1 & 1 & 0 & 0 \\ & 1 & 0 & 1 & 0 \\ \hline X_2 & 0 & c & d & A^2 \end{array}$$

Matrix $\hat{B}^2$ derived from B^2

Consider the 3-sum case involving $\tilde{B}^1$ and $\tilde{B}^2$. Assume the explicitly shown 1 in one or both of these matrices requires signing. The modifications are as follows, where $\hat{\tilde{B}}^1$ (resp. $\hat{\tilde{B}}^2$) corresponds to $\tilde{B}^1$ (resp. $\tilde{B}^2$).

(11.5.12)

$$\hat{\tilde{B}}^1 = \begin{array}{c|cccc} & \multicolumn{4}{c}{Y_1} \\ \hline X_1 & A^1 & f & f & 0 \\ \hline & e & 0 & 0 & 1 \\ \hline & 0 & 0 & 1 & 1 \end{array} \qquad \hat{\tilde{B}}^2 = \begin{array}{c|cccc} & \multicolumn{4}{c}{Y_2} \\ \hline & 1 & 1 & 0 & 0 \\ & 1 & 0 & 0 & h \\ \hline X_2 & 0 & g & g & A^2 \end{array}$$

Matrix $\hat{\tilde{B}}^1$ and $\hat{\tilde{B}}^2$ derived from $\tilde{B}^1$ and $\tilde{B}^2$

We leave it to the reader to verify by IR-pivots that the modifications are appropriate. Indeed, $\hat{\tilde{B}}^1$ (resp. $\hat{\tilde{B}}^2$) of (11.5.12) represents a 3-sum; one component is represented by $\tilde{B}^1$ (resp. $\tilde{B}^2$) of (11.5.6), and the second component is isomorphic to $M(W_4)$.

The reader likely anticipates that the analogue of Theorem (11.5.9) holds for $\{0,1\}$ totally unimodular matrices, where this time we permit for 3-sums the cases of (11.5.10)–(11.5.12) besides the ones of (11.5.6). That guess is correct. But the proof requires some care. For instance, we must show that the matrices of (11.5.10)–(11.5.12) are always smaller than the matrix of (11.5.5) produced by them. We must also establish that any one of the matrices of (11.5.10)–(11.5.12) is graphic or cographic if the related matrix of (11.5.6) is graphic or cographic. The latter task may seem very difficult. Indeed, in general it cannot be accomplished, as we find out in the proof of the next theorem.

(11.5.13) Theorem. *Any connected $\{0,1\}$ totally unimodular matrix is up to row and column indices a $\{0,1\}$ network matrix, or the transpose of such a matrix, or the matrix $B^{10.1}$ of (11.5.1), or may be constructed recursively by matrix 2-sums and 3-sums. The 2-sum rule is given by (11.5.4). The 3-sums are specified by the 3-sum rule (11.5.7), except that we also allow $\hat{B}^1$, $\hat{B}^2$, $\hat{\tilde{B}}^1$, $\hat{\tilde{B}}^2$ of (11.5.10)–(11.5.12) as component matrices. The building blocks are $\{0,1\}$ network matrices, their transposes, and the matrix $B^{10.1}$ of (11.5.1).*

Proof. We invoke Theorem (11.5.9) and the above derivation of $\hat{B}^1$, $\hat{B}^2$, $\tilde{\hat{B}}^1$, $\tilde{\hat{B}}^2$. The case of $B^{10.2}$ is not possible since that matrix cannot be scaled to become a $\{0, 1\}$ matrix. Thus, we are done, unless $\hat{B}^1$, $\hat{B}^2$, $\tilde{\hat{B}}^1$, or $\tilde{\hat{B}}^2$ is needed, and is of the same size as B or is not graphic or cographic. Because of pivots, symmetry, and duality, we may select any one of the above matrices, say $\hat{B}^1$, to eliminate these concerns.

First, we prove that $\hat{B}^1$ is smaller than B. By the proof of Theorem (11.3.16), the 3-separation $(X_1 \cup Y_1, X_2 \cup Y_2)$ of B inducing the 3-sum that in turn produces $\hat{B}^1$ plus B^2 or $\hat{B}^2$ satisfies $|X_1 \cup Y_1|, |X_2 \cup Y_2| \geq 6$. But $\hat{B}^1$ has length $|X_1 \cup Y_1| + 5$ and thus is smaller than B.

The second part is more complicated. We want to show that B^1 graphic or cographic implies $\hat{B}^1$ graphic or cographic. This goal turns out to be attainable in all cases but one. That exception we handle by a switch to a different 3-separation of B.

Suppose B^1 is graphic. As stated above, $\hat{B}^1$ represents a 3-sum. One component is the matroid of B^1. The second component is isomorphic to $M(W_4)$. We claim that 3-sum to be graphic. Indeed, any graph G for B^1 can be converted to one for $\hat{B}^1$ as follows. By (11.5.6), the edges of G not indexed by $X_1 \cup Y_1$ form a triangle C, say with vertices u, v, w. Then on an appropriately selected edge of C, say with endpoints u and v, we place a midpoint and connect that new node with w by a new edge. One easily verifies that this construction does produce a graph for $\hat{B}^1$.

We examine the second possibility, where B^1 is cographic and not graphic. Thus, B^1 is nonplanar. Let H be any graph for the transpose of B^1. By (11.5.6), the edges of H corresponding to the unlabeled rows and columns of B^1 form a cocircuit C^* of H of cardinality 3.

Assume C^* to be a 3-star of H. Analogously to the earlier case, one confirms $\hat{B}^1$ to be cographic, as desired.

Assume C^* to be not a 3-star of H. Thus, removal of the three edges of C^* transforms H to two connected nonempty graphs H_1 and H_2. Accordingly, the minor $H \backslash C^*$ is not 2-connected. That minor corresponds to the transpose of the submatrix A^1 of B^1. Thus, the transpose of A^1, and hence A^1, are not connected. Indeed, A^1 is a 1-sum of two matrices A^{11} and A^{12} where A^{11} corresponds to H_1, and A^{12} to H_2. We derive from H a graph $\tilde{H}_1$ (resp. $\tilde{H}_2$) by contracting the edges of H_2 (resp. H_1). In $\tilde{H}_1$ and $\tilde{H}_2$, the set C^* is a 3-star. If both $\tilde{H}_1$ and $\tilde{H}_2$ are planar, then one easily verifies H to be planar as well, a contradiction of the fact that B^1 is cographic and not graphic. Thus, we may assume $\tilde{H}_1$ to be nonplanar, and hence to have at least nine edges. We now redefine the 3-separation of B by shifting the submatrix A^{12} of A^1 to A^2. For the new 3-separation, the new B^1 corresponds to $\tilde{H}_1$, and thus is still cographic and not graphic. But C^* is now a 3-star of $\tilde{H}_1$, so the earlier case applies. □

So far, we have seen two applications that involve a rather obvious translation of Theorem (11.3.16). The next application makes a more sophisticated use of that theorem.

Characterization of Cycle Polytope

Given is a connected binary matroid M with groundset E. To each element $e \in E$, a real weight w_e is assigned. We want to find a disjoint union C of circuits of M so that $\sum_{e \in C} w_e$ is maximized. This problem occurs in a number of settings. We describe one, and reference others in Section 11.6.

Let G be a connected graph with edge set E and real weights w_e, $e \in E$. We want to partition the vertex set V of G into sets V_1 and V_2 such that the sum of the weights of the set C^* of edges connecting V_1 and V_2 is maximized. The set C^* is readily seen to be a disjoint union of the cocircuits of the graphic matroid $M(G)$ of G. Conversely, any such disjoint union of cocircuits of $M(G)$ is the set C^* for some partition of V. Thus, the graph problem, which usually is called the *max cut problem*, becomes the earlier matroid problem with $M = M(G)^*$.

The matroid problem can be easy or difficult, depending on M and on the sign of the weights. In general, the problem is known to be $\mathcal{NP}$-hard. We sketch a polyhedral approach for a special subclass. Let x^C be the characteristic vector for a disjoint union C of circuits of M. Thus, x^C is indexed by E, and the entry in position $e \in E$ of x^C is equal to 1 if $e \in C$, and equal to 0 otherwise. We view the x^C vectors to be in $\mathbb{R}^E$. Define $P(M)$ to be the convex hull of the x^C vectors. Usually, $P(M)$ is called the *cycle polytope* of M. By results of polyhedral combinatorics, we can solve the given problem efficiently if we can determine whether or not a vector $x \in \mathbb{R}^E$ is in $P(M)$, and if for $x \notin P(M)$ we can provide a hyperplane that separates x from $P(M)$. As a first step, we thus strive to find linear inequalities that are satisfied by all points of $P(M)$. We say that such inequalities are *valid* for $P(M)$. Notationally, for any subset $\overline{E} \subseteq E$, let $x(\overline{E}) = \sum_{e \in \overline{E}} x_e$.

Obviously, the inequalities

$$(11.5.14) \qquad\qquad 0 \leq x_e \leq 1;\ e \in E$$

are valid for $P(M)$. We derive additional valid inequalities as follows. By Lemma (3.3.6), each circuit of M intersects each cocircuit in an even number of elements. Let C be a disjoint union of circuits, C^* be a cocircuit, and F be a subset of C^* of odd cardinality. We claim that the following inequality is valid for $P(M)$.

$$(11.5.15) \quad x(F) - x(C^* \setminus F) \leq |F| - 1;\ F \subseteq C^* = \text{cocircuit};\ |F|\ \text{odd}.$$

Since $P(M)$ is the convex hull of the x^C vectors, we may establish validity of (11.5.15) by proving it for $x = x^C$, C being an arbitrary disjoint union

of circuits of M. If F is not contained in C, then $x(F) \leq |F| - 1$, and
(11.5.15) clearly holds. So assume $F \subseteq C$. We know that $F \subseteq C^*$. Since
$|F|$ is odd and $|C \cap C^*|$ is even, C has at least one element in $C^* \setminus F$. Thus,
once more (11.5.15) holds. We conclude that (11.5.15) is valid.

Define $Q(M)$ to be the subset of $\mathbb{R}^E$ defined by the inequalities of
(11.5.14) and (11.5.15). When is $P(M) = Q(M)$? We sketch the answer to
this question and its proof. First, one shows that $P(M) \neq Q(M)$ if M has
a minor isomorphic to F_7^*, $M(K_5)^*$, or R_{10}. Second, one establishes that
$P(M) = Q(M)$ if M is graphic or isomorphic to F_7, $M(K_{3,3})^*$, or $M(G_8)^*$,
where G_8 is given by (10.5.9). Third, one proves that $P(M) = Q(M)$ if M
is a 2-sum or Y-sum of two matroids M_1 and M_2 for which $P(M_1) = Q(M_1)$
and $P(M_2) = Q(M_2)$. Fourth and last, one concatenates the dual version
of Theorem (10.5.15) and Theorem (11.3.16) to the following result.

(11.5.16) Theorem.   *Let M be a connected binary matroid having no
F_7^* or $M(K_5)^*$ minors. Then M is graphic or isomorphic to F_7, $M(G_8)^*$,
$M(K_{3,3})^*$ or R_{10}, or may be constructed recursively by 2-sums and Y-sums.
The building blocks are graphic matroids and matroids isomorphic to F_7,
$M(G_8)^*$, $M(K_{3,3})^*$, and R_{10}.*

The four ingredients clearly imply the following conclusion.

(11.5.17) Theorem.   *Let M be a connected binary matroid. Then
$P(M) = Q(M)$ if and only if M has no F_7^*, $M(K_5)^*$, or R_{10} minors.*

References for additional material about $P(M)$ and $Q(M)$ are included
in Section 11.6.

We describe some additional applications of the regular matroid decomposition theorem without proofs. The first one concerns the number
of nonzeros in the rows of totally unimodular matrices having fewer rows
than columns.

Number of Nonzeros in Totally Unimodular Matrices

Let A be a totally unimodular matrix, say of size $m \times n$ with $m \leq n$. For
$i = 1, 2, \ldots, m$, let p_i be the number of nonzeros in row i of A. Define
$p^* = \min p_i$. Trivially, p^* is bounded from above by n, the number of
columns of A. That upper bound is attained by the $m \times n$ matrix A
containing only 1s. The example matrix has parallel columns. Thus, one
may conjecture that a tighter upper bound on p^* may exist in the absence
of parallel columns. The next theorem confirms this conjecture by proving
m, the number of rows of A, to be an upper bound on p^* when A has no
parallel columns.

(11.5.18) Theorem. *Let A be a totally unimodular matrix of size $m \times n$ with $m \leq n$. For $i = 1, 2, \ldots, m$, let p_i be the number of nonzeros in row i of A, and define $p^* = \min p_i$. If A has no parallel columns, then $p^* \leq m$.*

Triples in Circuits

The next theorem addresses the following question. Given are three elements e, f, g of a binary matroid M. When does some circuit of M include them all?

It is not difficult to reduce the problem to the case where the matroid is sufficiently connected. For that situation the answer is as follows.

(11.5.19) Theorem. *Let e, f, g be distinct elements of a 3-connected binary matroid M. Assume that M does not have a 3-separation with at least four elements on each side. Then there is no circuit of M containing e, f, and g if and only if $\{e, f, g\}$ is a cocircuit of M or the following condition holds. M is graphic, and in the corresponding graph, the edges e, f, and g are edges with a common endpoint.*

Odd Cycles

Given is an undirected graph G. What is the structure of G if every two cycles of G, each of odd length, have a node in common? Here, too, it is not difficult to reduce the problem to the situation where the graph is sufficiently connected. The following theorem provides the answer for that case.

(11.5.20) Theorem. *Let G be a 3-connected graph on at least six vertices and without parallel edges. Assume that G does not have a 3-separation with at least four edges on each side. Then every two cycles of G, each with an odd number of edges, have a common vertex if and only if G observes* (i) *or* (ii) *below.*

(i) *Deletion of some vertex or deletion of the edges of a triangle from G results in a bipartite graph.*
(ii) *G can be drawn in the projective plane so that every region is bounded by an even number of edges.*

Another application concerns the construction of the connected, undirected, and signed graphs without so-called odd-K_4 minors. Relevant definitions and details of the construction are included in Chapter 13.

In the last section, we point out some extensions and identify relevant references.

11.6   Extensions and References

The main reference for the entire chapter is Seymour (1980b), which contains Seymour's decomposition theorem for regular matroids.

Lemmas (11.2.1) and (11.2.9) constitute the easy part of the regular matroid decomposition theorem. They are due to Brylawski (1975). For $k \geq 4$, k-sums with regular components are not necessarily regular. An example for $k = 4$ is given in Truemper (1985b). Intuitively, one is tempted to argue that regularity of k-sums must be assured when the connecting matroid is sufficiently large and structurally rich. Indeed, this notion can be made precise by sufficient conditions that, for any $k \geq 4$, assure regularity of k-sums with regular components. Because of space limitations, we omit a detailed treatment.

All decomposition results of Section 11.3 either are taken directly from Seymour (1980b), or are implied by that reference. We have described them using 3-, Δ-, and Y-sums. Seymour (1980b) relies on Δ-sums.

A very efficient version of the regularity/total unimodularity test of Section 11.4 is presented in Truemper (1990). Indeed, the algorithm described there has the currently best worst-case bound of all known schemes. Other polynomial algorithms, using different ideas, are given in Cunningham and Edmonds (1978), Bixby, Cunningham, and Rajan (1986), and Rajan (1986). We should mention that Seymour (1980b) already contains implicitly a polynomial, though not very efficient, testing algorithm. A polynomial test for deciding regularity of matroids *a priori* not known to be binary is given in Truemper (1982a).

Theorems (11.5.9) and (11.5.13), which are matrix versions of Theorem (11.3.16), are given without proof in Seymour (1985a), and Nemhauser and Wolsey (1988). Seymour (1985a) uses the hypergraph terminology of Berge (1973). The polytope question $P(M) \stackrel{?}{=} Q(M)$ answered by Theorem (11.5.17) is just one example of numerous questions concerning flows, circuits, and cutsets in matroids. Seymour (1981a) contains a wide-ranging investigation of these issues. One of them concerns the *sum of circuits property* first defined in Seymour (1979b). It is shown in Seymour (1981a) that this property holds for a matroid M if and only if M is binary and has no F_7^*, $M(K_5)^*$, or R_{10} minors. That result, Theorems (10.5.15) and (11.3.16), and an amazing symmetry of $P(M)$ allow one to completely resolve the $P(M) \stackrel{?}{=} Q(M)$ question by Theorem (11.5.17), which is due to Barahona and Grötschel (1986). The proof sketched here is from Grötschel and Truemper (1989b), which contains additional results about $P(M)$ and treats computational aspects. Earlier results for special matroid classes and applications are described in Orlova and Dorfman (1972), Edmonds and Johnson (1973), Hadlock (1975), Barahona (1983), Barahona and Mahjoub (1986), and Barahona, Grötschel, Jünger, and Reinelt (1988). Additional results for the cycle polytope are given in Grötschel and Truemper (1989a). Theorem (11.5.18) is proved in Bixby and Cunningham (1987). Theorem (11.5.19) is from Seymour (1986a). Theorem (11.5.20) is due to Lovász and is included in Gerards, Lovász, Schrijver, Seymour, and Truemper (1991). References for the odd-K_4 result are given in Chapter 13. Additional applications of the regular matroid decomposition theorem are described in Seymour (1981d), (1981f). Bland and Edmonds (1978) have used the decomposition to reduce linear programs with totally unimodular constraint matrix to a sequence of maximum flow and shortest route problems.

Chapter 12

Almost Regular Matroids

12.1 Overview

So far in this book, we have always used matrices to understand matroids. We have selected a base of the matroid to be investigated. Then we have constructed for that base a representation matrix over some field or even an abstract matrix. Finally, we have analyzed the matrix structure to deduce matroid results. In the third step, we have employed pivots, in particular in the path shortening technique, to modify the matrix in agreement with some change of the matroid base. We have also deleted or added rows and columns to represent matroid reductions or extensions.

Occasionally, we have reversed the just-described roles of matroids and matrices. An example is the test of total unimodularity in Section 11.4 via a test of matroid regularity. A second example is the analysis of the structure of totally unimodular matrices in Section 11.5 via the structure of regular matroids. But generally, it seems to be difficult to answer matrix questions with matroid techniques. In particular, the taking of matroid minors, a most useful matroid operation, has a cumbersome translation into matrix language unless one permits pivots. But the pivot operation almost always changes the matrix structure rather drastically. Thus, pivots often impede the understanding of matrix structure.

The preceding arguments seem to lead to the inescapable conclusion that matrix properties generally cannot be conveniently analyzed with matroids. In particular, one is inclined to accept that conclusion in the following setting. The matroid property in question, say $\mathcal{P}$, is inherited under submatrix-taking. One wants to understand the minimal matrices that do not have $\mathcal{P}$. We call any such matrix a *minimal violation matrix* of $\mathcal{P}$.

In this chapter, we show that the above reasoning, flawless as it may seem, is not valid for the investigation of the minimal violation matrices of certain properties. Indeed, we describe a general matroid technique for a class of such problems where $\mathcal{P}$ observes certain conditions. We demonstrate the technique for the case where $\mathcal{P}$ is the property of regularity. The method deduces a matroid formulation that involves a class of binary matroids called almost regular. An analysis of that class produces the ΔY matroid construction stated earlier in Theorem (4.4.16). Additional analysis leads to a number of matrix constructions of surprising simplicity. Two of the constructions generate the minimal violation matrices for the following two properties: regularity and total unimodularity.

At the outset, the sections of this chapter may appear to be like the pieces of a puzzle: diverse and seemingly unrelated. Only toward the end of the chapter do the relationships and functions of the puzzle pieces emerge. We sketch now the content of each section.

In Section 12.2, we determine for undirected graphs G the minimal subgraphs that rule out success for a certain signing of the edges of G. The signing process is supposed to convert G to a graph G' with edge labels "+1" and "−1." The labels are to be such that in every chordless circuit of G', the edge labels sum (in $\mathbb{R}$) to prescribed values specified in a given vector α. If such signing is possible, then we call the resulting graph G' α-balanced. The signing condition is then linked to the set $\mathcal{N}$ of the binary minimal violation matrices of regularity. That way, we determine complete characterizations for two proper subsets of $\mathcal{N}$. We also find an admittedly weak characterization of the remaining matrices of $\mathcal{N}$.

In Section 12.3, we shift our attention from graphs and the set $\mathcal{N}$ to several matrix classes. In particular, we define the class $\mathcal{U}$ of real complement totally unimodular matrices, the classes $\mathcal{A}$ and $\mathcal{B}$ of almost representative matrices over GF(3) and GF(2), respectively, and the class $\mathcal{V}$ of real minimal violation matrices of total unimodularity. The definitions of these classes will give the impression that they are quite unrelated. But in a way, these classes as well as $\mathcal{N}$ are different manifestations of one and the same phenomenon: the absence of matroid regularity plus certain minimality conditions.

Another puzzle piece is introduced in Section 12.4. There we describe the previously mentioned technique for the investigation of the minimal violation matrices of certain matrix properties $\mathcal{P}$. We specialize the general method to the case where $\mathcal{P}$ is the property of regularity. As a result, we define the class of almost regular matroids. We state and sketch a proof of the construction of the almost regular matroids via ΔY extension sequences. The latter result should be familiar since a summary is included in Section 4.4.

The odd and unrelated puzzle pieces of Sections 12.2–12.4 are merged to a rather beautiful picture in Section 12.5. The almost regular matroids are seen to be the matroid manifestation of the matrices of the classes of $\mathcal{A}$, $\mathcal{B}$, $\mathcal{N}$, $\mathcal{U}$, and $\mathcal{V}$. In particular, the construction of the almost regular matroids via ΔY extension sequences produces an elementary construction of the class $\mathcal{U}$ of complement totally unimodular matrices. With $\mathcal{U}$ in hand, we very easily construct the remaining classes $\mathcal{A}$, $\mathcal{B}$, $\mathcal{N}$, and $\mathcal{V}$. We thus have a construction for the binary minimal violation matrices of regularity, and for the real minimal violation matrices of total unimodularity. In the final section, 12.6, we sketch applications and extensions, and cite references.

The chapter assumes knowledge of Chapters 2, 3, 4, 9, and 10.

12.2   Characterization of Alpha-Balanced Graphs

We are given an undirected graph G. For each chordless circuit C of G, we are given an integer number $\alpha_C = 0, 1, 2$, or 3. We collect the α_C in a vector α. To each edge of G, we want to assign the label $+1$ or -1 so that in the resulting graph G' the real sum of the edge labels of each chordless cycle C is congruent (mod 4) to α_C. When such labels can be found, we say that G' is α-balanced. Suppose an α-balanced G' cannot be produced. Then G apparently contains a configuration of chordless cycles with conflicting requirements. In this section, we identify the possible sources of such conflicts. Specifically, we pinpoint three subgraph configurations that collectively represent the minimal obstacles to α-balancedness. In the last portion of the section, we specialize that result for the case when α is the zero vector. That specialization leads to a characterization of the binary matrices that may be signed to become so-called balanced $\{0, \pm 1\}$ real matrices. From the latter characterization, we deduce a partition of the class $\mathcal{N}$ of the binary minimal violation matrices of regularity. The partition consist of three subclasses. Two of them are well described, and indeed are readily constructed. But the third subclass has a rather unsatisfactory characterization. One could say that the subsequent sections of this chapter are devoted to replacing that unsatisfactory description with a mathematically appealing and practically useful one.

We begin with the detailed technical discussion. We use a number of terms that we define in the next few paragraphs.

It is convenient for us to consider every graph to be undirected and to have a priori a $+1$ or -1 label on each edge. If nothing is said about a graph, then all labels are assumed to be $+1$. Thus, the previously described assignment of $\{\pm 1\}$ labels to G becomes a change of some edge labels. That change we call a *signing* of G. We *scale* a star of G by multiplying the edge labels of that star by -1. We *scale* G by a sequence of scaling steps, each one involving some star of G. The *label sum* of a subgraph $\overline{G}$ is the real sum of the labels of the $\overline{G}$ edges. That sum is denoted by $L(\overline{G})$.

The emphasis on nodes in this section demands that we temporarily abandon our usual viewpoint where nodes are edge subsets. Thus, we view nodes as points, and consider edges to be unordered node pairs. That approach is reasonable since we never have to deal with the contraction operation or with parallel edges. We still view trees, paths, and cycles to be given by their edge sets. Most subgraphs will be induced by some node subset. We use *n-subgraph* to specify that case. Special instances are *n-path* and *n-cycle*. If $\overline{G}$ is a subgraph of G but not an n-subgraph, then G has an edge that is not in $\overline{G}$, but both of whose endpoints are in $\overline{G}$. Such an edge is a *G-chord* for $\overline{G}$.

Let α be an integer vector whose entries are in one-to-one correspondence with the n-cycles of a graph G. Throughout, it is assumed that each entry of α is $0, 1, 2,$ or 3. Then G is *α-balanced* if for each n-cycle C the label sum $L(C)$ satisfies $L(C) \equiv \alpha_C \pmod 4$. Note that scaling in G changes $L(C)$ by a multiple of 4, and thus does not affect α-balancedness. Suppose a graph G has only $+1$s as edge labels and is not α-balanced. Then possibly an α-balanced graph G' can be deduced from G by signing. The label changes of edges in an n-cycle C, say producing C', modify $L(C)$ for some integer k to $L(C') = L(C) + 2k = |C| + 2k$. Thus, if $L(C') \equiv \alpha_C \pmod 4$ is to be achieved at all, then necessarily $|C| \equiv \alpha_C \pmod 2$. From now on, we assume that any α satisfies this necessary condition.

The relation "is an n-subgraph of" is transitive. In particular, every n-cycle of an n-subgraph of G is an n-cycle of G. Let α be given for G. By the above observation, it makes sense to apply the term "α-balanced" not just to G, but also to n-subgraphs of G. In particular, α-balancedness of an n-cycle C of G means $L(C) \equiv \alpha_C \pmod 4$. We also say in that case that C *agrees with* α.

One may suspect that signing to achieve α-balancedness is essentially unique. The next lemma confirms this notion. The proof is almost identical to that of Lemma (9.2.6).

(12.2.1) Lemma.   *Let G and G' be connected α-balanced graphs that are identical up to edge labels. Then G' may be obtained from G by scaling.*

Proof. Let T be a tree of G, and T' be the corresponding tree of G'. Because of scaling, we may suppose that the labels of G and G' agree on the edges of T and T'. Suppose the labels of G and G' differ, say on an edge e of G and on the corresponding edge e' of G'. The edges e and e' form fundamental cycles C and C' with T and T', respectively. Select T and e, and thus T' and e', so that the cardinality of the cycles is minimum.

Suppose C and C' have chords. By the minimality condition, the label of
any chord of C must agree with that of the corresponding chord of C'. But
then C and C' do not have minimum cardinality, a contradiction. Thus,
C and C' are chordless cycles with labels in agreement except for e and e'.
But then $L(C)$ and $L(C')$ differ by ± 2, and necessarily $L(C) \not\equiv \alpha_C \pmod 4$
or $L(C') \not\equiv \alpha_C \pmod 4$, a contradiction.   □

In this section, the drawings of graphs follow special rules. A solid
straight line connecting two nodes represents an edge, while a solid line with
a short zigzag segment indicates a path where all intermediate nodes have
degree 2. A broken line represents a path connecting the two endpoints
of the broken line, and two or more such paths may have one or more
intermediate nodes in common. However, the path of a broken line has no
intermediate node in common with any node explicitly shown. The labels
on edges are always omitted.

Of particular interest are the following graphs.

(12.2.2)

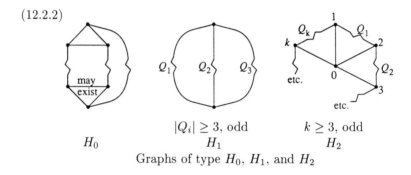

$$|Q_i| \geq 3, \text{ odd} \qquad k \geq 3, \text{ odd}$$
$$H_0 \qquad\qquad H_1 \qquad\qquad H_2$$

Graphs of type H_0, H_1, and H_2

With the aid of the following lemma, it is easy to check whether or not a
graph G of type H_0, H_1, or H_2 may be signed to become α-balanced for a
given α.

(12.2.3) Lemma. *The following statements are equivalent for a graph
G of type H_0, H_1, or H_2, and a given α.*

  (i) *G can be signed so that an even (resp. odd) number of n-cycles do not
       agree with α.*
  (ii) *G can be signed so that every n-cycle (resp. every n-cycle except a
       designated one) agrees with α.*

Proof. It is easily seen that one can always sign H_0, H_1, and H_2 such that
at most one designated n-cycle does not agree with α. Now every edge of
H_0, H_1, and H_2 is part of exactly two n-cycles, and hence every signing of
G produces the same number $\pmod 2$ of n-cycles that do not agree with
α.   □

The main theorem of this section follows. It establishes the central role of n-subgraphs of type H_0, H_1, and H_2 when one wants to achieve α-balancedness by signing.

(12.2.4) Theorem. *For a given vector α, a graph G may be signed to become α-balanced if and only if every n-subgraph of type H_0, H_1, and H_2 can be so signed and $\alpha_C \equiv |C| (\mathrm{mod}\, 2)$ for every n-cycle C of G.*

We prove the theorem in a moment. As an aside, we mention a corollary that rephrases the theorem in terms of edge subsets.

(12.2.5) Corollary. *Let β be a $\{0, 1\}$ vector whose entries are in one-to-one correspondence with the n-cycles of a graph G. Then there exists a subset F of the edge set of G such that $|F \cap C| \equiv \beta_C (\mathrm{mod}\, 2)$, for all n-cycles C of G, if and only if the latter condition is true for all n-subgraphs of type H_0, H_1, and H_2 of G.*

Proof. For each n-cycle C of G, define $\alpha_C = 2\beta_C - |C| (\mathrm{mod}\, 4)$. We call the requirement $|F \cap C| \equiv \beta_C (\mathrm{mod}\, 2)$ the β-*condition*. Obviously, we only need to prove the "if" part of the corollary. Thus, we assume that the β-condition can be satisfied for each n-subgraph H of type H_0, H_1, or H_2, say by edge subset F_H of H. Sign H so that precisely the edges of F_H receive a $+1$. By the definition of α and by the β-condition, for any n-cycle C of H there are integral l_C and k_C such that $|F_H \cap C| + 2l = \beta_C = (\alpha_C + 4k_C + |C|)/2$. Because of this equation and the signing rule for H, $L(C) = 2|F_H \cap C| - |C| \equiv \alpha_C (\mathrm{mod}\, 4)$. Thus, H is α-balanced. By Theorem (12.2.4), G can be signed to become α-balanced. Consider G to be so signed. Let F be the subset of edges of G with $+1$ label. Since for any n-cycle C we have $L(C) \equiv \alpha_C (\mathrm{mod}\, 4)$, there are integers k_C and l_C so that $|F \cap C| = (\alpha_C + 4k_C + |C|)/2 = \beta_C + 2l_C$. Thus, F satisfies the β-condition. $\square$

We accomplish the proof of Theorem (12.2.4) in two steps. First we prove a rather technical lemma. Let G be a graph without parallel edges, and P_0 be a path of G where all intermediate nodes have degree 2. For a given vector α, the graph G is *almost α-balanced with respect to P_0* if all n-cycles of G that do not contain P_0 agree with α.

(12.2.6) Lemma. *Let G be a graph without parallel edges, and let vector α be given. Suppose that every n-subgraph of type H_0, H_1, or H_2 of G can be signed to become α-balanced, and that G is almost α-balanced with respect to a given P_0. Assume C_1 and C_2 are n-cycles of G that include P_0, and let $P_3 \subseteq C_1 \cap C_2$ be a path of maximal cardinality satisfying $P_0 \subseteq P_3$. If one of the n-cycles C_1, C_2 agrees with α, then the other n-cycle agrees with α as well, provided one of the following conditions is satisfied:*

(a) $|P_3| \geq 2$;
(b) $|P_3| = 1$ *and* $(C_1 \cup C_2) - P_3$ *is not an n-cycle of* $G - P_3$.

Proof. The proof is by induction on $|P_3|$. The lemma holds trivially for the maximal value that $|P_3|$ may take on since then $C_1 = C_2$. Hence, we will prove validity for $|P_3| = l$ assuming that the lemma holds whenever $|P_3| \geq l+1$. In the nontrivial case, we have $C_1 \neq C_2$. Thus, the endpoints u and v of P_3 have degree 3 in $\overline{G} = C_1 \cup C_2$. The graph $\overline{G}$ is depicted below. P_1, P_2, and P_3 are the three paths u to v, and for $i = 1, 2$, $C_i = P_i \cup P_3$.

(12.2.7)

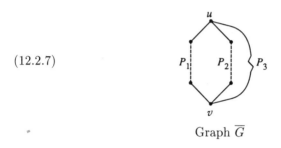

Graph $\overline{G}$

Note that $(C_1 \cup C_2) - P_3$ is the set $P_1 \cup P_2$. Furthermore, $|P_1| \geq 2$ since C_2 is an n-cycle of G. Similarly, $|P_2| \geq 2$. It will be convenient to consider two cases.

(1) $|P_1|$ or $|P_2| = 2$.

Without loss of generality, suppose $|P_1| = 2$. Since C_2 is an n-cycle, the intermediate node of P_1, say w, cannot be a node of P_2. Addition of all edges of G from w to intermediate nodes of P_2 produces the following n-subgraph $\overline{\overline{G}}$ of G.

(12.2.8)

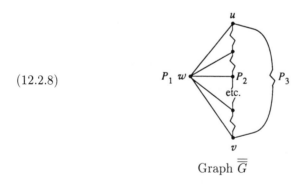

Graph $\overline{\overline{G}}$

If $|P_3| = 1$, then w of $\overline{\overline{G}}$ has degree of at least 3 since otherwise $(C_1 \cup C_2) - P_3$ is an n-cycle of $G - P_3$. Thus, $\overline{\overline{G}}$ is a graph of type H_1 or H_2, and all n-cycles of $\overline{\overline{G}}$ are α-balanced except possibly for C_1 and C_2, since G is almost α-balanced.

By assumption, $\overline{\overline{G}}$ can be signed to become α-balanced. So by Lemma (12.2.3), C_2 must agree with α if C_1 does, and vice versa.

(2) Both $|P_1|, |P_2| \geq 3$.

For $i = 1, 2$, let a_i, b_i be the vertices of P_i adjacent to u, v, respectively. Thus, $\overline{G}$ is as follows.

(12.2.9)

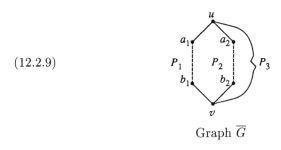

Graph $\overline{G}$

Clearly, u, v, a_1, a_2, b_1, b_2 are all distinct. Suppose there is a path in $G - P_3$ from u to v using only vertices of $P_1 \cup P_2$ and also avoiding a_2 and b_1. Then there exists an n-path P with these properties, and $P \cup P_3$ is an n-cycle C of G. Now $C \cap C_1$ contains a path that in turn properly contains P_3. Thus, by induction and part (a), C agrees with α if and only if C_1 does. The same conclusion holds for C and C_2, so C_1 and C_2 both agree with α or both do not. Hence, we may suppose that there is no such P. Then P_1 and P_2 have no vertex in common except for u, v. Also, every $(G - P_3)$-chord for the cycle $P_1 \cup P_2$ is incident with a_2 or b_1. If there is no such chord, we must be in case (a) of the lemma, and $C_1 \cup C_2$ is a graph of type H_1. Since $P_1 \cup P_2$ agrees with α, we again have the desired result by Lemma (12.2.3). Hence, suppose there exists at least one such chord. Repeat the above argument, but this time try to find a P avoiding a_1 and b_2. If again we are unsuccessful, then the $(G - P_3)$-chords for $P_1 \cup P_2$ are found only at a_1 or b_2. Thus, there are at most two such chords, one from a_1 to a_2, the other from b_1 to b_2, and $C_1 \cup C_2$ must be a graph of type H_0. All n-cycles agree with α except at most C_1 and C_2, so Lemma (12.2.3) produces the desired conclusion. $\qquad \Box$

Proof of Theorem (12.2.4). For proof of the nontrivial "if" part, it is sufficient to consider the case where all edges but those incident at some node m have been signed such that $G - \{m\}$ is α-balanced. We want to sign the edges of G incident at m so that G becomes α-balanced. Let C_1 and C_2 be two n-cycles of G, each containing edges (i, m) and (j, m) for some i and j. Derive G_1 from G by deleting all neighbors of m not equal to i or j. Clearly, C_1 and C_2 are contained in G_1, and G_1 is almost α-balanced with respect to $P_0 = \{(i, m), (j, m)\}$. Since G_1 is an n-subgraph of G, every n-subgraph of type H_0, H_1, or H_2 of G_1 can be signed to become α-balanced. So by Lemma (12.2.6), C_1 and C_2 agree with α or they both do not.

Define a graph J, also with $\{\pm 1\}$ arc labels, from G as follows. Each edge of G incident at node m becomes a node of J. An arc connects two nodes of J if the corresponding edges of G are part of at least one n-cycle C of G. This arc is signed $+1$ if C agrees with α, and -1 otherwise. By the previous argument, the classification of each arc of J is well-defined. For clarity we use "arc" (resp. "edge") in connection with J (resp. G). Suppose we change the sign of edge (i, m) in G. In J, we must change the sign on all arcs incident at node (i, m), so this is a scaling step. Conversely, scaling in J leads to signing of edges in G incident at m. Clearly, it is possible to scale J such that all arcs of a given forest have $+1$ labels. Furthermore, note that a given cycle of J has an even number of -1 arcs if and only if that is true after scaling.

It is claimed that every cycle of J has an even number of -1 arcs. It is sufficient to prove the claim for an n-cycle C_J of J. We will consider two cases depending on $k = |C_J|$.

(1) $k \geq 4$.

By definition of J, the graph G contains the following subgraph $\overline{G}$, where each path Q_i forms an n-cycle C_i with edges (i, m) and $(i + 1, m)$ ($k + 1$ is interpreted as 1).

(12.2.10)

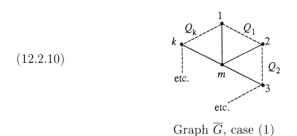

Graph $\overline{G}$, case (1)

It is claimed that $C_G = \bigcup_{i=1}^{k} Q_i$ is an n-cycle of G. We first show that C_G is a cycle. Suppose Q_1 and Q_j have a node u in common, where $u \neq 2$ if $j = 2$, and where without loss of generality $j \neq k$. Since C_i is an n-cycle, for any i we have $v \notin Q_i$, where $v \neq i, i + 1$ is a neighbor of m in G. Hence, u is not equal to any endpoint of Q_1 or Q_j. If $j \neq 2$, define $\overline{P}$ to be composed of the paths from 1 to u on Q_1 and u to j on Q_j. Clearly, $\overline{P}$ contains no neighbor v of m except for nodes 1 and j, and we may replace it by an n-path P observing the same condition. But P and the edges $(1, m)$, (j, m) form an n-cycle of G. But then, since $j \neq k$, C_J cannot be an n-cycle of J. If $j = 2$, $\overline{P}$ is composed of paths 1 to u on Q_1 and u to 3 on Q_2 $(= Q_j)$. Again, we conclude that C_J is not an n-cycle of J. Hence, C_G must be a cycle of G. Similar arguments prove that C_G is indeed an n-cycle of G. Since $\overline{G}$ is an n-subgraph of G of type H_2, it can be signed to become α-balanced. By Lemma (12.2.3), an even number of n-cycles of

$\overline{G}$ do not agree with α. C_G cannot be one of these, since it is an n-cycle of $G - \{m\}$. The C_i, $i = 1, 2, \ldots, k$, constitute the remaining n-cycles of $\overline{G}$. The fact that an even number of the C_i do not agree with α, results in an even number of -1 arcs in the n-cycle C_J of J.

(2) $k = 3$.

Again, G has $\overline{G}$ of (12.2.10) as a subgraph. If $C_G = \bigcup_{i=1}^3 Q_i$ is an n-cycle of G, then arguments as for the case $k \geq 4$ yield the desired conclusion. So suppose Q_1 and Q_2 prevent C_G from being an n-cycle, i.e., Q_1 and Q_2 have a node $i \neq 1, 2, 3$ in common, or there exists a G-chord for $Q_1 \cup Q_2$. Define $P_0 = \{(2, m)\}$, $P_1 = Q_1 \cup \{(1, m)\}$, and $P_2 = Q_2 \cup \{(3, m)\}$. Graph $\overline{G} = \bigcup_{i=0}^2 P_i$ is shown below.

(12.2.11)

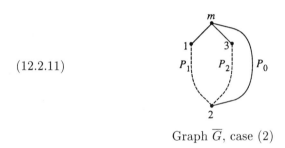

Graph $\overline{G}$, case (2)

Derive G_1 from G by deleting all neighbors of m not equal to 1, 2, or 3. Clearly, $C_i = P_0 \cup P_i$, $i = 1, 2$, is contained in G_1. Because of scaling in J, we may suppose that the arc of J connecting nodes $(1, m)$ and $(3, m)$ has a $+1$ label. This implies that every n-cycle C of G containing edges $(1, m)$ and $(3, m)$ agrees with α, and G_1 is therefore almost α-balanced with respect to P_0. Furthermore, every n-subgraph of G_1 of type H_0, H_1, or H_2 can be signed to become α-balanced. Since $P_1 \cup P_2$ is not an n-cycle of $G_1 - P_0$, either part (a) or part (b) of Lemma (12.2.6) holds, and C_1 agrees with α if and only if C_2 does. But this implies that C_J has an even number of -1 arcs.

The remainder is simple. We scale J so that all arcs of an arbitrarily selected forest T of J receive $+1$ labels. Then all out-of-forest arcs must also have $+1$ labels, since otherwise a cycle with an odd number of -1 arcs has been found. Related signing in G results in an α-balanced graph. $\square$

One application of Theorem (12.2.4) is as follows. Call a binary matrix B *balancedness-inducing* if its 1s can be replaced by ± 1s so that the resulting real matrix A satisfies the following requirement. For each $k \geq 2$, each $k \times k$ submatrix of A of the form given by (12.2.12) below must have real determinant 0. By Lemma (9.2.4), the determinant condition is equivalent to the demand that the entries of the submatrix sum to $0 \pmod 4$.

(12.2.12)

$$\begin{bmatrix} \pm1 & & & \pm1 \\ \pm1\,\pm1 & & & \\ & \pm1 \cdot & & \\ & & \cdot\,\pm1 & \\ & & \cdot\,\pm1\,\pm1 & \end{bmatrix}$$

Matrix whose bipartite graph
is a chordless cycle

The next theorem characterizes the class of balancedness-inducing binary matrices. As we shall see, the proof requires one easy application of Theorem (12.2.4).

(12.2.13) Theorem. *A binary matrix B is balancedness-inducing if and only if B does not have a submatrix $\overline{B}$ such that $BG(\overline{B})$ is one of the graphs H_1 or H_2 below.*

(12.2.14)

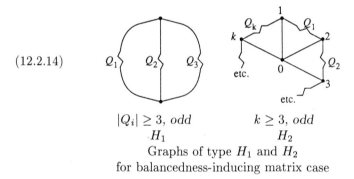

$|Q_i| \geq 3$, *odd* $k \geq 3$, *odd*

H_1 H_2

Graphs of type H_1 and H_2
for balancedness-inducing matrix case

Proof. Let B be a binary matrix. Define G to be the bipartite graph $BG(B)$ with additional $+1$ labels on the edges. Each n-cycle of G corresponds precisely to a $\{0,1\}$ submatrix of B that looks like the matrix of (12.2.12) except for the signs of the entries. Signing B to produce an A with the desired property is equivalent to signing G so that each n-cycle C of the resulting graph has $L(C) \equiv 0 \pmod{4}$. Put differently, an appropriate signing of B is possible if and only if G can be signed to become a 0-balanced graph, say G'. We call G' as well as the related $\{0, \pm1\}$ signed real version of B *balanced*. By Lemma (12.2.1), G' is unique up to scaling. The latter operation corresponds to scaling of some rows and columns of A by -1 factors. Thus, up to such scaling, the matrix A is unique. The proof of Lemma (12.2.1) implies a simple scheme to effect the signing if it is possible at all.

By Theorem (12.2.4), G' and thus A exist if and only if every n-subgraph of G of type H_0, H_1, or H_2 (see (12.2.2)) can be signed to become balanced. The graph H_0 is not bipartite, so it cannot be present in the bipartite G. For the same reason, the paths Q_1, Q_2, and Q_3 of any H_1 n-subgraph must have the same parity. One readily sees that H_1 with a $+1$ label on each edge has an odd number of circuits C with $L(C) \equiv 0(\mathrm{mod}\,4)$ if and only if for $i = 1, 2, 3$, $|Q_i|$ is odd and at least 3. By Lemma (12.2.3), the latter condition characterizes the H_1 graphs that cannot be signed to become balanced. Corresponding arguments for H_2 show that H_2 cannot be signed to become balanced if and only if k, which is the number of spokes of the wheel-like graph, is odd and at least 3. $\square$

Balancedness-inducing matrices are related to regular ones as follows.

(12.2.15) Lemma.   *Let B be a regular matrix. Then B is balancedness-inducing. Furthermore, any balanced $\{0, \pm 1\}$ real matrix derived from B by signing is totally unimodular.*

Proof. Regularity implies that a totally unimodular matrix A can be deduced from B by signing. By Lemma (9.2.4), every $k \times k$, $k \geq 2$, submatrix A equal to the matrix of (12.2.12) must have its entries sum to $0(\mathrm{mod}\,4)$. Thus, A is balanced. As observed above, any balanced matrix A' derived from B by signing must be obtainable from A by scaling. Thus, A' is totally unimodular. $\square$

Suppose we want to characterize the class $\mathcal{N}$ of binary minimal violation matrices of regularity. Theorem (12.2.13) and Lemma (12.2.15) bring us quite close to that goal, as follows.

(12.2.16) Theorem.   *Let $\mathcal{N}$ be the class of binary minimal violation matrices of regularity. Then $\mathcal{N}$ has a partition into the following three subclasses $\mathcal{N}_1, \mathcal{N}_2, \mathcal{N}_3$.*

(a) *$\mathcal{N}_1$ (resp. $\mathcal{N}_2$) is the set of binary matrices B for which $\mathrm{BG}(B)$ is a graph of type H_1 (resp. H_2) of (12.2.14).*
(b) *$\mathcal{N}_3$ is the set of binary balancedness-inducing matrices B satisfying the following condition. Any balanced $\{0, \pm 1\}$ real matrix A derived from B by signing is a minimal violation matrix of total unimodularity with at least three ± 1s in some row and in some column.*

Proof. Let B be any matrix in $\mathcal{N}$. Thus, B is not regular, but every proper submatrix of B is regular. Suppose B is not balancedness-inducing. By Theorem (12.2.13), $\mathrm{BG}(B)$ has as n-subgraph a graph H of type H_1 or H_2 of (12.2.14). Since H cannot be signed to become balanced, by Lemma (12.2.15) we must have $\mathrm{BG}(B) = H$. Thus, $B \in (\mathcal{N}_1 \cup \mathcal{N}_2)$.

Now suppose B to be balancedness-inducing. Let A be any real $\{0, \pm 1\}$ balanced matrix obtained from B by signing. Since B is not regular, A cannot be totally unimodular. Since every proper submatrix of B is regular, by Lemma (12.2.15) every proper submatrix of A is totally unimodular. Thus, A is a minimal violation matrix of total unimodularity. The nonregularity of B implies that B is not graphic or cographic. Thus, B has a row and a column with at least three 1s. Then A has at least three ±1s in some row or column. We conclude that $B \in \mathcal{N}_3$.

So far we have shown that $\mathcal{N} \subseteq \mathcal{N}_1 \cup \mathcal{N}_2 \cup \mathcal{N}_3$. It is easy to see by (12.2.14) that any $B \in (\mathcal{N}_1 \cup \mathcal{N}_2)$ is a minimal nonregular matrix. Thus, $\mathcal{N}_1 \cup \mathcal{N}_2 \subseteq \mathcal{N}$. Let $B \in \mathcal{N}_3$. By definition, every proper submatrix of B is regular. If B is regular, then by Lemma (12.2.15) any balanced matrix A induced by B is totally unimodular. But this violates the definition of $\mathcal{N}_3$. Thus, $\mathcal{N}_3 \subseteq \mathcal{N}$.

At this point, we know $\mathcal{N} = \mathcal{N}_1 \cup \mathcal{N}_2 \cup \mathcal{N}_3$. By Theorem (12.2.13), $\mathcal{N}_1 \cup \mathcal{N}_2$ consist of the minimal binary matrices that are not balancedness-inducing. The matrices of $\mathcal{N}_3$ are by definition balancedness-inducing. Thus, $\mathcal{N}_1 \cup \mathcal{N}_2$ and $\mathcal{N}_3$ are disjoint. By (12.2.14), $\mathcal{N}_1$ and $\mathcal{N}_2$ are disjoint as well. Thus, $\mathcal{N}_1$, $\mathcal{N}_2$, and $\mathcal{N}_3$ partition $\mathcal{N}$ as claimed. □

Theorem (12.2.16) is a substantial step toward the goal of understanding the class $\mathcal{N}$ of binary minimal violation matrices of regularity. Indeed, the subclasses $\mathcal{N}_1$ and $\mathcal{N}_2$ of such matrices have a simple and appealing description. But the characterization of the remaining class $\mathcal{N}_3$ in terms of minimal violation matrices of total unimodularity has little value unless we understand the latter matrices. In the next section, we take small but nevertheless useful steps toward understanding those matrices.

12.3 Several Matrix Classes

We define four classes of matrices. Each of them is connected in some way with the binary minimal violation matrices of regularity. We point out the relationships as the classes are introduced one by one. Their significance will become apparent in Sections 12.4 and 12.5.

Complement Totally Unimodular Matrices

We begin with $\{0, 1\}$ real matrices. For such a matrix U, we may define the following complement operations. Let k index a row of U. Then the *row k complement* of U is the real $\{0, 1\}$ matrix U' derived from U as follows. For all column indices j with $U_{kj} = 1$ and for all row indices $i \neq k$, replace the entry U_{ij} by its complement, i.e., 1 by 0, and 0 by 1. Let l index a column of U. Then the *column l complement* of U is the transpose of the row l complement of U^t.

For example, if U is the 4×4 identity with indices k and l as shown,

(12.3.1)

$$U = \begin{array}{c} \\ k \end{array} \begin{array}{c} l \\ \left[\begin{array}{cccc} 1 & 0 & 0 & 0 \\ 0 & 1 & 0 & 0 \\ 0 & 0 & 1 & 0 \\ 0 & 0 & 0 & 1 \end{array}\right] \end{array}$$

Matrix U

then the row k complement of U is the matrix U' below.

(12.3.2)

$$U' = \left[\begin{array}{cccc} 1 & 0 & 0 & 0 \\ 1 & 1 & 0 & 0 \\ 1 & 0 & 1 & 0 \\ 1 & 0 & 0 & 1 \end{array}\right]$$

Row k complement of U

The column l complement of U' is the following matrix U''.

(12.3.3)

$$U'' = \left[\begin{array}{cccc} 1 & 1 & 1 & 1 \\ 1 & 0 & 1 & 1 \\ 1 & 1 & 0 & 1 \\ 1 & 1 & 1 & 0 \end{array}\right]$$

Column l complement of U'

How many different matrices, up to indices, may be derived from an $m \times n$ $\{0,1\}$ matrix U in a sequence of complement operations? The answer relies on two readily verified claims. For $k \neq l$, let U' be the row k complement of U, and let U'' be the row l complement of U. Then up to a change of indices, the row l complement of U can be seen to be U''. Suppose one obtains a matrix from U by a row complement step followed by a column complement step. Then the same matrix results if one reverses the order of the two steps. By these two claims, any matrix obtainable from a given $m \times n$ $\{0,1\}$ matrix U in a sequence of complement operations may up to indices be produced by at most one row complement step and/or one column complement step. Thus, at most $(m+1)(n+1)$ numerically different matrices may be deduced from U by repeated complement steps.

With respect to total unimodularity, the complement operation can be very destructive. For example, U of (12.3.1) is the 4×4 identity, and thus totally unimodular. Consider the matrix U'' of (12.3.3) deduced from U by two complement steps. In the right-hand corner, U'' has a 3×3 submatrix with real determinant 2. Thus, U'' is not totally unimodular.

We define a real $\{0,1\}$ matrix U to be *complement totally unimodular* if U and all matrices derivable from U by possibly repeated complement operations are totally unimodular. Clearly, complement total unimodularity is maintained under submatrix-taking and complement operations. We collect the complement totally unimodular matrices in a set $\mathcal{U}$. By the above observation and examination of a few additional examples, one quickly sees that complement total unimodularity is a very demanding property. In fact, one is inclined to believe that there are only a few structurally different complement totally unimodular matrices. But that notion turns out to be mistaken, as we shall see in Section 12.5.

Complement total unimodularity of U implies that certain extensions of U are totally unimodular as follows.

(12.3.4) Lemma. *Let U be a real $\{0, 1\}$ matrix that is complement totally unimodular. Enlarge U to a matrix U' by adjoining a row or column containing only 1s. Then U' is totally unimodular.*

Proof. Let $\overline{U}$ be any square submatrix of U'. We must show that $\det_{\mathbb{R}} = 0$ or ± 1. In the nontrivial case, $\overline{U}$ intersects the vector of 1s adjoined to U. A real pivot on any 1 of that vector followed by deletion of the pivot row and column produces a matrix $\overline{\overline{U}}$. One readily confirms that, up to scaling by $\{\pm 1\}$ factors, $\overline{\overline{U}}$ is a submatrix of a row or column complement of U. Thus, $\overline{\overline{U}}$ is totally unimodular, and hence $\det_{\mathbb{R}} \overline{U} = 0$ or ± 1 as desired.   □

Almost Representative Matrices

We change fields and consider matrices over GF(3) or GF(2). Let A be a matrix over GF(3), and B its support matrix. View B to be over GF(2). Let M be the ternary matroid represented by A over GF(3), and N be the binary matroid represented by B over GF(2). By the derivation of B from A, certain sets are bases of both M and N. In fact, the two matroids may be identical. Suppose they are not. If there is just one set that is a base of one of the matroids and not of the other one, then we say that A *almost represents N* over GF(3), and that B *almost represents M* over GF(2). One may re-express the assumption as follows. The GF(3)-determinant of each square submatrix of A is nonzero if and only if the GF(2)-determinant of the corresponding square submatrix of B is nonzero, except for one square submatrix $\overline{A}$ of A and for the corresponding submatrix $\overline{B}$ of B. By Corollary (3.5.3), we know that $\det_3 \overline{A} \neq 0$ and $\det_2 \overline{B} = 0$. The arguments to come will confirm this fact.

Suppose we perform a GF(3)-pivot on a ± 1 of $\overline{A}$ in A. We also carry out the corresponding GF(2)-pivot on the related 1 of $\overline{B}$ in B. In the matrices so deduced from A and B, the difference between M and N manifests itself by submatrices $\overline{\overline{A}}$ and $\overline{\overline{B}}$ that are smaller than $\overline{A}$ and $\overline{B}$. Indeed, had we carried out the respective pivots just in $\overline{A}$ and $\overline{B}$ and deleted the pivot row and column, we would have obtained $\overline{\overline{A}}$ and $\overline{\overline{B}}$. Because of this reduction possibility, we may assume $\overline{A}$ and $\overline{B}$ to be of order 2×2. A simple case analysis of the 2×2 matrices proves that the difference in determinants between $\overline{A}$ and $\overline{B}$ can be produced in essentially one way. That is, we must have, up to scaling in $\overline{A}$,

(12.3.5)
$$\overline{A} = \begin{bmatrix} \text{-}1 & 1 \\ 1 & 1 \end{bmatrix}; \qquad \overline{B} = \begin{bmatrix} 1 & 1 \\ 1 & 1 \end{bmatrix}$$

Submatrices $\overline{A}$ and $\overline{B}$ proving $M \neq N$

Thus, $\det_3 \overline{A} \neq 0$ and $\det_2 \overline{B} = 0$, as predicted by Corollary (3.5.3). Evidently, the same relationship must have held prior to any pivots. The reader who has covered Section 3.5 surely recognizes the similarity of the above arguments to those of the proof of Theorem (3.5.2) and Corollary (3.5.3). Here we have a GF(3)-matrix A instead of the abstract matrix of Section 3.5.

We continue with $\overline{A}$ and $\overline{B}$ given by (12.3.5). We perform one more GF(3)-pivot on a ± 1 of $\overline{A}$ in A, and also carry out the related GF(2)-pivot in B. This change effectively reduces $\overline{A}$ to a 1×1 matrix $\overline{\overline{A}}$, and $\overline{B}$ to a 1×1 matrix $\overline{\overline{B}}$, for which $\det_3 \overline{\overline{A}} \neq 0$ and $\det_2 \overline{\overline{B}} = 0$. Thus, $\overline{\overline{B}} = [\,0\,]$. Because of scaling of A, we may assume $\overline{\overline{A}} = [-1]$.

Before going on, we record the insight attained so far in the following lemma.

(12.3.6) Lemma. *Let A be a matrix over GF(3). View the support matrix B of A to be over GF(2). Assume that the GF(3)-determinant of each square submatrix of A is nonzero if and only if the GF(2)-determinant of the corresponding submatrix of B is nonzero, with the exception of just one submatrix $\overline{A}$ in A and the related submatrix $\overline{B}$ in B, both of order $k \geq 2$. Then by GF(3)-pivots within the submatrix $\overline{A}$ and scaling, and by corresponding GF(2)-pivots in B, the matrices A and B can be transformed to matrices with determinants agreeing analogously to A and B, except for a submatrix $\overline{\overline{A}} = [-1]$ and $\overline{\overline{B}} = [\,0\,]$.*

For notational convenience, we now redefine A and B to be the matrices produced by the pivots. Thus, $\overline{\overline{A}} = [-1]$ is a submatrix of A, and $\overline{\overline{B}} = [\,0\,]$ is the related submatrix of B. We want to analyze the structure of A and B. To this end, let y be the row index of $\overline{\overline{A}}$ and of $\overline{\overline{B}}$, and let x be the column index. Assume that X (resp. Y) is the index set of the remaining rows (resp. columns) of A and B. Also assume that A has no zero rows or columns. Recall that $\underline{1}$ denotes a column vector containing only 1s. We claim that up to row and column scaling of A, the matrices A and B are of the form given by (12.3.7) below, where U is a $\{0, 1\}$ matrix viewed to be over GF(3) or GF(2) as needed.

(12.3.7)

$$
A = \begin{array}{c|c|c|} & x & Y \\ \hline y & -1 & 1^t \\ \hline X & 1 & U \end{array} \quad ; \quad
B = \begin{array}{c|c|c|} & x & Y \\ \hline y & 0 & 1^t \\ \hline X & 1 & U \end{array}
$$

Matrix A over GF(3) for M,
and matrix B over GF(2) for N

The claim plus additional facts make up the next theorem.

(12.3.8) Theorem.

(a) *Let A be a matrix over GF(3) without zero rows or columns, and let B be a matrix over GF(2) of the same size and with the same row and column indices. Assume that the GF(3)-determinant of each square submatrix of A is nonzero if and only if the GF(2)-determinant of the corresponding submatrix of B is nonzero, except for one 1×1 submatrix $\overline{\overline{A}} = [-1]$ of A and the corresponding submatrix $\overline{\overline{B}} = [0]$ of B, say with row index y and column index x. Let the remaining rows of A be indexed by X, and the remaining columns by Y. Then up to a scaling of rows and columns of A, the matrices A and B are given by (12.3.7), where U is a $\{0,1\}$ matrix to be viewed over GF(3) or GF(2) as needed. When U is considered to be real, then it is complement totally unimodular.*

(b) *Let A and B be the matrices of (12.3.7). The column x and the row y must be present, but X or Y may be empty. Assume that the submatrix U of either matrix is a $\{0,1\}$ matrix that is complement totally unimodular when considered real. View A to be over GF(3), and B to be over GF(2). Then the GF(3)-determinant of each square submatrix of A is nonzero if and only if the GF(2)-determinant of the corresponding submatrix of B is nonzero, except for the 1×1 submatrix $\overline{\overline{A}} = [-1]$ of A indexed by x and y, and the corresponding submatrix $\overline{\overline{B}} = [0]$ of B.*

Proof. We establish part (a). A and B of (12.3.7) correctly display $\overline{\overline{A}}$ and $\overline{\overline{B}}$. Suppose column x of A contains a 0, say in row $i \in X$. Since A has no zero rows, there is a $j \in Y$ with $A_{ij} = \pm 1$. From the rows x, i and from the columns y, j of A and B, we extract the submatrices

(12.3.9)

$$
\overline{A} = \begin{array}{c|c|c|} & y & j \\ \hline x & 1 & \gamma \\ \hline i & 0 & \pm 1 \end{array} \quad ; \quad
\overline{B} = \begin{array}{c|c|c|} & y & j \\ \hline x & 0 & \delta \\ \hline i & 0 & 1 \end{array}
$$

Submatrices $\overline{A}$ and $\overline{B}$ of counterexample

We ignore the entries γ and δ. Indeed, for any γ and δ, we have $\det_3 \overline{A} \neq 0$ and $\det_2 \overline{B} = 0$, which contradicts the presumed agreement of determinants. Thus, by scaling in A, we may assume that column x of A and column x of B contain only 1s except for the entry of $\overline{\overline{A}}$ or $\overline{\overline{B}}$ in row y. By symmetry, we also may assume that the row y of A and the row y of B contain only 1s except for the entry in column x.

It remains for us to prove that U is a $\{0, 1\}$ matrix that occurs in both A and B, and that U, when viewed as real, is complement totally unimodular. If U of A contains a -1, say in row i and column j, then rows x, i and columns y, j of A define a 2×2 GF(3)-singular submatrix of A. But in B, these rows and columns specify a GF(2)-nonsingular submatrix, a contradiction. Thus, U in A is a $\{0, 1\}$ matrix. Because of the agreement of determinants of A and B, the matrix U must also occur in B, this time considered over GF(2), of course. By Theorem (9.2.9), the agreement of determinants on U when viewed over GF(3) and GF(2) implies that U as real matrix is totally unimodular.

Suppose we perform a GF(2)-pivot in column x, row i of B of (12.3.7). The latter matrix is like B except that the indices x and i have traded places and U has been replaced by its row i complement U'. We perform the analogous GF(3)-pivot in A. That pivot plus some scaling with $\{\pm 1\}$ factors produces a matrix A' that is like A except that the indices x and i have switched and U has become U'. The determinants of A' and B' agree analogously to A and B, except for the $[-1]$ submatrix in row y and column i of A' and the corresponding $[0]$ submatrix of B. By the preceding discussion, U' must be totally unimodular. Using additional pivots, we see that U as real matrix is complement totally unimodular.

We turn to part (b). Let C be a square submatrix of A, and let D be the corresponding submatrix of B. We must show that $\det_3 C$ is nonzero if and only if $\det_2 D$ is nonzero, with the single exception of $C = \overline{\overline{A}}$ and $D = \overline{\overline{B}}$. Suppose C intersects at most one of column x and row y of A. By the complement total unimodularity of U and Lemma (12.3.4), the matrix C is totally unimodular. By Theorem (9.2.9), the determinants of C and D agree as desired. This leaves the case where C and D properly include $\overline{\overline{A}}$ and $\overline{\overline{B}}$, respectively. Carry out a GF(3)-pivot on any 1 in column x of C. Then delete the pivot row and column. Let a matrix C' result. Correspondingly, reduce D by a GF(2)-pivot to D'. Up to scaling by $\{\pm 1\}$ factors in C', both reduction steps produce a submatrix of a row complement of U with an additional row of 1s adjoined. By the above discussion, the determinants of C' and D', and hence of C and D, agree.   □

The above proof contains an observation that we want to record as a lemma for future reference.

(12.3.10) Lemma. *Let B be the matrix over* GF(2) *of* (12.3.7). *Then* a GF(2)-pivot in column x or row y of B transforms B to a matrix B' structured like B, except that a row index and a column index have traded places and U has been replaced by a row or column complement of U.

Collect in sets $\mathcal{A}$ and $\mathcal{B}$ the possible cases of A and B, respectively, of part (b) of Theorem (12.3.8). That is, each $A \in \mathcal{A}$ and each $B \in \mathcal{B}$ is given by (12.3.7), and the $\{0,1\}$ submatrix U of either matrix when viewed as real is complement totally unimodular. We permit the extreme cases where U is trivial or empty. Thus, one of the index sets X and Y, or even both of them, may be empty.

Let N be the matroid represented by some $B \in \mathcal{B}$ over GF(2). By part (b) of Theorem (12.3.8), the corresponding $A \in \mathcal{A}$ almost represents N over GF(3). For this reason, we call $\mathcal{A}$ a collection of *almost representative matrices over* GF(3). Analogously, the matroid M represented by an $A \in \mathcal{A}$ over GF(3) is almost represented by the corresponding $B \in \mathcal{B}$ over GF(2). Thus, $\mathcal{B}$ is a collection of *almost representative matrices over* GF(2).

Minimal Violation Matrices of Total Unimodularity

So far in this section, we have defined three matrix classes: $\mathcal{U}$, $\mathcal{A}$, and $\mathcal{B}$. We need one additional class $\mathcal{V}$, which contains the real minimal violation matrices of total unimodularity. Thus, every $V \in \mathcal{V}$ is not totally unimodular, but this is so for every proper submatrix of V. Clearly, each $V \in \mathcal{V}$ is square. To avoid uninteresting instances, we exclude from $\mathcal{V}$ the cases V of order 1. Thus, each $V \in \mathcal{V}$ is for some $k \geq 2$, a $k \times k$ $\{0, \pm 1\}$ matrix. Let W be the support matrix of V. Consider V to be real or over GF(3) as needed below, and W to be over GF(2). We have the following result.

(12.3.11) Theorem.   *Let V be any matrix of $\mathcal{V}$, and let W be its binary support matrix. Then by GF(3)-pivots in V and scaling with $\{\pm 1\}$ factors, and by corresponding GF(2)-pivots in W, the matrices V and W can be transformed to matrices $A \in \mathcal{A}$ and $B \in \mathcal{B}$, respectively, of order at least 2.*

Conversely, suppose in $A \in \mathcal{A}$ and $B \in \mathcal{B}$ of order at least 2, we perform GF(3)-pivots and related GF(2)-pivots, respectively, so that A' and B' result that satisfy the following condition. Let $\overline{X}$ be the subset of $X \cup \{x\}$ indexing columns of A' and B', and $\overline{Y}$ be the subset of $Y \cup \{y\}$ indexing rows. The condition is that $|\overline{X}| \geq 2$ or $|\overline{Y}| \geq 2$. Then the submatrix $\overline{A}$ of A' indexed by $\overline{X}$ and $\overline{Y}$ is in $\mathcal{V}$, and the corresponding submatrix $\overline{B}$ of B' is the support of $\overline{A}$.

Proof. We start with the first part, where V and W are given. Let $\overline{V}$ be a proper submatrix of V. Then $\overline{V}$ as real matrix is totally unimodular. Thus, by Theorem (9.2.9), $\overline{V}$ as GF(3) matrix has $\det_3 \overline{V}$ nonzero if and only if the corresponding $\overline{W}$ of W has $\det_2 \overline{W}$ nonzero. By analogous arguments, exactly one of $\det_3 V$ and $\det_2 W$ is nonzero. Thus, V and W constitute a pair of matrices to which Lemma (12.3.6) can be applied. Accordingly, V and W can by pivots and scaling be transformed to A over GF(3) and B over GF(2), respectively, with agreeing determinants, except for a submatrix $\overline{\overline{A}} = [-1]$ of A and $\overline{\overline{B}} = [0]$ of B. By part (a) of Theorem (12.3.8), we have $A \in \mathcal{A}$ and $B \in \mathcal{B}$.

Reversal of the above arguments essentially proves the converse part. We only need to show that the matrices $\overline{A}$ of A' and $\overline{B}$ of B' deduced from $A \in \mathcal{A}$ and $B \in \mathcal{B}$ are the ones with disagreeing determinants. Let M (resp. N) be the matroid represented by A over GF(3) (resp. B over GF(2)). In the matroid M, the set $X \cup \{x\}$ is a base. But that set is not a base in N. By assumption, the set $\overline{X}$ with $|\overline{X}| \geq 2$ is the subset of $X \cup \{x\}$ indexing columns of A', and $\overline{Y}$ is the subset of $Y \cup \{y\}$ indexing rows. Since $X \cup \{x\}$ is a base of M but not of N, the submatrix $\overline{A}$ of A' indexed by $\overline{X}$ and $\overline{Y}$ has $\det_3 \overline{A}$ nonzero, while the corresponding submatrix $\overline{B}$ of B' has $\det_2 \overline{B} = 0$. $\qquad \Box$

Several simple but useful results follow from Theorem (12.3.11). Recall from Section 2.3 that a $\{0, \pm 1\}$ matrix is *Eulerian* if in every row and every column the entries sum to $0 \pmod 2$. Equivalently, each row and column must have an even number of nonzeros.

(12.3.12) Corollary.
(a) *A square $\{0, \pm 1\}$ matrix of order at least 2 is in $\mathcal{V}$ if and only if the following holds. V can be scaled with $\{\pm 1\}$ factors to become, for some square, nonsingular, complement totally unimodular matrix U, the matrix*

(12.3.13)

$$
\begin{array}{|c|c|}
\hline
\alpha & a \\
\hline
b & U^{-1} \\
\hline
\end{array}
\quad ; \quad
\begin{array}{l}
a = \underline{1}^t \cdot U^{-1} \\
b = U^{-1} \cdot \underline{1} \\
\alpha = \underline{1}^t \cdot U^{-1} \cdot \underline{1} - 2
\end{array}
$$

Matrix V up to scaling

(b) *For every matrix $V \in \mathcal{V}$, the real inverse of V contains only $\frac{1}{2}$ entries, $|\det_{\mathbb{R}} V| = 2$, V is Eulerian, and the real sum of the entries of V in $\mathbb{R}$ is congruent to $2 \pmod 4$.*

Proof. We start with part (a). By Theorem (12.3.11), we may deduce V, up to scaling with $\{\pm 1\}$ factors, from a square GF(3)-nonsingular $A \in \mathcal{A}$ by GF(3)-pivots. Redefine V so that it is the appropriately scaled version of the original V. The matrix A is given by (12.3.7). We may recreate A from V by performing the GF(3)-pivots in reverse order. Suppose we use real pivots instead. All intermediate real matrices must be numerically identical to the GF(3)-matrices in the original GF(3)-pivot sequence, by virtue of the fact that each 2×2 submatrix of each such real matrix is totally unimodular. A different conclusion applies to the last pivot. If performed in GF(3), it would produce A. But carried out in $\mathbb{R}$, it produces a matrix $\tilde{A}$ containing $\det_{\mathbb{R}} V$ as entry where A has the -1. All other entries of $\tilde{A}$ must agree numerically with those of A. Now $\det_{\mathbb{R}} V$ is the real determinant of a 2×2 submatrix of the predecessor matrix of the real pivot sequence. Since $\det_{\mathbb{R}} V \neq 0, \pm 1$, we must have $|\det_{\mathbb{R}} V| = 2$. We know that the final pivot, when done in GF(3), produces the -1 of A instead of $\det_{\mathbb{R}} V$. Thus, $\det_{\mathbb{R}} V$ is congruent to $-1 (\bmod 3)$, and accordingly $\det_{\mathbb{R}} V = 2$.

We just have proved that $\tilde{A}$ is the matrix

$$(12.3.14) \qquad \tilde{A} = \begin{array}{c|c|c} & x & Y \\ \hline y & 2 & 1' \\ \hline x & 1 & U \end{array}$$

Matrix $\tilde{A}$ derived by real pivots from V

The submatrix U is complement totally unimodular. By the existence of the real pivot sequence, $\det_{\mathbb{R}} U$ must be nonzero. Thus, $|\det_{\mathbb{R}} U| = 1$.

Suppose we employ the following well-known method for computing the inverse of V. We begin with the real matrix $[I \mid V]$. Then we carry out elementary row operations until the submatrix V of $[I \mid V]$ has become an identity matrix. At that time, the submatrix I of $[I \mid V]$ has become V^{-1}. We claim that the matrix $\tilde{A}$ of (12.3.14) contains in compact form the results of most of these row operations. For a proof, we first note that each one of the real pivots deducing $\tilde{A}$ of (12.3.14) from a scaled version of V involves a ± 1 as pivot element. Then by the relationship between elementary row operations and pivots described in Section 2.3, the pivots producing $\tilde{A}$ correspond to row operations and scaling steps with $\{\pm 1\}$ factors in $[I \mid V]$ that convert the submatrix I of $[I \mid V]$ to the matrix

$$(12.3.15) \qquad \tilde{V}^1 = \begin{array}{|c|c|} \hline 1 & 1' \\ \hline 0 & U \\ \hline \end{array}$$

Matrix $\tilde{V}^1$ derived from $[I \mid V]$
by row operations and scaling

and that change the submatrix V of $[I \mid V]$ to the matrix

$$(12.3.16) \qquad \tilde{V}^2 = \begin{array}{|c|c|} \hline 2 & 0 \\ \hline 1 & 1 \ddots 1 \\ \hline \end{array}$$

Matrix $\tilde{V}^2$ derived from V
by row operations and scaling

Observe that the columns of $\tilde{A}$ save the first one occur in $\tilde{V}^1$, and that the first column of $\tilde{A}$ is also the first column of $\tilde{V}^2$, as is implied by the discussion of Section 2.3.

We now perform row operations in $[\tilde{V}^1 \mid \tilde{V}^2]$ that convert $\tilde{V}^2$ to an identity matrix. Correspondingly, $\tilde{V}^1$ becomes the inverse matrix of a scaled version of V. Up to scaling, that inverse matrix is

$$(12.3.17) \qquad \tilde{V} = \frac{1}{2} \begin{array}{|c|c|} \hline -1 & \mathbf{1}^t \\ \hline \mathbf{1} & \tilde{U} \\ \hline \end{array}; \qquad \tilde{U} = 2U - \mathbf{1} \cdot \mathbf{1}^t$$

Inverse matrix $\tilde{V}$ of scaled version of V

A multiplication check verifies that the matrix of (12.3.13) is $\tilde{V}^{-1}$, and thus is V up to scaling. This fact proves part (a).

Part (b) is now easily shown. We already know $|\det_{\mathbb{R}} V| = 2$. That V^{-1} is a $\{\pm\frac{1}{2}\}$ matrix is evident from the scaled version $\tilde{V}$ of V^{-1} given by (12.3.17). The matrix V is Eulerian, since this is clearly so for the scaled version given by (12.3.13). The latter matrix has its entries sum to $2 \pmod 4$. Scaling does not affect that result, so the same conclusion applies to V. $\qquad\qquad\qquad\qquad\qquad\qquad\qquad\qquad\qquad\square$

Let us summarize the main results of this section for the matrix classes $\mathcal{U}$, $\mathcal{A}$, $\mathcal{B}$, and $\mathcal{V}$. The class $\mathcal{U}$ contains the real $\{0,1\}$ complement totally unimodular matrices. $\mathcal{A}$ and $\mathcal{B}$ contain the almost representative matrices over GF(3) and GF(2), respectively, as depicted by (12.3.7). From any one of the three classes, the remaining two are obtained by trivial operations, as is evident from (12.3.7). The class $\mathcal{V}$ contains the real minimal violation matrices of total unimodularity of order at least 2. By (12.3.13), each $V \in \mathcal{V}$ can be readily computed from a square $\mathbb{R}$-nonsingular $U \in \mathcal{U}$. With equal ease, we can derive $\mathcal{V}$ from $\mathcal{A}$ or $\mathcal{B}$. But note that we do not know how to deduce the entire class $\mathcal{U}$, or $\mathcal{A}$ or $\mathcal{B}$, from $\mathcal{V}$. It turns out that this is not possible by the matrix operations of scaling by $\{\pm 1\}$ factors, pivots, submatrix-taking, and change of fields from $\mathbb{R}$ to GF(2) or GF(3). We prove this fact in Section 12.5. There we construct $\mathcal{U}$, $\mathcal{A}$, $\mathcal{B}$, and $\mathcal{V}$, taking the following viewpoint. First we construct $\mathcal{U}$ by a process yet to be described. Then we deduce $\mathcal{A}$, $\mathcal{B}$, and $\mathcal{V}$ as just mentioned.

Recall that Theorem (12.2.16) establishes a partition of the class $\mathcal{N}$ of binary minimal violation matrices of regularity. Two subclasses labeled $\mathcal{N}_1$ and $\mathcal{N}_2$ are well described by that theorem. But the third class $\mathcal{N}_3$ is not well explained. Indeed, each $B \in \mathcal{N}_3$ is the support matrix of a minimal violation matrix V of total unimodularity with at least three nonzeros in some row and some column. Thus, $V \in \mathcal{V}$. As an aside, the just-mentioned bound of 3 on the number of nonzeros in some row and some column of V can now be strengthened to 4 by part (b) of Corollary (12.3.12). At any rate, the construction of $\mathcal{V}$ via $\mathcal{U}$ gives a construction of $\mathcal{N}_3$. So the yet-to-be-described construction process for $\mathcal{U}$ effectively brings to an end the quest for an understanding of the structure of $\mathcal{N}$.

The promised construction of $\mathcal{U}$ relies on some matroid results that we introduce in the next section.

12.4 Definition and Construction of Almost Regular Matroids

As argued in the introductory section of this chapter, one is tempted to claim that matroids are not suitable for investigations of minimal violation matrices. There it is also claimed that this argument is flawed. Here we show why, by providing a general method for a matroid-based investigation of the minimal violation matrices of certain matrix properties. We specialize the method to a particular instance. In doing so, we define and analyze the matroids that we called almost regular in Section 4.4. In particular, we establish the construction for almost regular matroids that was already listed in Section 4.4. We use that construction to obtain a construction for the class $\mathcal{U}$ of the complement totally unimodular matrices.

We begin with a general discussion about matrix properties and matroids. Let $\mathcal{P}$ be a property defined for the matrices over a field $\mathcal{F}$, where $\mathcal{F}$ must be GF(2) or GF(3). Technically, one may consider $\mathcal{P}$ to be a subset of the matrices over $\mathcal{F}$. The property is to be maintained under submatrix-taking, row and column permutations, scaling with $\{\pm 1\}$ factors, and $\mathcal{F}$-pivots, and when a row or column unit vector is adjoined.

Examples for $\mathcal{P}$ are regularity, graphicness, cographicness, graphic-or-cographicness, and planarity, all defined for $\mathcal{F} = \text{GF}(2)$. Also qualifying is the following property for $\mathcal{F} = \text{GF}(3)$. A $\{0, \pm 1\}$ matrix has the property when over $\mathbb{R}$ it is totally unimodular.

Suppose we want to understand the matrices over $\mathcal{F}$ that are minimal violation matrices of $\mathcal{P}$. For any matrix A over $\mathcal{F}$, define $M(A)$ to be the matroid represented by A over $\mathcal{F}$. $M(A)$ may be representable over $\mathcal{F}$ by a number of different matrices A'. If $\mathcal{F} = \text{GF}(2)$, then all such A' are obtainable from A by GF(2)-pivots. By assumption, both A' and A have $\mathcal{P}$ or they do not. Thus, it is well defined when we declare $M(A)$ to have $\mathcal{P}$ if A has $\mathcal{P}$. The same conclusion applies when $\mathcal{F} = \text{GF}(3)$. This time any A' representing $M(A)$ over GF(3) may, by a slightly modified proof of Lemma (9.2.6), be obtained from A by GF(3)-pivots and scaling with $\{\pm 1\}$ factors. By the assumptions on $\mathcal{P}$, thus A' and A have $\mathcal{P}$ or they both do not.

Note that series or parallel extensions of $M(A)$ maintain $\mathcal{P}$, for such an extension corresponds to at most one $\mathcal{F}$-pivot in A followed by adjoining of a row or column unit vector. By assumption, the latter operations maintain $\mathcal{P}$.

We describe a five-step process that leads to insight into the minimal violation matrices of $\mathcal{P}$ over $\mathcal{F}$.

Step 1. Let $\mathcal{W}$ be the class of minimal matrices over $\mathcal{F}$ that do not have $\mathcal{P}$. Assign to the indices of each matrix $W \in \mathcal{W}$ the following additional labels. If an index z labels a row (resp. column) of W, then assign the label "*con*" (resp. "*del*"). For example, if the rows of W are indexed by X and the columns by Y, then W with the labels is the matrix of (12.4.1) below. From now on, we assume each $W \in \mathcal{W}$ to be so labeled. We also assume that the elements of the matroid $M(W)$ are labeled correspondingly.

(12.4.1)

Minimal violation matrix W with labels

The unusual "*con*" and "*del*" labels of W and $M(W)$ tell the following. If element z of the matroid $M(W)$ has a "*con*" label, then $z \in X$. By the minimality of W, deletion of row $z \in X$ from W results in a matrix W' with $\mathcal{P}$. Correspondingly, $M(W')$, which is $M(W)/z$, has $\mathcal{P}$. The "*con*" label on z allows us to predict this outcome for $M(W)/z$. That is, "*con*"traction of an element of $M(W)$ with a "*con*" label produces a minor having $\mathcal{P}$. Similarly, "*del*"etion of an element with a "*del*" label results in a minor having $\mathcal{P}$. Note that we cannot tell from the labels whether for an element z with "*con*" label the minor $M\backslash z$ has $\mathcal{P}$. Similarly, when z has a "*del*" label, we are ignorant about M/z having or not having $\mathcal{P}$.

Collect in a set $\mathcal{M}_1$ the matroids $M(W)$ with $W \in \mathcal{W}$. For brevity, we call an element with "*con*" (resp. "*del*") label simply a "*con*" (resp. "*del*") element.

Step 2. Establish elementary facts about the matrices $W \in \mathcal{W}$. If possible, translate some of these facts into matroid language so that they apply to the matroids of $\mathcal{M}_1$. Let $\mathcal{E}$ be the collection of such matroid facts.

Step 3. At this time, we reverse the sequence of arguments. We use certain necessary conditions satisfied by the matroids of $\mathcal{M}_1$ to define a class $\mathcal{M}_2$ of matroids representable over $\mathcal{F}$. The conditions for membership in $\mathcal{M}_2$ are as follows. Each $M \in \mathcal{M}_2$ must not have $\mathcal{P}$. Each one of its elements must be labeled "*con*" or "*del*" in such a way that a "*con*" (resp. "*del*") label on an element z of M implies M/z (resp. $M\backslash z$) to have $\mathcal{P}$. Finally, M must satisfy the conditions of $\mathcal{E}$. Clearly, $\mathcal{M}_1$ is a subset of $\mathcal{M}_2$.

Step 4. Enlarge $\mathcal{M}_2$ by adding for each member M all proper minors. The elements of these minors are labeled in agreement with the labels of M. By definition, $\mathcal{M}_2$ is now closed under minor-taking.

Step 5. Analyze the structure of the matroids of $\mathcal{M}_2$. Specialize the conclusions for $\mathcal{M}_2$ to $\mathcal{M}_1$. Finally, translate the latter results to statements about the matrices $W \in \mathcal{W}$.

The above procedure is of course more of a recipe than an algorithm. Let us demonstrate its use by applying it to the case where $\mathcal{F}$ is GF(2) and $\mathcal{P}$ is regularity.

Step 1. By Theorem (12.2.16), we have a good understanding of a portion of the class $\mathcal{N}$ of minimal violation matrices of regularity. Indeed, at this point, only the subclass $\mathcal{N}_3$ is poorly characterized. That class consists of the binary support matrices of the minimal violation matrices of total unimodularity with at least three (by Corollary (12.3.12), at least four) nonzeros in some row and some column. Thus, we take $\mathcal{W}$ to be $\mathcal{N}_3$. We assign "con" and "del" labels to the rows and columns, respectively, of the matrices of $\mathcal{W}$. Then we define $\mathcal{M}_1$ to be the set of matroids $M(W)$ with $W \in \mathcal{W}$.

Step 2. By Corollary (12.3.12), each matrix of $\mathcal{N}_3$ is Eulerian. That is, each $W \in \mathcal{W}$ as given by (12.4.1) has an even number of 1s in each row and each column. This latter fact has a convenient translation into matroid language: Each $M \in \mathcal{M}_1$ has a base such that each fundamental circuit (resp. cocircuit) has an even number of "con" (resp. "del") elements. Now each circuit (resp. cocircuit) of M is the symmetric difference of some of these fundamental circuits (resp. cocircuits). We conclude that each circuit (resp. cocircuit) has an even number of "con" (resp. "del") elements. Another fact, trivial yet important, is that M has at least one "con" element and at least one "del" element. The preceding conditions on circuits, cocircuits, and labels we declare to be the collection $\mathcal{E}$ of matroid facts about $\mathcal{M}_1$.

Step 3. We reverse the arguments and define $\mathcal{M}_2$ from facts known for $\mathcal{M}_1$. Specifically, $\mathcal{M}_2$ is the class of binary matroids M satisfying the following conditions. First, M must be nonregular. Second, each element z of M must be labeled "con" or "del." The "con" (resp. "del") label must imply that M/z (resp. $M\backslash z$) is regular. Third, the circuits and cocircuits of M must obey the following *parity condition*: Each circuit (resp. cocircuit) is to have an even number of "con" (resp. "del") elements. Fourth and last, the following *existence condition* must be satisfied: There is to be at least one "con" element and at least one "del" element. The matroids of $\mathcal{M}_2$ so defined we call *almost regular*.

Step 4. We enlarge $\mathcal{M}_2$ by adding all possible minors. Each minor assumes the labels of the matroid producing it. We claim that each minor so added to $\mathcal{M}_2$ is regular or almost regular. By duality and induction, we only need to consider a 1-element deletion of an element z in an almost regular M. If $M\backslash z$ is regular, we are done. So assume $M\backslash z$ to be nonregular. A "con" (resp. "del") label on an element $w \neq z$ of M implies M/w (resp. $M\backslash w$) to be regular. Thus, for such w the minor $(M\backslash z)/w$ (resp. $(M\backslash z)\backslash w$) of $M\backslash z$ is regular as well. Since $M\backslash z$ is nonregular, z must have a "con" label. By (3.3.11), each circuit of $M\backslash z$ is a circuit of M, and each cocircuit of $M\backslash z$ is a minimal member of the collection $\{C^* - \{z\} \mid C^* = \text{cocircuit of } M\}$. By the parity condition, each circuit (resp. cocircuit) of M has an even number of "con" (resp. "del") elements. Since z has a "con" label, the same conclusion applies to the circuits and cocircuits of $M\backslash z$. The parity and existence conditions for M imply that M has at least two "con" elements and at least two "del" elements. Thus, $M\backslash z$ satisfies the existence condition. These arguments establish $M\backslash z$ to be almost regular. They also prove that the enlarged $\mathcal{M}_2$ is the class of almost regular matroids plus their regular minors. As desired, $\mathcal{M}_2$ is now closed under minor-taking.

Step 5. We must analyze $\mathcal{M}_2$. Then we must specialize the results to $\mathcal{M}_1$. Finally, we must express the latter results in matrix terminology to obtain conclusions about $\mathcal{W}$, and hence about $\mathcal{N}_3$.

In the remainder of this section, we carry out the tasks mandated by Step 5. We begin with some elementary facts about almost regular matroids. Let M be such a matroid. We dualize M in the usual way, but also switch "con" (resp. "del") labels to "del" (resp. "con") labels. As usual, we denote that dual matroid of M by M^*. The next lemma establishes M^* to be almost regular and provides some additional facts about M. A *hyperplane* of a matroid M is a maximal set of M with rank equal to the rank of M minus 1. A *cohyperplane* is a hyperplane in the dual matroid M^* of M.

(12.4.2) Lemma.   *Let M be an almost regular matroid. Then the following holds.*

(a) *The dual matroid M^* of M is almost regular.*
(b) *Every nonregular minor of M is almost regular.*
(c) *The set of "con" (resp. "del") elements of M is a cocircuit and a cohyperplane (resp. a circuit and a hyperplane) of M.*

Proof. (a) This is proved by routine checking of the definition of almost regular matroids given in Step 3 above.
(b) The proof is given under Step 4 above.
(c) Since M is nonregular, it has by Theorem (9.3.2) an F_7 or F_7^* minor. By duality and part (a), we may assume presence of F_7. The "con" and "del" labels for F_7 are unique up to isomorphism, as may be checked by a simple case analysis. Indeed, the matrix B^7 given by (12.4.3) below is a representation matrix of F_7 with correct "con" and "del" labels.

(12.4.3)

$$B^7 = \begin{array}{c} \\ con \\ del \\ del \end{array} \begin{array}{c} \overset{\substack{d\\e\\f}}{} \overset{\substack{c\\o\\n}}{} \overset{\substack{c\\o\\n}}{} \overset{\substack{c\\o\\n}}{} \\ \begin{bmatrix} 0 & 1 & 1 & 1 \\ 1 & 1 & 0 & 1 \\ 1 & 0 & 1 & 1 \end{bmatrix} \end{array}$$

Matrix B^7 for matroid F_7
with "*con*" and "*del*" labels

Take B to be any representation matrix of M that displays an F_7 minor in agreement with B^7 of (12.4.3). We claim that B can be partitioned as

(12.4.4)

$$B = \begin{array}{c} \\ \\ con \\ X_1 \; del \\ del \\ \\ X_2 \; del \end{array} \begin{array}{cccc|c} \multicolumn{4}{c}{Y_1} & Y_2 \\ \substack{d\\e\\f} \; \substack{c\\o\\n} \; \substack{c\\o\\n} \; \substack{c\\o\\n} & \substack{c\\o\\n} \\ \hline 0 \; 1 \; 1 \; 1 & a \\ 1 \; 1 \; 0 \; 1 & \\ 1 \; 0 \; 1 \; 1 & \\ \hline b & 0/1 \end{array}$$

Matrix B displaying F_7 minor

Note that up to indices, B^7 is displayed in the upper left corner of B. For each $x \in X_2$, the minor M/x is nonregular. Thus, x must have a "*del*" label. Similarly, for each $y \in Y_2$, nonregularity of $M\backslash y$ implies a "*con*" label for y. So far, we have justified all labels of B.

We claim that the subvectors a and b of B contain only 1s. If the subvector b has a 0, say in row $x \in X_2$, then the 1s in that row of B correspond to a cocircuit of M with exactly one "*del*" element, in violation of the parity condition. Similarly, a 0 in the subvector a contradicts the parity condition for a fundamental circuit of M.

At this point, we have shown that the first column and the first row of B contain only 1s except for the 0 in the $(1,1)$ position. This fact, plus the given assignments of labels to B, implies that the elements with "*con*" (resp. "*del*") labels form a cocircuit and cohyperplane (resp. circuit and hyperplane) of M, as is easily confirmed by direct checking. □

We now review the construction of the almost regular matroids as described in Section 4.4. The starting point for the construction of an almost regular matroid is either F_7 or a 1-element extension of R_{10}, both with appropriate "*con*" and "*del*" labels. The representation matrix of F_7 we have already seen. It is B^7 of (12.4.3). The matrix for the extension of R_{10} we call B^{11}. It is derived from the matrix of (10.2.8) as follows. We permute the rows of the matrix of (10.2.8) so that the last row becomes the first one. To the resulting matrix, we adjoin a new leftmost column. Then we assign appropriate "*con*" and "*del*" labels. The matrices B^7 and B^{11} have already been listed under (4.4.13). We repeat them here for ready reference.

$$(12.4.5)$$

$$B^7 = \begin{array}{c} \\ con \\ del \\ del \end{array} \begin{array}{cccc} d & c & c & c \\ e & o & o & o \\ f & n & n & n \\ \hline \left[\begin{array}{cccc} 0 & 1 & 1 & 1 \\ 1 & 1 & 0 & 1 \\ 1 & 0 & 1 & 1 \end{array}\right] \end{array} \; ; \qquad B^{11} = \begin{array}{c} \\ \\ con \\ del \\ del \\ del \\ del \end{array} \begin{array}{cccccc} d & c & c & c & c & c \\ e & o & o & o & o & o \\ f & n & n & n & n & n \\ \hline \left[\begin{array}{cccccc} 0 & 1 & 1 & 1 & 1 & 1 \\ 1 & 1 & 0 & 0 & 1 & 1 \\ 1 & 1 & 1 & 0 & 0 & 1 \\ 1 & 0 & 1 & 1 & 0 & 1 \\ 1 & 0 & 0 & 1 & 1 & 1 \end{array}\right] \end{array}$$

Labeled matrices B^7 and B^{11}

By now, the reader has acquired enough machinery, in particular that of graphs plus T sets of Section 10.2, that he/she can quickly verify the matroid of B^{11} to be almost regular. Thus, we omit details of that check.

As stated above, one of the two matroids represented by B^7 or B^{11} is the starting point for the construction of an almost regular matroid. The construction itself consists of a sequence of series or parallel extension steps and of triangle-to-triad and triad-to-triangle exchanges. The extension steps we call SP *steps,* and the exchanges steps, ΔY *exchanges.* These operations are controlled by rules that may be summarized as follows. A parallel (resp. series) extension is permitted only if the involved element z has a "*con*" (resp. "*del*") label. The new element receives the same label as z. The ΔY exchanges are depicted in terms of representation matrices by (4.4.5)–(4.4.7). The conditions on labels are summarized by (4.4.14). Instead of covering all possibilities for the ΔY exchange as done in Section 4.4, here we show just one instance based on (4.4.6). All other cases can be reduced by GF(2)-pivots to the one displayed. In (12.4.6) below, the matrix on the left represents the matroid with the triangle $\{e, f, g\}$, and the one on the right the matroid with the triad $\{x, y, z\}$.

$$(12.4.6)$$

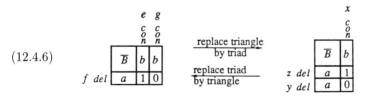

ΔY exchange rule for almost regular matroids

In Section 4.4, we define a sequence of SP extensions and ΔY exchanges under the preceding rules to be a *restricted ΔY extension sequence.* The next result justifies our reliance on such sequences.

(12.4.7) Lemma. *Let a matroid M' be created from an almost regular matroid by a restricted ΔY extension sequence. Then M' is almost regular.*

Proof. By induction, we may assume M' to be derived from M in a single SP extension or ΔY exchange step. Consider the first case. Clearly, M' has a minor isomorphic to M. Thus, M' is nonregular since M is almost regular. The parity condition for M' is readily verified because of the restriction that a parallel (resp. series) extension is only permitted for a "*con*" (resp. "*del*") element of M. For the same reason, for each "*con*" (resp. "*del*") element of M', we have M'/z (resp. $M'\backslash z$) regular. Clearly, both "*con*" and "*del*" labels occur in M'. Thus, M' is almost regular.

The ΔY exchange is almost as easily proved. By Theorem (11.2.11), such an exchange maintains regularity. Thus, by contradiction, M' is nonregular. The remaining conditions are readily verified with the matrices of (12.4.6). Thus, M' is almost regular. $\Box$

The surprising fact is that restricted ΔY extension sequences create all almost regular matroids from the two matroids given by B^7 and B^{11} of (12.4.5). The precise statement is given in Theorem (4.4.16), which we repeat here.

(12.4.8) Theorem. *The class of almost regular matroids has a partition into two subclasses. One of the subclasses consists of the almost regular matroids producible by ΔY extension sequences from the matroid represented by B^7 of (12.4.5). The other subclass is analogously generated by B^{11} of (12.4.5). There is a polynomial algorithm that obtains an appropriate ΔY extension sequence for creating any almost regular matroid from the matroid of B^7 or B^{11}, whichever applies.*

Proof. The existing proof is so long that we cannot include it here. Nevertheless, we sketch the main arguments, since they involve interesting matroid decomposition ideas involving the matroids R_{10} and R_{12} of Sections 10.2 and 11.3.

We take M to be a minimal almost regular matroid that cannot be produced by restricted ΔY extension sequences from any one of the two matroids given by (12.4.5). Put differently, M cannot be reduced to one of the two matroids by restricted ΔY extension sequences. For short, we say that M is not reducible.

By the minimality, M obviously cannot have series or parallel elements. In addition, it turns out that M cannot have a 3-separation with at least four elements on each side. Indeed, such a 3-separation implies a 3-sum decomposition of M where one component is a wheel. That 3-sum is readily proved to be reducible, a contradiction of the minimality of M.

Suppose M does not have R_{10} or R_{12} minors. First, one can show that M must have a "*del*" element z such that $M\backslash z$ is graphic. Thus, M can be represented by a graph plus T set. Rather complicated arguments then prove M to be isomorphic to F_7 or F_7^* with appropriate labels. A single ΔY exchange transforms the F_7^* case to F_7.

Next, we assume that M has an R_{10} minor. Rather easily, we reach the conclusion that M is isomorphic to the matroid of B^{11} or its dual. The second case can be transformed to the first one by one ΔY exchange. The proof relies on the graph plus T set representation of Section 10.2. That approach also produces the insight, at present irrelevant but later useful, that any almost regular matroid with an R_{10} minor must be labeled in such a way that the "con" elements do not form a base and the "del" elements do not form a cobase.

One case remains, where M has an R_{12} minor but no R_{10} minors. By duality and the symmetry of R_{12}, we may assume that M has a "del" element z such that $M \backslash z$ has an R_{12} minor. To investigate this case, we apply the recursive decomposition algorithm of Section 10.5, starting with $\mathcal{H} = \{R_{12}\}$. The class of matroids under consideration is $\mathcal{M}_2$, defined in Steps 3 and 4 at the beginning of this section. In the first iteration of the decomposition algorithm, we use the 3-separation of R_{12} given by (11.3.11). We find exactly two matroids that prevent induced 3-separations. They are duals of each other. Let V_{13} be one of them. Thus, after the first iteration, we have $\mathcal{H}' = \{V_{13}, V_{13}^*\}$, which is the set $\mathcal{H}$ for the next iteration. There we see that both V_{13} and V_{13}^* induce certain 4-sum decompositions. We conclude the second iteration with $\mathcal{H}' = \emptyset$, and stop the decomposition algorithm.

We utilize the output of the decomposition algorithm as follows. As argued at the beginning of the proof, M cannot have a 3-separation with at least four elements on each side. In particular, the 3-separation of any R_{12} minor, as given by (11.3.11), cannot induce a 3-separation of M. By the results of the decomposition algorithm, M then has a V_{13} or V_{13}^* minor.

By duality, we only need to pursue the case where M has a V_{13} minor. We prove that one component of a certain 4-sum induced by V_{13} is almost regular and is represented by a graph plus T set with special structure. In fact, that graph can be created by a particular ΔY extension sequence. For that sequence, the last SP extension step plus the subsequent ΔY exchanges can be viewed as final steps of a construction of M, thus contradicting the minimality of M. $\Box$

Later we need the following observation made in the preceding proof.

(12.4.9) Lemma. *Let M be an almost regular matroid. Then M has an R_{10} minor if and only if it is produced from the almost regular matroid of B^{11} by some restricted ΔY extension sequence. Furthermore, if M has an R_{10} minor, then the set of "con" elements does not form a base of M, and the set of "del" elements does not form a cobase.*

At this point, it probably is not apparent how the matrix classes $\mathcal{U}$,

$\mathcal{A}$, $\mathcal{B}$, $\mathcal{V}$ of Section 12.3 are related to the class of almost regular matroids. The next section will quickly change that situation.

12.5 Matrix Constructions

Let us pause for a moment to assess our position. In Section 12.3, we have defined the matrix classes $\mathcal{U}$, $\mathcal{A}$, $\mathcal{B}$, and $\mathcal{V}$. We have seen that knowledge of just one of the classes $\mathcal{U}$, $\mathcal{A}$, or $\mathcal{B}$ permits easy construction of all others. In Section 12.4, we have established an elementary procedure for creating the almost regular matroids from two initial matroids given by B^7 and B^{11} of (12.4.5). In this section, we use these results to determine elementary constructions for $\mathcal{U}$, $\mathcal{A}$, $\mathcal{B}$, and $\mathcal{V}$. From $\mathcal{B}$ or $\mathcal{V}$, we obtain the class $\mathcal{N}_3$ of Theorem (12.2.16). Thus, we solve the characterization problem of the minimal binary nonregular matrices.

We could proceed in several ways. Particularly appealing appears to be the following route. We first tie a subclass of $\mathcal{U}$ to the almost regular matroids. Then we identify the remaining members of $\mathcal{U}$. Finally, we construct $\mathcal{U}$, $\mathcal{A}$, $\mathcal{B}$, and $\mathcal{V}$, in that order. The first step is accomplished by the following lemma.

(12.5.1) Lemma. *A binary matroid M is almost regular if and only if M is represented by a binary nonregular matrix of the form*

(12.5.2)

$$
\tilde{B} = \begin{array}{c|c|c}
 & \begin{matrix} x \\ d \\ e \\ f \end{matrix} & \begin{matrix} Y \\ c \\ o \\ n \end{matrix} \\
\hline
y \ \ con & 0 & 1' \\
\hline
X \ \ del & 1 & U \\
\end{array}
$$

Matrix $\tilde{B}$ for almost regular matroid M

where U when viewed over $\mathbb{R}$ is complement totally unimodular.

Proof. We start with the "only if" part. Let M be almost regular. According to some arbitrary choice, partition the set of "*del*" elements of M into a singleton set $\{x\}$ and a remainder X. Similarly, partition the set of "*con*" elements into $\{y\}$ and Y. By Lemma (12.4.2), $X \cup \{x\}$ is a circuit and hyperplane of M, and $Y \cup \{y\}$ is a cocircuit and cohyperplane. These facts imply $X \cup \{y\}$ to be a base of M, and the corresponding representation matrix to be $\tilde{B}$ of (12.5.2) for some $\{0, 1\}$ matrix U.

We must show U to be complement totally unimodular. Since element x of M has a "*del*" label, $M \backslash x$ is regular. Thus, the column submatrix of $\tilde{B}$ indexed by Y is regular. Indeed, because of the 1s in row y of that submatrix, the matrix U when viewed as real must be totally unimodular.

By GF(2)-pivots in column x or row y of $\tilde{B}$, we can transform U to all matrices obtainable from U by complement operations. As just argued, each such matrix when viewed over $\mathbb{R}$ must be totally unimodular. Thus, U is complement totally unimodular.

For proof of the "if" part, we reverse the above arguments. Thus, according to the nonregular matrix $\tilde{B}$ and the complement totally unimodular matrix U, the matroid M is nonregular, and each of its "con" (resp. "del") labels indicates regularity upon a contraction (resp. deletion). The parity condition is easily confirmed for the fundamental circuits and cocircuits displayed by $\tilde{B}$. Thus, that condition holds for all circuits and cocircuits. Since $\tilde{B}$ is nonregular, it must have at least three rows and at least three columns. Thus, the existence condition on "con" and "del" labels is satisfied. We conclude that M is almost regular.   □

Recall from Section 12.3 that the class $\mathcal{B}$ consists of the binary matrices B of the form

(12.5.3)

$$B = \begin{array}{c|c|c} & x & Y \\ \hline y & 0 & 1' \\ \hline X & 1 & U \end{array}$$

Matrix B of class $\mathcal{B}$

where U when viewed over $\mathbb{R}$ is complement totally unimodular. Lemma (12.5.1) thus implies the following result for $\mathcal{B}$.

(12.5.4) Corollary. *The representation matrices $\tilde{B}$ of (12.5.2) of the almost regular matroids become upon removal of labels precisely the nonregular matrices of $\mathcal{B}$.*

Proof. This follows directly from Lemma (12.5.1) and a comparison of $\tilde{B}$ of (12.5.2) with B of (12.5.3).   □

By Lemma (12.5.1) and Corollary (12.5.4), we have a characterization of the subclass of $\mathcal{U}$ whose members produce the nonregular matrices of $\mathcal{B}$. The next lemma establishes the structure of the remaining members of $\mathcal{U}$, i.e., those generating the regular matrices of $\mathcal{B}$. To this end, we define any nonempty square triangular matrix U satisfying for all $i \geq j$, $U_{ij} = 1$, to be *solid triangular*. When we add parallel or zero vectors any number of times to such a matrix, we get a *solid staircase* matrix. A typical example of such a matrix is given below.

(12.5.5)

$$\begin{array}{|c|} \hline \quad 0 \\ 1s \\ \hline \end{array}$$

Solid staircase matrix

We need an auxiliary result about solid staircase matrices.

(12.5.6) Lemma. *A $\{0, 1\}$ matrix is a solid staircase matrix if and only if it has no 2×2 identity as submatrix.*

Proof. The "only if" part is elementary. The "if" part is proved by a straightforward inductive argument. One removes a row with maximum number of 1s, invokes induction, then adds that row again for the desired conclusion.   □

Here is the promised characterization of the regular matrices of $\mathcal{B}$.

(12.5.7) Lemma. *A matrix B of $\mathcal{B}$ as given by (12.5.3) is regular if and only if the submatrix U of B is a zero matrix or solid staircase matrix, or becomes a matrix of the latter type by some complement steps.*

Proof. We start with the "if" part. We must show that any B of (12.5.3) with U as specified is regular. Evidently, this is so when U is a zero matrix. So assume that by some complement steps, U becomes a solid staircase matrix. According to the proof of Lemma (12.5.1), any such steps correspond to GF(2)-pivots in B. Thus, they maintain regularity. Series and parallel extensions of a binary matroid also retain regularity. Thus, we may further assume that B has no parallel or unit vectors. By (12.5.3), U then has no parallel or zero vectors. Thus, U is solid triangular, and B is

(12.5.8)

$$
B = \begin{array}{c|c|c}
 & x & Y \\
\hline
y & 0 & 1^t \\
\hline
X & 1 & \begin{matrix} 1 & . \\ & . & . \\ 1s & & 1 \end{matrix}
\end{array}
$$

Matrix B with solid triangular U

We claim that B represents the graphic matroid of a wheel, with $X \cup \{x\}$ as set of rim edges, and $Y \cup \{y\}$ as set of spokes. The edges x and y are such that $X \cup \{y\}$ is a path. The claim is easily verified. One only checks that the fundamental cycles for $X \cup \{y\}$ are displayed by B of (12.5.8). We conclude that B is regular.

For proof of the "only if" part, we assume B of (12.5.3) to be regular. Since B is in $\mathcal{B}$, the submatrix U when viewed over $\mathbb{R}$ is complement totally unimodular. We are done if U is a zero matrix. So assume U to be nonzero. We may assume that U has no zero or parallel vectors. Thus, we must show U or a matrix obtained from U by complement steps, to be solid triangular.

If U has a 3×3 identity submatrix, say indexed by $\overline{X} \subseteq X$ and $\overline{Y} \subseteq Y$, then B contains the matrix of (12.5.9) below. A pivot on any 1 of the 3×3 identity submatrix, plus a pivot in the resulting matrix on the 1 in the (x, y) position, produces a matrix that displays an F_7 minor. But then B

is not regular, a contradiction.

(12.5.9)

$$
\begin{array}{c|c|ccc}
 & x & \multicolumn{3}{c}{\overline{Y}} \\
\hline
y & 0 & 1 & 1 & 1 \\
\hline
 & 1 & 1 & 0 & 0 \\
\overline{X} & 1 & 0 & 1 & 0 \\
 & 1 & 0 & 0 & 1 \\
\end{array}
$$

Nonregular submatrix of B induced
by a 3×3 identity submatrix of U

Suppose U has a 2×2 identity submatrix. If U is connected, then U or its transpose contains the submatrix $\begin{bmatrix} 1 & 0 & 1 \\ 0 & 1 & 1 \end{bmatrix}$, say indexed by $\overline{X} \subseteq X$ and $\overline{Y} \subseteq Y$. By duality, we may assume the former case. Then B contains the matrix

(12.5.10)

$$
\begin{array}{c|c|ccc}
 & x & \multicolumn{3}{c}{\overline{Y}} \\
\hline
y & 0 & 1 & 1 & 1 \\
\hline
\overline{x} & 1 & 1 & 0 & 1 \\
 & 1 & 0 & 1 & 1 \\
\end{array}
$$

Nonregular submatrix of B induced
by a certain 2×3 submatrix of U

which represents an F^7 minor, and once more we have a contradiction.

Still assume that U has a 2×2 identity submatrix. As argued above, U does not contain a 3×3 identity submatrix and is not connected. Thus, U has exactly two connected blocks, neither of which contains a 2×2 identity submatrix. By Lemma (12.5.6) and by the absence of zero or parallel vectors from U, each block must be solid triangular. Let z index a row of U with maximum number of 1s. The row z complement of U is then solid triangular, as desired.

Finally, assume that U has no 2×2 identity submatrix. By Lemma (12.5.6) and by the absence of zero or parallel vectors, U is then solid triangular. □

We are ready to state and validate the following construction of the complement totally unimodular matrices. The construction relies on four seemingly strange matrices. Their origin will become clear shortly.

(12.5.11) Construction of $\mathcal{U}$. *Define real $\{0,1\}$ matrices U^0, U^1, U^7, and U^{11} as follows.*

(12.5.12) $U^0 = \boxed{0}$; $U^1 = \boxed{1}$; $U^7 = \begin{bmatrix} 1 & 0 & 1 \\ 0 & 1 & 1 \end{bmatrix}$; $U^{11} = \begin{bmatrix} 1 & 0 & 0 & 1 & 1 \\ 1 & 1 & 0 & 0 & 1 \\ 0 & 1 & 1 & 0 & 1 \\ 0 & 0 & 1 & 1 & 1 \end{bmatrix}$

Complement totally unimodular
matrices U^0, U^1, U^7, and U^{11}

Starting with $U = U^0, U^1, U^7$, or U^{11}, apply a sequence of operations each of which is one of (i), (ii), or (iii) below.

(i) *Perform a row or column complement operation.*
(ii) *Add a zero or parallel row or column vector.*
(iii) *(ΔY exchange) If U is one of the two matrices below, replace U by the other matrix.*

(12.5.13)

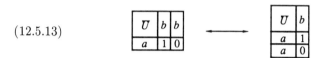

ΔY exchange for complement
totally unimodular matrices

Collect in sets $\mathcal{U}_0$, $\mathcal{U}_1$, $\mathcal{U}_7$, and $\mathcal{U}_{11}$ the matrices that can be so deduced from U^0, U^1, U^7, and U^{11}, respectively. These sets form a partition of $\mathcal{U}$.

An example matrix constructed from U^7 is given by (12.5.25) below.

Proof of Construction (12.5.11).   It is easy to see that the steps (i)–(iii) produce from $U^0 = [\,0\,]$ the class of zero matrices, which thus constitutes $\mathcal{U}_0$. Indeed, steps (i) and (iii) never apply. For validation of the remaining cases, we assign a "*con*" (resp. "*del*") label to each column (resp. row) of any U produced by Construction (12.5.11) from U^1, U^7, or U^{11}. For the moment, these labels are purely formal. Next, we embed U into the matrix $\tilde{B}$ of (12.5.2), which we repeat here.

(12.5.14)

$$\tilde{B} = \begin{array}{c|c|c} & \begin{matrix} x \\ d \\ e \\ f \end{matrix} & \begin{matrix} Y \\ c \\ o \\ n \end{matrix} \\ \hline y \;\; con & 0 & 1' \\ \hline X \;\; del & 1 & U \end{array}$$

Matrix $\tilde{B}$ derived from matrix U

Steps (i)–(iii) of Construction (12.5.11) are then equivalent to the following operations on $\tilde{B}$. The complement-taking of step (i) corresponds to a GF(2)-pivot in row y or column x of $\tilde{B}$. The addition of zero or parallel vectors of step (ii) is the addition of parallel vectors or of unit vectors with 1 in row y or column x. Finally, the exchange of step (iii) is the ΔY exchange depicted by (12.4.6).

We take these observations one step further. Let M be the labeled matroid represented by $\tilde{B}$. We claim that the operations just specified for $\tilde{B}$ are equivalent to the operations of restricted ΔY extension sequences in M. Indeed, the addition of parallel vectors or of unit vectors with 1 in row y or column x of $\tilde{B}$ becomes the restricted SP extension for M. The GF(2)-pivot in $\tilde{B}$ plus the exchange given by (12.4.6) are equivalent to a restricted ΔY exchange in M.

We interpret the matrices U of $\mathcal{U}_1$, $\mathcal{U}_7$, and $\mathcal{U}_{11}$ in terms of the related matroid M. We start with $\mathcal{U}_1$. The matrix $\tilde{B}$ of (12.5.14) produced by $U = U^1 = [\,1\,]$ is

(12.5.15)

$$
\begin{array}{c}
\begin{array}{cc} x & Y \\ d & c \\ e & o \\ f & n \end{array} \\
\begin{array}{r} y\ con \\ X\ del \end{array}
\begin{array}{|cc|} \hline 0 & 1 \\ 1 & 1 \\ \hline \end{array}
\end{array}
$$

Matrix $\tilde{B}$ derived from $U = U^1 = [\,1\,]$

The corresponding M is the graphic matroid of the wheel with two spokes. Indeed, the spokes are labeled "con" and form the set $Y \cup \{y\}$. The rim edges are labeled "del" and constitute the set $X \cup \{x\}$. We interpret restricted ΔY extension sequences for M as sequences of graph operations applied to the preceding graph. Each such operation is either a subdivision of a "del" edge into two "del" edges, or an addition of a "con" edge parallel to a "con" edge, or an exchange of a triangle by a 3-star, or an exchange of a 3-star by a triangle. In the latter two operations, the labels are assigned according to (4.4.14), which we repeat below.

(12.5.16)

Triangle and 3-star with labels

We leave it to the reader to verify that any graph produced by these operations is obtainable from some wheel graph with at least two spokes by subdivision of rim edges and addition of edges parallel to spokes. In terms of M, the rim edges of such an extended wheel form the set $X \cup \{x\}$, and the spokes constitute $Y \cup \{y\}$. We interpret this result in terms of U as in the proof of Lemma (12.5.7). Thus, we see that U or some matrix obtained from U by complement operations, is a solid staircase matrix.

We turn to the cases of U^7 and U^{11}. The matrix U^7 (resp. U^{11}) produces as $\tilde{B}$ the matrix B^7 (resp. B^{11}) of (12.4.5). In either case, $\tilde{B}$ is nonregular, and M is an almost regular matroid. By Theorem (12.4.8), restricted ΔY extension sequences produce all almost regular matroids from the two matroids represented by B^7 and B^{11}. By Lemma (12.5.1) and Corollary (12.5.4), the sets $\mathcal{U}_7$ and $\mathcal{U}_{11}$ thus contain the complement totally unimodular matrices that produce the nonregular members of the class $\mathcal{B}$.

By Lemma (12.5.7) and the above characterization of $\mathcal{U}_0$ and $\mathcal{U}_1$, the class $\mathcal{U}_0 \cup \mathcal{U}_1$ contains precisely the complement totally unimodular matrices U that produce the regular matrices B of $\mathcal{B}$. We have also seen that the class $\mathcal{U}_7 \cup \mathcal{U}_{11}$ generates the nonregular matrices B of $\mathcal{B}$. Thus, $\mathcal{U}_0 \cup \mathcal{U}_1 \cup \mathcal{U}_7 \cup \mathcal{U}_{11}$ is equal to $\mathcal{U}$. The sets $\mathcal{U}_0 \cup \mathcal{U}_1$ and $\mathcal{U}_7 \cup \mathcal{U}_{11}$ are necessarily disjoint. Evidently, this is also so for $\mathcal{U}_0$ and $\mathcal{U}_1$. Theorem (12.4.8) says that any almost regular matroid can be generated from exactly one of the matroids represented by B^7 and B^{11}. Thus, $\mathcal{U}_7$ and $\mathcal{U}_{11}$ are disjoint. We conclude that $\mathcal{U}_0$, $\mathcal{U}_1$, $\mathcal{U}_7$, and $\mathcal{U}_{11}$ form a partition of $\mathcal{U}$ as claimed by Construction (12.5.11). □

Construction (12.5.11) has two interesting corollaries.

(12.5.17) Corollary. *Let U be a nonempty complement totally unimodular matrix without zero or parallel vectors. Then U, or some matrix obtainable from U by complement operations, contains a unit vector.*

Proof. We use the notation of the proof of Construction (12.5.11). By the assumptions, U is in $\mathcal{U}_1 \cup \mathcal{U}_7 \cup \mathcal{U}_{11}$. Let U define $\tilde{B}$ of (12.5.14), and M be the labeled matroid represented by $\tilde{B}$. Now U has no zero or parallel vectors. Thus, M has no series or parallel elements. M is obtained from the graphic matroid of (12.5.15) or from the almost regular matroid of B^7 or B^{11} by a restricted ΔY extension sequence. In all three cases, the initial matroid has a triangle with two "*con*" labels and one "*del*" label, and the final matroid has a triangle with two "*con*" labels or triad with two "*del*" labels. By duality, we may assume M to have a triangle C with two "*con*" labels. In the notation for $\tilde{B}$ of (12.5.14), assume $x \notin C$ and $y \in C$. Thus, y is one of the two "*con*" elements of C. The second "*con*" element of C must be in Y. The third element of C, with "*del*" label, must be in X. Then column $z \in Y$ of $\tilde{B}$ has exactly two 1s, one of which is in row y. Thus, column z of U is a unit vector. If $x \in C$ or $y \notin C$, we can achieve the desired configuration by GF(2)-pivots in column x or row y of $\tilde{B}$. The pivots correspond to complement operations for U. Thus, the lemma holds in all cases. □

(12.5.18) Corollary. *If $U \in \mathcal{U}$ is $\mathbb{R}$-nonsingular, then U is in $\mathcal{U}_1$ or $\mathcal{U}_7$, but not in $\mathcal{U}_0$ or $\mathcal{U}_{11}$.*

Proof. Candidate classes are the claimed ones and $\mathcal{U}_{11}$. Let $U \in \mathcal{U}_{11}$, and let M be the matroid defined via $\tilde{B}$ of (12.5.14). By the proof of Construction (12.5.11), M is an almost regular matroid produced by some restricted ΔY extension sequence from the matroid represented by B^{11}. Lemma (12.4.9) says that the set $Y \cup \{y\}$ of "*con*" elements of M does not form a base, and the set $X \cup \{x\}$ of "*del*" elements does not form a cobase. By assumption, U is $\mathbb{R}$-nonsingular. Since U is complement totally unimodular, U when viewed as binary is $\mathrm{GF}(2)$-nonsingular as well. But then, by $\tilde{B}$ of (12.5.14), the set $Y \cup \{y\}$ is a base of M, a contradiction.   □

At this point, it is a simple matter to construct the classes $\mathcal{A}$, $\mathcal{B}$, and $\mathcal{V}$. For completeness, we include details. We start with $\mathcal{A}$ and $\mathcal{B}$. Recall that $\mathcal{A}$ (resp. $\mathcal{B}$) is the class of almost representative matrices over $\mathrm{GF}(3)$ (resp. $\mathrm{GF}(2)$).

(12.5.19) Construction of $\mathcal{A}$ and $\mathcal{B}$.   *The classes $\mathcal{A}$ and $\mathcal{B}$ are deduced from $\mathcal{U}$ as follows. Each matrix of $\mathcal{A}$ (resp. $\mathcal{B}$) is precisely a matrix of the form*

(12.5.20)

$$
\begin{array}{|c|c|}
\hline
\alpha & 1^t \\
\hline
1 & U \\
\hline
\end{array}
$$

Matrix of $\mathcal{A}$ or $\mathcal{B}$

For $\mathcal{A}$ (resp. $\mathcal{B}$), the matrix is over $\mathrm{GF}(3)$ (resp. $\mathrm{GF}(2)$), with $\alpha = -1$ (resp. $\alpha = 0$). The submatrix U is a $\{0,1\}$ matrix that, when considered over $\mathbb{R}$, is in $\mathcal{U}$.

Proof. Validity of the construction follows directly from the definition of $\mathcal{A}$ and $\mathcal{B}$ following Lemma (12.3.10).   □

Next we give a construction for $\mathcal{V}$, the class of minimal violation matrices of total unimodularity.

(12.5.21) Construction of $\mathcal{V}$.   *The class $\mathcal{V}$ is deduced from $\mathcal{U}$ as follows. Each matrix V of $\mathcal{V}$ is up to scaling by $\{\pm 1\}$ factors of the form*

(12.5.22)

$$
\begin{array}{|c|c|}
\hline
\alpha & a \\
\hline
b & U^{-1} \\
\hline
\end{array}
\quad ; \quad
\begin{array}{l}
a = 1^t \cdot U^{-1} \\
b = U^{-1} \cdot 1 \\
\alpha = 1^t \cdot U^{-1} \cdot 1 - 2
\end{array}
$$

Matrix V up to scaling

The matrix U is a square $\mathbb{R}$-nonsingular matrix of $\mathcal{U}_1$ or $\mathcal{U}_7$. If $U \in \mathcal{U}_1$, then V has exactly two nonzeros in each row and in each column; indeed, the bipartite graph $\mathrm{BG}(V)$ is a cycle. If $U \in \mathcal{U}_7$, then V has at least four nonzeros in some row and in some column.

Proof. Validity of the construction follows from Corollaries (12.3.12) and (12.5.18).   □

Corollary (12.5.17) and Construction (12.5.21) yield an interesting result.

(12.5.23) Corollary.   *Every matrix $V \in \mathcal{V}$ has a row or column with exactly two nonzeros.*

Proof. The inverse of V is, up to scaling by $\{\pm 1\}$ factors, given by $\tilde{V}$ of (12.3.17). The U^{-1} occurring in V of (12.5.22) is the inverse of the U defining $\tilde{V}$ of (12.3.17). A simple check confirms that replacement of U by a matrix deduced from U by complement operations effectively may be viewed as scaling in $\tilde{V}$. Thus, up to scaling, U and all matrices obtainable from U by complement operations produce the same V. Since U is nonsingular, it cannot contain zero or parallel vectors. By Corollary (12.5.17), U, or some matrix deduced by complement operations from U, contains a unit vector. Let U' be that matrix. The inverse of U' contains a unit vector as well. By (12.5.22) and the above discussion, V must have a row or column with exactly two nonzeros. $\square$

At long last, we can fill the gap left by Theorem (12.2.16) and complete the characterization of minimal nonregular submatrices. The desired theorem is as follows.

(12.5.24) Theorem.   *Let $\mathcal{N}$ be the class of binary minimal violation matrices of regularity. Then $\mathcal{N}$ has a partition into three subclasses $\mathcal{N}_1$, $\mathcal{N}_2$, $\mathcal{N}_3$ as follows.*

(a) *$\mathcal{N}_1$ (resp. $\mathcal{N}_2$) is the set of binary matrices B for which $\mathrm{BG}(B)$ is a graph of type H_1 (resp. H_2) of (12.2.14).*
(b) *$\mathcal{N}_3$ is the set of binary support matrices of $V \in \mathcal{V}$ produced via (12.5.22) from the $\mathbb{R}$-nonsingular matrices $U \in \mathcal{U}_7$.*

Proof. Part (a) is taken from Theorem (12.2.16). Part (b) follows from that theorem and Construction (12.5.21) of $\mathcal{V}$. $\square$

The above constructions are readily performed by hand. That way one can rapidly produce structurally interesting matrices. An example of a complement totally unimodular matrix U and a minimal violation matrix V of total unimodularity generated that way is as follows. We first obtain by Construction (12.5.11) the following complement totally unimodular matrix U.

(12.5.25)

$$U = \begin{array}{c c} & \begin{array}{cccccc} z & i & b & f & e & c \end{array} \\ \begin{array}{c} a \\ g \\ h \\ j \\ k \\ d \end{array} & \left[\begin{array}{ccc|ccc} 1 & 1 & 1 & 1 & 1 & 1 \\ 0 & 1 & 0 & 1 & 1 & 1 \\ 0 & 0 & 1 & 1 & 1 & 1 \\ \hline 1 & 0 & 1 & 1 & 0 & 1 \\ 0 & 0 & 0 & 0 & 1 & 1 \\ 1 & 0 & 1 & 0 & 0 & 1 \end{array}\right] \end{array}$$

Complement totally unimodular matrix U
representing R_{12}

The labels may be used to verify that U represents the regular matroid R_{12} of (10.2.9). One only needs to check via the graph plus T set of (10.2.9) the fundamental circuits for the base $\{a, d, g, h, j, k\}$, which is the index set of the rows of U. The indicated partition of U corresponds to the 3-separation $(\{a, b, g, h, i, z\}, \{c, d, e, f, j, k\})$ of R_{12}. Straightforward computations confirm that U is $\mathbb{R}$-nonsingular, and that up to scaling by $\{\pm 1\}$ factors and row and column exchanges, U is its own inverse. Thus, for this special case, we are tempted to use U instead of U^{-1} in the formula for V of (12.5.22). But U^{-1} is a scaled and permuted version of U. Because of the scaling, the formula (12.5.22) cannot be used. But we may rely on part (b) of Corollary (12.3.12). According to that result, V is Eulerian and its entries sum in $\mathbb{R}$ to $2(\bmod 4)$. Using these two facts, we determine the following V from U.

(12.5.26)

$$V = \begin{array}{c|c|cccccc} & & z & i & b & f & e & c \\ \hline & 1 & 1 & 0 & 0 & 0 & 0 & 0 \\ \hline a & 0 & 1 & 1 & 1 & 1 & 1 & 1 \\ g & 0 & 0 & 1 & 0 & 1 & 1 & 1 \\ h & 0 & 0 & 0 & 1 & 1 & 1 & 1 \\ j & 0 & 1 & 0 & 1 & 1 & 0 & 1 \\ k & 0 & 0 & 0 & 0 & 0 & 1 & 1 \\ d & 1 & 1 & 0 & 1 & 0 & 0 & 1 \end{array}$$

Matrix V deduced from U of (12.5.25)

The above example supports the following result.

(12.5.27) Lemma. *The regular matroid R_{12} has a real nonsingular $\{0, 1\}$ representation matrix that is complement totally unimodular.*

Construction (12.5.11) and the matrix V of (12.5.26) permit us to relate $\mathcal{V}$ to the regular matroids R_{10} and R_{12} of (10.2.8) and (10.2.9), as follows.

(12.5.28) Lemma. *Let M be the binary matroid represented by the binary support matrix W of a minimal non-totally unimodular matrix $V \in \mathcal{V}$. Then M does not have R_{10} minors, but may have an R_{12} minor.*

Proof. By Theorem (12.3.11), M is represented not just by W as stated, but also by some $B \in \mathcal{B}$ given by (12.5.20). Indeed, the submatrix U^{-1} of V of (12.5.22) is the inverse of the matrix U defining B. By Construction (12.5.21) for $\mathcal{V}$, we have $U \in \mathcal{U}_1 \cup \mathcal{U}_7$. Thus, $U \notin \mathcal{U}_{11}$. By the proof of Construction (12.5.11) for $\mathcal{U}$, $U \notin \mathcal{U}_{11}$ implies that M has no R_{10} minor. The matrix V of (12.5.26) demonstrates that M may have an R_{12} minor. $\square$

Lemma (12.5.28) implies the claim made toward the end of Section 12.3 that none of the classes $\mathcal{U}$, $\mathcal{A}$, and $\mathcal{B}$ can be produced from $\mathcal{V}$ by the matrix operations of scaling by $\{\pm 1\}$ factors, pivots, submatrix taking, and change of fields from $\mathbb{R}$ to GF(2) or GF(3). In particular, the matrix U^{11} of (12.5.12) cannot be obtained that way.

In the last section, we discuss applications and extensions, and include references.

12.6 Applications, Extensions, and References

The entire Section 12.2 is taken from Truemper (1982b). Variations of the graph signing problem of Section 12.2 are treated in Zaslavsky (1981a), (1981b), (1982), (1987), (1989), (1991). The notion of α-balancedness extends the concept of $\{0,1\}$ balanced matrices covered extensively elsewhere (see Berge (1972), (1973), (1980), Fulkerson, Hoffman, and Oppenheim (1974), and Anstee and Farber (1984)). Profound decomposition results for balanced matrices are proved in Conforti and Rao (1989), (1991a)–(1991d), Conforti and Cornuéjols (1990), and Conforti, Cornuéjols, and Rao (1991). In Truemper and Chandrasekaran (1978), balancedness of $\{0,1\}$ matrices is tied to total unimodularity by the exclusion of certain minimal violation matrices of total unimodularity. That result motivated the search for a construction of $\mathcal{V}$.

A $\{0,1\}$ matrix is *perfect* if the polyhedron $\{x \mid A \cdot x \leq \underline{1}, x \geq 0\}$ has only integer vertices. A graph is *perfect* if its clique/node incidence matrix is perfect. Any $\{0,1\}$ balanced matrix is perfect, but the converse does not hold. The so-called strong perfect graph conjecture says that the minimal nonperfect graphs, and thus the minimal nonperfect matrices, have a certain simple structure (see, e.g., Padberg (1974)). There are numerous partial results with regard to that conjecture. Approaches based on graph decomposition have been cited in Section 10.7. Padberg (1974) contains a very interesting matrix-based attack on the problem.

The concept of complement total unimodularity of Section 12.3 is defined in Truemper (1980b). Almost representative matrices are characterized in Truemper (1982b). A number of papers explicitly or implicitly include properties of the class $\mathcal{V}$ (see Ghouila-Houri (1962), Camion (1963a), (1963b), (1965), Chandrasekaran (1969), Gondran (1973), Padberg (1975), (1976), Tamir (1976), Kress and Tamir (1980), Truemper (1977), (1978), (1980b), (1982b), and de Werra (1981)). But none of the results captures the complexity of $\mathcal{V}$. Indeed, the cited results do not contain a single clue to how one might construct even a small subset of structurally different matrices of $\mathcal{V}$.

Section 12.4 relies on Truemper (1991), (1992). The analysis technique applies not just to the cited properties, but also to certain representability questions. An important case is treated in Truemper (1982b), where the minimal violation matrices of abstract matrices not representable over GF(2) or GF(3) are characterized.

Section 12.5 is entirely based on Truemper (1992). That reference also contains alternate constructions using pivots. The constructions do not have descriptions as brief as the ones included here, but they are well suited for hand calculations. Truemper (1992) also includes a proof that every 3-connected almost regular matroid different from F_7 and F_7^* has a binary representation matrix that is balanced. An example is the matrix V of (12.5.26). The proof relies on an extension of an efficient algorithm of Fonlupt and Raco (1984) that proves existence of a balanced binary representation matrix for any regular matroid. The algorithm is a modification of a scheme due to Camion (1968) where the latter existence result was first established.

Chapter 13

Max-Flow Min-Cut Matroids

13.1 Overview

Concepts, theorems, or algorithms in one area of mathematics often inspire new approaches in another area. In turn, the ensuing new developments in the latter area may lead to new ideas in the former one. In this chapter, we describe an interesting instance of this cyclic process.

We start with the *max flow problem* for undirected graphs, which may be defined as follows. Given is a connected and undirected graph G. One of the edges of G, say l, is declared to be *special*. To each edge e of G other than l, a nonnegative integer h_e is assigned and called the *capacity* of e. Define G to have *flow value* F if there are F cycles, not necessarily distinct, that satisfy the following two conditions. Each cycle of the collection must contain the special edge l, and any other edge e of G is allowed to occur altogether in at most h_e of the cycles. The max flow problem demands that one solve the problem max F. The solution value must be accompanied by a collection of cycles producing that value.

A companion of the max flow problem is the following *min cut problem*. For any cocycle D of G containing the special edge l, define the *capacity* of D to be the sum of the capacities of the edges in D other than l. Denote the capacity of D by $h(D)$. The min cut problem asks one to solve min $h(D)$. The solution value must be accompanied by a cocycle D containing the edge l and having the solution value as capacity.

The famous max-flow min-cut theorem for graphs says that max $F =$ min $h(D)$ no matter how the nonnegative integral edge capacities are selected.

The max flow and min cut problems for graphs have obvious matroid translations. In the above description, one replaces the graph G by a matroid M and specifies elements instead of edges, and circuits and cocircuits instead of cycles and cocycles. One might conjecture that the max-flow min-cut theorem still holds in the expanded setting. But this is not so in general. Counterexamples can be produced with small matroids, in particular with U_4^2, the rank 2 uniform matroid on four elements, and with F_7^*, the Fano dual matroid. Indeed, because of the symmetry of U_4^2 as well as of F_7^*, the conjecture is false for these matroids no matter which element is declared to be special. In general, there are matroids where the equality $\max F = \min h(D)$ does or does not hold for all capacity vectors h, depending on the selection of the special element. Thus, we are motivated to define a matroid M with a special element l to have the *max-flow min-cut property*, or to be a *max-flow min-cut matroid*, if $\max F = \min h(D)$ no matter which nonnegative integral values are assigned as capacities. Absence or presence of the max-flow min-cut property for M is evidently governed by the connected component of M containing the element l. Thus, for the purposes of characterizing the max-flow min-cut matroids, we might as well restrict ourselves to connected matroids.

Using an ingenious but complicated induction hypothesis, Seymour proved in a long paper that the connected max-flow min-cut matroids are precisely the connected binary matroids where the special element l is not contained in any F_7^* minor. In this chapter, we prove Seymour's result using a quite different approach. Let us define $\mathcal{M}$ to be the class of connected binary matroids where each matroid has a special element l such that l is not contained in any F_7^* minor. In Section 13.2, we establish 2- and Δ-sum decomposition theorems for the matroids of $\mathcal{M}$. In Section 13.3, we use these theorems to prove Seymour's result that $\mathcal{M}$ is precisely the class of connected max-flow min-cut matroids. In Section 13.4, we use a part of the proof to validate a construction for $\mathcal{M}$ that involves certain 2- and Δ-sums. We show that the construction can be determined for any matroid of $\mathcal{M}$ in polynomial time, and thus conclude that one can test for the max-flow min-cut property of binary matroids in polynomial time. We also describe polynomial algorithms for the following problems involving the matroids of $\mathcal{M}$: the max flow problem, the min cut problem, and a certain shortest circuit problem.

In Section 13.5, we examine an interesting graph application of the above results for $\mathcal{M}$. Let H be an undirected graph each of whose edges is declared to be odd or even. Recall that K_4 is the complete graph on four vertices. Declare a K_4 minor of H to be an odd-K_4 minor if it has a certain property that is defined via the relative position of its even and odd edges. Then let $\mathcal{G}$ be the class of 2-connected graphs without odd-K_4 minors. The graphs of $\mathcal{G}$ have pleasant properties, so an understanding of their structure is desirable. It turns out that the graphs of $\mathcal{G}$ can be linked to the class $\mathcal{M}$. As a result, we can apply the cited construction for $\mathcal{M}$ to understand the structure of the graphs of $\mathcal{G}$. In that way, we obtain a construction for the graphs of $\mathcal{G}$. At the same time, we produce a polynomial test for membership in $\mathcal{G}$. Evidently, we have moved from the max-flow min-cut property of graphs to the max-flow min-cut matroids, and then to the graphs without odd-K_4 minors. Thus, we have an instance of the cyclic process mentioned in the introductory paragraph.

The final section, 13.6, contains additional applications, extensions, and references.

The chapter makes use of Chapters 2, 3, and 5–11. It is also assumed that the reader has a basic knowledge of linear programming. Relevant references are included in Section 13.6.

13.2   2-Sum and Delta-Sum Decompositions

Recall that $\mathcal{M}$ is the class of connected binary matroids where each matroid has a special element l such that l is not contained in any F_7^* minor. In this section, we first show that any 2-separable matroid of $\mathcal{M}$ has a certain 2-sum decomposition where both component matroids are also in $\mathcal{M}$. Then we establish that any 3-connected nonregular matroid of $\mathcal{M}$ has a particular Δ-sum decomposition where again the components are in $\mathcal{M}$. These two results will be used in the next section to show that $\mathcal{M}$ is precisely the class of max-flow min-cut matroids. Before proceeding, the reader may want to review briefly the results for 2- and Δ-sums in Sections 8.2 and 8.5, respectively.

We begin with some definitions. By symmetry, there is essentially just one way to declare an element of the Fano matroid F_7 or of its dual F_7^* to be the special element l. When this is done, we get the matroid F_7 *with l* or F_7^* *with l*. Suppose two binary matroids M and M' contain the element l. If an isomorphism exists between M and M' that takes the element l of one of the matroids to l of the other one, then the two matroids are *l-isomorphic*. Finally, we emphasize that l is always the special element of any matroid in $\mathcal{M}$. Note that F_7 with l is in $\mathcal{M}$, while F_7^* with l is not.

We are ready for the detailed discussion of the 2- and Δ-sum results for $\mathcal{M}$.

2-Sum Decomposition

The structure theorem for the 2-sum case is as follows.

(13.2.1) Theorem.   *Any 2-separable matroid $M \in \mathcal{M}$ on a set E has a 2-sum decomposition where both components M_1 and M_2 are connected* minors of M, contain l, and thus are in $\mathcal{M}$. In addition, M_2 has an element $y \notin l$ so that any set $C \subseteq E$ is a circuit of M with l if and only if (i) or (ii) below holds.

(i) C is a circuit of M_2 with l but not y.
(ii) $C = (C_1 - \{l\}) \cup (C_2 - \{y\})$ where C_1 is a circuit of M_1 with l, and where C_2 is a circuit of M_2 with both l and y.

Proof. Using the results of Section 8.2 for 2-separations and 2-sums, as well as the path shortening technique of Chapter 5, one readily shows that any 2-separable $M \in \mathcal{M}$ has a representation matrix B of the form

(13.2.2)

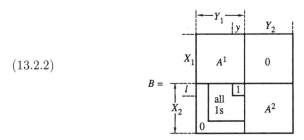

Matrix B of $M \in \mathcal{M}$ with exact 2-separation

Note the position of the element l in X_2, and the indicated element $y \in Y_1$. By Section 8.2, M is a 2-sum where both components M_1 and M_2 are connected minors of M, have the element l, and are represented by the matrices B^1 and B^2 of (13.2.3) below. Routine arguments using B, B^1 and B^2 confirm the claims of the theorem concerning the circuits of M, M_1, and M_2. $\square$

(13.2.3)

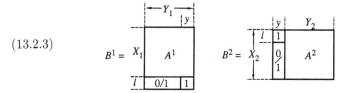

Matrices B^1 and B^2 of 2-sum decomposition of M

Delta-Sum Decomposition

We turn to the much more challenging case where M is 3-connected. In Section 13.3, it will be shown that any regular matroid of $\mathcal{M}$ has the max-flow min-cut property. Thus, we concentrate here on the situation where M is not regular. For that case, one can prove M either to be isomorphic to F_7, or to have a particular Δ-sum decomposition. The precise statement is as follows.

(13.2.4) Theorem.   *Any 3-connected nonregular matroid $M \in \mathcal{M}$ on a set E is isomorphic to the Fano matroid F_7, or has a Δ-sum decomposition where the components M_1 and M_2 are connected minors of M, contain l, and thus are in $\mathcal{M}$. In the Δ-sum decomposition, both connecting triangles of M_1 and M_2 contain l. Let the remaining elements of the connecting triangle of M_1 (resp. M_2) be a and b (resp. v and w). Then any set $C \subseteq E$ is a circuit of M with l if and only if (i), (ii), (iii), or (iv) below holds.*

(i) *C is a circuit C_1 of M_1 with l but without a and b.*

(ii) *C is a circuit C_2 of M_2 with l but without v and w.*

(iii) *$C = (C_a - \{a\}) \cup (C_v - \{v\})$ where C_a is a circuit of M_1 with l and a but without b, and where C_v is a circuit of M_2 with l and v but without w.*

(iv) *$C = (C_b - \{b\}) \cup (C_w - \{w\})$ where C_b is a circuit of M_1 with l and b but without a, and where C_w is a circuit of M_2 with l and w but without v.*

The proof of Theorem (13.2.4) takes up the remainder of this section. We proceed as follows. First, we show that any 3-connected nonregular $M \in \mathcal{M}$ is isomorphic to F_7 or has a minor that is l-isomorphic to the matroid N_8 of $\mathcal{M}$ defined by the matrix B^8 below.

(13.2.5)

$$B^8 = \begin{array}{c c} & \begin{array}{c c c | c} & Y_1 & & Y_2 \\ & u\ \ v\ \ w & & l \end{array} \\ \begin{array}{c} X_1 \begin{array}{c} d \\ \end{array} \\ \begin{array}{c} a \\ X_2\ b \\ c \end{array} \end{array} & \begin{array}{|c c c | c|} \hline 1 & 1 & 1 & 0 \\ \hline 1 & 0 & 1 & 1 \\ 1 & 1 & 0 & 1 \\ 0 & 1 & 1 & 0 \\ \hline \end{array} \end{array}$$

Matrix B^8 for the matroid N_8

Evidently, $(X_1 \cup Y_1, X_2 \cup Y_2)$ is a 3-separation of N_8. Note that Y_2 contains just the element l. It is easy to confirm that N_8 is indeed in $\mathcal{M}$.

Next, we establish that M has a minor N with the following properties. N contains l and is l-isomorphic to N_8, and a 3-separation of N corresponding to $(X_1 \cup Y_1, X_2 \cup Y_2)$ of N_8 under one of the l-isomorphisms induces a 3-separation of M. From that induced 3-separation of M, we finally derive the Δ-sum decomposition claimed in Theorem (13.2.4).

Induced 3-Separation

We begin the detailed discussion. The first lemma deals with the presence of N_8 minors in M.

(13.2.6) Lemma.   *Let M be a 3-connected nonregular matroid of $\mathcal{M}$. Then M is isomorphic to the Fano matroid F_7, or has a minor that is l-isomorphic to N_8.*

Proof. If M has seven elements, then it must be isomorphic to F_7 since membership in $\mathcal{M}$ rules out F_7^*. So assume that M has at least eight elements. If M does not have any F_7^* minors, then by the splitter result of Lemma (11.3.19), M has an F_7 minor and is 2-separable, or is regular. The assumptions made here rule out these cases. Thus, M has an F_7^* minor, say N. Obviously, the minor N does not contain the element l.

According to Lemma (5.2.4), M has a connected N' minor that is a 1-element extension of N by l. If in N' the element l is parallel to or in series with some other element, then clearly l is part of an F_7^* minor of N' and thus of M, a contradiction. Thus, N' is 3-connected. Now F_7^* has no 3-connected 1-element expansion, so N' is obtained from N by addition of l. Routine calculations confirm that up to l-isomorphism, there is just one addition case that does not have an F_7^* minor with l. That case is represented by the matrix of (13.2.5). □

From now on, we assume that M is a 3-connected nonregular matroid of $\mathcal{M}$ with at least eight elements. By Lemma (13.2.6), M has a minor that is l-isomorphic to N_8. For one such minor, we will exhibit a 3-separation that corresponds to the 3-separation $(X_1 \cup Y_1, X_2 \cup Y_2)$ of N_8 under one of the l-isomorphisms, and that induces a 3-separation of M. For this task, we invoke Corollary (6.3.25) and Theorem (6.3.28). Recall that Corollary (6.3.25) contains sufficient conditions for induced k-separations, while Theorem (6.3.28) extends those conditions to the case where M has a special subset L of elements. For the case at hand, we define L to be the set containing just l. We combine Corollary (6.3.25) and Theorem (6.3.28) for this special set L to the following theorem.

(13.2.7) Theorem. *Suppose a 3-connected N given by the matrix B^N*

(13.2.8)

$$
B^N = \begin{array}{c|c|c|}
 & Y_1 & Y_2 \\
\hline
X_1 & A^1 & 0 \\
\hline
X_2 & D & A^2 \\
\hline
\end{array}
$$

Partitioned matrix B^N for N

is in $\mathcal{M}$ and has $l \in (X_2 \cup Y_2)$. Assume that $(X_1 \cup Y_1, X_2 \cup Y_2)$ is a k-separation of N. Furthermore, assume that $N/(X_2 \cup Y_2)$ has no loops and that $N \backslash (X_2 \cup Y_2)$ has no coloops. Finally, assume for every 3-connected 1-element extension of N in $\mathcal{M}$, say by an element z, that the pair $(X_1 \cup Y_1, X_2 \cup Y_2 \cup \{z\})$ is a k-separation of that extension. Then for any 3-connected matroid $M \in \mathcal{M}$ with a minor l-isomorphic to N, the following holds. Any k-separation of any such minor that corresponds to $(X_1 \cup Y_1, X_2 \cup Y_2)$ of N under one of the l-isomorphisms induces a k-separation of M.

We use Theorem (13.2.7) plus the recursive decomposition scheme explained in Section 10.5 to establish 3-separations for the nonregular 3-connected matroids of $\mathcal{M}$. We do not repeat here details of the decomposition scheme, so the reader may want to review that material before proceeding. The first iteration of that scheme is effectively accomplished by the following lemma.

(13.2.9) Lemma. *Let M be a 3-connected matroid of $\mathcal{M}$ with N_8 as minor. Then the 3-separation $(X_1 \cup Y_1, X_2 \cup Y_2)$ of N_8 as given by (13.2.5) induces a 3-separation of M, or M has a minor that is l-isomorphic to the matroid N_9 represented by B^9 below.*

(13.2.10)

$$B^9 = \begin{array}{c|ccc|cc} & \multicolumn{3}{c|}{Y_1} & \multicolumn{2}{c}{Y_2} \\ & u & v & w & l & z \\ \hline X_1\ d & 1 & 1 & 1 & 0 & 1 \\ \hline & a & 1 & 0 & 1 & 1 & 1 \\ X_2\ b & 1 & 1 & 0 & 1 & 0 \\ & c & 0 & 1 & 1 & 0 & 0 \\ \end{array}$$

Matrix B^9 for the matroid N_9

Proof. We assume absence of the specified N_9 minors in M, and apply Theorem (13.2.7). The matroid N_8 plays the role of N of that theorem. Since in N_8 we have $Y_2 = \{l\}$, we have $l \in (X_2 \cup Y_2)$. It also is easily checked that N_8 satisfies the conditions of Theorem (13.2.7) involving $N/(X_2 \cup Y_2)$ and $N \backslash (X_1 \cup Y_1)$. For verification of the remaining conditions of the theorem, we must compute the 3-connected 1-element extensions N' of N_8, say by an element z, such that in N' the pair $(X_1 \cup Y_1, X_2 \cup Y_2 \cup \{z\})$ is not a 3-separation. We are done by contradiction once we show that any such case of N' is l-isomorphic to N_9 or is not in $\mathcal{M}$.

We first consider the addition of z. For this, we adjoin to B^8 of (13.2.5) a column $[g/h]$ representing the element z. The partition of $[g/h]$ corresponds to that of B^8. By assumption, $(X_1 \cup Y_1, X_2 \cup Y_2 \cup \{z\})$ is not a 3-separation of N'. Thus, we must have $g = 1$. By the 3-connectedness of N', the vector h must have one or three 1s. Routine checking confirms that all such instances are l-isomorphic to N_9. We turn to the expansion by z. We adjoin to B^8 of (13.2.5) a row $[e \mid f]$ representing the element z. The partition of $[e \mid f]$ corresponds to that of B^8. The conditions imposed on z demand that e has one or three 1s and that $f = 1$. In each case, the element l of N' can be placed into an F_7^* minor, and thus is not in $\mathcal{M}$. $\qquad\square$

In the terminology of the decomposition scheme of Section 10.5, we have just completed the first iteration. We begin the second iteration by analyzing N_9 for 3-separations. That matroid has several such separations, some of them useful for our purposes, and others not. Indeed, in the initial investigation into the class $\mathcal{M}$, a 3-separation of N_9 was selected that led to a large number of iterations of the decomposition scheme. But eventually a better choice was found. It is the 3-separation $(\{a, b, d, u\}, \{c, l, v, w, z\})$. It can be displayed in the accustomed format once we compute the representation matrix for N_9 corresponding to the base $\{b, c, v, w\}$. That matrix for N_9 is as follows.

(13.2.11)

$$
\begin{array}{c}
\quad\quad a\; d\; u\; l\; z \\
\begin{array}{c} b \\ w \\ v \\ c \end{array}
\left[
\begin{array}{ccc|cc}
1 & 1 & 1 & 0 & 0 \\
1 & 0 & 1 & 1 & 1 \\
1 & 1 & 0 & 1 & 0 \\
0 & 1 & 1 & 0 & 1
\end{array}
\right]
\end{array}
$$

Matrix for N_9 corresponding to base $\{b, c, v, w\}$

The second iteration of the decomposition scheme is accomplished by the following lemma.

(13.2.12) Lemma. *Let M be a 3-connected matroid of $\mathcal{M}$ with N_9 as minor. Then the 3-separation $(\{a, b, d, u\}, \{c, l, v, w, z\})$ of N_9 induces a 3-separation of M.*

Proof. The arguments are virtually identical to those for Lemma (13.2.9), except that this time each 3-connected 1-element extension satisfies the conditions of Theorem (13.2.7) or is not in $\mathcal{M}$. We leave it to the reader to fill in the details. ☐

Lemmas (13.2.6), (13.2.9), and (13.2.12) imply the following theorem, which thus is the result of two iterations of the decomposition scheme.

(13.2.13) Theorem. *Let M be a 3-connected nonregular matroid of $\mathcal{M}$ with at least eight elements. Then M has a minor N with the following properties. N contains the element l and is l-isomorphic to N_8, and a 3-separation of N corresponding to the 3-separation $(X_1 \cup Y_1, X_2 \cup Y_2)$ of N_8 under one of the l-isomorphisms induces a 3-separation of M.*

Proof. By Lemma (13.2.6), M has a minor that contains l and that is l-isomorphic to the matroid N_8. We may suppose that N_8 itself is that minor. Then by Lemma (13.2.9), M has a 3-separation induced by the 3-separation $(X_1 \cup Y_1, X_2 \cup Y_2)$ of N_8, or M has a minor with l that is l-isomorphic to the matroid N_9. In the former case, we are done. In the latter case, we apply Lemma (13.2.12). Accordingly, M has a 3-separation induced by the 3-separation $(\{a, b, d, u\}, \{c, l, v, w, z\})$ of N_9. From the matrix of (13.2.11) for N_9, we now delete the last column. Evidently, up to indices other than l, a matrix for N_8 results. The matroid N represented by that matrix is thus l-isomorphic to N_8. Furthermore, by the derivation of N from N_9, the 3-separation $(\{a, b, d, u\}, \{c, l, v, w\})$ of N also induces the 3-separation of M derived earlier from N_9. But that 3-separation of N corresponds to the 3-separation $(X_1 \cup Y_1, X_2 \cup Y_2)$ of N_8. Thus, the case involving N_9 also leads to the desired conclusion.   □

From 3-Separation to Delta-Sum

The reader probably anticipates that the conversion of the just-proved 3-separation of M to the Δ-sum decomposition claimed in Theorem (13.2.4) is straightforward. Unfortunately, the situation is not quite as simple. In particular, we must obtain some insight into the position of l relative to the 3-separation of M before we can proceed to the Δ-sum decomposition. We obtain this insight next.

By (13.2.5), the minor $N = N_8/c$ of N_8 is l-isomorphic to the Fano matroid F_7 with l. Indeed, the groundset of N is $\{a, b, d, l, u, v, w\}$, and the 3-separation $(\{d, u, v, w\}, \{a, b, l\})$ of N induces the same 3-separation in M that N_8 induces.

From now on, we do not need the matroid N_8 any more. Thus, we can switch notation, and can utilize the sets X_1, X_2, Y_1, and Y_2 so far employed for N_8 to denote induced 3-separations of M. We do this next in a representation matrix B of M. That matrix displays the just-defined F_7 minor N, with indices a, b, d, u, v, w, l and 3-separation $(\{d, u, v, w\}, \{a, b, l\})$.

(13.2.14)

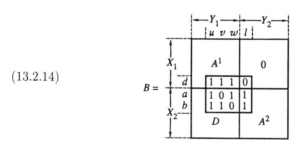

Matrix B of M displaying N
and induced 3-separation $(X_1 \cup Y_1, X_2 \cup Y_2)$

To capture the role of l in the 3-separation $(X_1 \cup Y_1, X_2 \cup Y_2)$ of M, we define l to *straddle* the 3-separation if a shift of l from Y_2 to Y_1 results in another 3-separation of M. The next theorem says that l must straddle the 3-separation. This fact will be essential for the Δ-sum decomposition to come.

(13.2.15) Theorem. *Let a 3-connected matroid $M \in \mathcal{M}$ be represented by the matrix B of (13.2.14), where $(X_1 \cup Y_1, X_2 \cup Y_2)$ is a 3-separation of M. Then the element l straddles that 3-separation.*

Proof. Evidently, the element l straddles the 3-separation of M if and only if column l of the submatrix A^2 of B and column u of the submatrix

D of B are parallel. The former column vector we call g^l, and the latter one g^u. Note that g^u is the sum (in GF(2)) of the columns v and w of D.

Assume that l does not straddle the 3-separation. Thus, g^l and g^u are not parallel. Equivalently, one of these vectors contains a 1 in a row $x \in X_2$ where the other one has a 0. Thus, two cases are possible.

In the first case, $g^l_x = 0$ and $g^u_x = 1$. Then the submatrix of B indexed by a, b, d, x and l, u, v, w is either

(13.2.16)

$$
\begin{array}{c|ccc|c}
 & u & v & w & l \\
\hline
d & 1 & 1 & 1 & 0 \\
a & 1 & 0 & 1 & 1 \\
b & 1 & 1 & 0 & 1 \\
x & 1 & 0 & 1 & 0 \\
\end{array}
$$

Submatrix of B

or is obtained from the matrix of (13.2.16) by exchanging the row indices a and b, and the column indices v and w. Thus, we may assume that the situation depicted by (13.2.16) is at hand. Let $\overline{M}$ be the minor represented by the matrix of (13.2.16). The minor $\overline{M} \backslash b$ of $\overline{M}$ turns out to be an F_7^* minor with l, and thus $M \notin \mathcal{M}$, a contradiction.

In the second case, $g^l_x = 1$ and $g^u_x = 0$. If row x of D is nonzero, or equivalently, if row x has 1s in both columns v and w, then rows a, b, d, x and columns l, v, w prove l to be in an F_7^* minor, a contradiction. Thus, row x of D is zero. By this fact and by the completion of the first case, we can narrow down the second case as follows. There is a nonempty subset $\overline{\overline{X}}_2 \subseteq X_2$ such that for all $x \in \overline{\overline{X}}_2$, we have $g^l_x = 1$ while row x of D is equal to 0. Furthermore, for all $x \in (X_2 - \overline{\overline{X}}_2)$, we have $g^l_x = g^u_x$.

Let $\overline{X}_2$ be the index set of the nonzero rows of D. By the above definition of $\overline{\overline{X}}_2$, we know $\overline{X}_2 \subseteq (X_2 - \overline{\overline{X}}_2)$. We now redraw B of (13.2.14) by enlarging and repartitioning the submatrices A^2 and D. The revised B is as follows.

(13.2.17)

Repartitioned matrix B

The partition of the rows of B of (13.2.17) is induced by the sets $\overline{X}_2$ and $\overline{\overline{X}}_2$ introduced above. The partition of Y_2 into the sets l, $\overline{Y}_2$, and $\overline{\overline{Y}}_2$ is based on the information given in the submatrices indexed by $\overline{X}_2$, $\overline{Y}_2$ and $\overline{X}_2$, $\overline{\overline{Y}}_2$. Thus, each column of the former submatrix is zero or parallel to the subvector $\overline{g}$ of column l. By the definition of $\overline{X}_2$, the subvector $\overline{g}$ also occurs in D, as shown.

Suppose $\overline{\overline{Y}}_2 = \emptyset$. Then it is readily checked via B of (13.2.17) that $X_1 \cup Y_1 \cup \overline{X}_2$ is one side of a 2-separation of the 3-connected M, a contradiction. Thus, $\overline{\overline{Y}}_2 \neq \emptyset$.

Derive a matrix $\overline{B}$ from B by deleting all 1s in positions (x, y) where $x \in \overline{X}_2$ and $y \in (\{l\} \cup \overline{Y}_2)$. Consider the bipartite graph $\mathrm{BG}(\overline{B})$. If that graph does not have a path from l to $\overline{\overline{Y}}_2$, then arguments analogous to those of the proof of Lemma (5.2.11), prove M to be 2-separable. Thus, a path exists from l to some $y \in \overline{\overline{Y}}_2$. Because of path-shortening pivots, we may assume that the path has exactly two arcs. Thus, we can extract from $\overline{B}$ the following submatrix $\overline{\overline{B}}$,

(13.2.18)

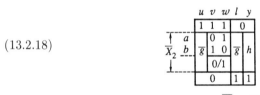

Submatrix $\overline{\overline{B}}$ of $\overline{B}$

where $h \neq 0$ and $h \neq \overline{g}$. Reduce $\overline{X}_2$ to a minimal set containing a and b such that the just-mentioned condition is still satisfied by the correspondingly reduced $\overline{g}$ and h. Thus, $|\overline{X}_2| = 2$ or 3. Since $\overline{g}$ has two 1s in rows a and b, we have $|\overline{X}_2| = 3$ if and only if h has either two or no 1's in rows a and b. If h is a unit vector, then a pivot on the 1 of h plus deletion of the pivot column produces a previously resolved instance where $g_x^l \neq g_x^u$ and where a row x of D is nonzero. Otherwise, $|\overline{X}_2| = 3$, and h has two 1s in rows a and b. In each one of the three possible cases, a pivot in h plus deletion of the pivot column and of one row proves that l is in an F_7^* minor, a contradiction. $\quad\Box$

As described in Section 8.5, a Δ-sum decomposition is derived from a 3-sum decomposition via a certain ΔY exchange. We first carry out the 3-sum decomposition of M using the matrix of (13.2.14), except that this time we ignore the column index u and the details of that column. We are guided in that decomposition by the matrices of (8.3.10) and (8.3.11). Below, we list the matrix B for M, as well as the representation matrices B^1 and $\tilde{B}^2$ for the components M_1 and $\tilde{M}_2$ of the 3-sum decomposition of M.

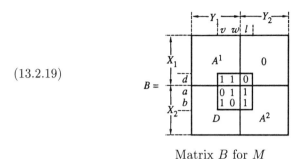

(13.2.19)

Matrix B for M

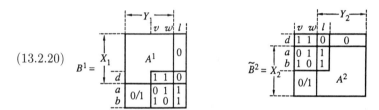

(13.2.20)

Matrices B^1 and $\tilde{B}^2$ of 3-sum decomposition of M

The Δ-sum decomposition has M_1 and a second matroid, say M_2, as components, where M_2 is derived from $\tilde{M}_2$ by the exchange of the triad $\{d, v, w\}$ with a triangle, say Z, according to the rule of (4.4.5). For the particular case at hand, we modify that exchange using the following arguments.

By Theorem (13.2.15), the element l straddles the 3-separation $(X_1 \cup Y_1, X_2 \cup Y_2)$ of M. Thus, in B, the columns v and w of the submatrix D span the column l of the submatrix A^2. This implies that one element of M_2 in the triangle Z is parallel to l. Thus, we can delete that element from M_2, and still have enough information to compose M_1 and the reduced M_2 to M again. Accordingly, we redefine M_2 to be that reduced matroid. From $\tilde{B}^2$ of (13.2.20), we see that the just-defined matroid M_2 may be taken to be $\tilde{M}_2/d$. Note that M_2 may have just six elements, in which case it is isomorphic to the wheel matroid $M(W_3)$.

For later reference, we include the matrices B^1 and B^2 of the Δ-sum decomposition of M into M_1 and M_2 in (13.2.21) below. We have simplified the matrices by omitting indices that are of no consequence for the subsequent discussion. With B^1 and B^2 of (13.2.21), one readily confirms the statements (i)–(iv) of Theorem (13.2.4) about certain circuits of M, M_1 and M_2. We leave the simple calculations to the reader. Thus, we have completed the proof of that theorem.

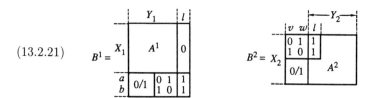

$(13.2.21)$

Matrices B^1 and B^2 of Δ-sum decomposition of M

In the next section, we use Theorems (13.2.1) and (13.2.4) to characterize the max-flow min-cut matroids with special element l by the exclusion of U_4^2 minors and of F_7^* minors with l.

13.3 Characterization of Max-Flow Min-Cut Matroids

Recall from the introduction that a max-flow min-cut matroid is a connected matroid with a special element l such that for any nonnegative integral edge capacity vector h, we have $\max F = \min h(D)$. Here $\max F$ is the optimal value of the max flow problem, where one must find a maximum number of cycles with l such that each element $e \neq l$ is contained in at most h_e of these cycles. Then $\min h(D)$ is the optimal value of the min cut problem, where one must determine a cocircuit D with l such that $h(D)$, defined to be the sum of the values h_e of the elements $e \in (D - l)$, is minimum.

In this section, we show that a connected matroid with a special element l has the max-flow min-cut property if and only if M has no U_4^2 minors and has no F_7^* minors with l. By Theorem (3.5.2), a matroid is binary if and only if it has no U_4^2 minors. Thus, the claimed characterization is equivalent to the statement that $\mathcal{M}$, the class of connected binary matroids having no F_7^* minors with l, is the class of connected max-flow min-cut matroids.

The characterization is summarized by the following theorem and corollary.

(13.3.1) Theorem. *A connected matroid M with a special element l is a max-flow min-cut matroid if and only if M has no U_4^2 minors and has no F_7^* minors with l.*

(13.3.2) Corollary. *$\mathcal{M}$ is the class of connected max-flow min-cut matroids.*

We begin the proof of Theorem (13.3.1) by noting that the result is trivially true for matroids whose groundset contains only the special element l. Thus, we assume from now on that all matroids M examined below have at least one additional element besides l. We first formulate the max flow problem and the min cut problem as linear programs with an integrality condition.

Max Flow Problem

Let M be a connected but not necessarily binary matroid whose groundset E contains a special element l. We construct the following matrix H from M. The rows of H correspond to the elements of E other than l, and the columns to the circuits C of M containing l. The entry of H in row e and column C is then 1 if e occurs in circuit C, and 0 otherwise. Consider the following linear program involving H and an arbitrary nonnegative integral vector h. Recall that $\underline{1}$ is a vector containing only 1s. All vectors are assumed to be column vectors of appropriate dimension. Below, the abbreviation "s. t." stands for "subject to."

$$
(13.3.3) \qquad
\begin{aligned}
\max \quad & \underline{1}^t \cdot r \\
\text{s. t.} \quad & H \cdot r \le h \\
& r \ge 0
\end{aligned}
$$

We call this problem the *fractional max flow problem*, since it becomes the max flow problem with capacity vector h when we require the solution vector r to be integral. Indeed, for any integral solution vector r, the entry in position C specifies the number of times the circuit C is to be selected.

Min Cut Problem

The linear programming dual of (13.3.3) is

$$
(13.3.4) \qquad
\begin{aligned}
\min \quad & h^t \cdot s \\
\text{s. t.} \quad & H^t \cdot s \ge \underline{1} \\
& s \ge 0
\end{aligned}
$$

We call (13.3.4) the *fractional min cut problem*. We justify the term next. Suppose that we require the solution vector s of (13.3.4) to be integral. According to the constraints of (13.3.4), any optimal solution s can then be assumed to be a $\{0, 1\}$ vector. We do this from now on when we impose the integrality condition on s of (13.3.4). By the constraint $H^t \cdot s \ge \underline{1}$ of (13.3.4), the vector s is thus the incidence vector of a subset Z of E that intersects every circuit encoded by a column of H. Let $C^* = Z \cup \{l\}$. We claim that C^* contains a cocircuit of M with l. For a proof, select a base X of M that contains l and that avoids the set Z as much as possible. Collect in a set $\overline{Y}$ each element $y \in (E - X)$ whose fundamental circuit C_y with X contains l. By this definition, $\overline{Y} \cup \{l\}$ is the fundamental cocircuit of M that l forms with $(E - X)$. Furthermore, by the derivation of Z from the vector s, Z intersects, for each $y \in \overline{Y}$, the fundamental circuit C_y in some element z different from l. Suppose no such z is equal to y. Then for some $z \in Z$, $(X - \{z\}) \cup \{y\}$ is a base, which proves that X does not avoid Z as much as possible, a contradiction. Thus, $\overline{Y} \subseteq Z$, and C^* contains a cocircuit of M with l as claimed. Since the vector h is nonnegative, we may assume C^* to be that cocircuit.

By Lemma (3.4.25), any circuit of M with l and any cocircuit of M with l cannot intersect just on the element l. Thus, any cocircuit of M with l is a candidate for producing an integral solution for (13.3.4), and the best such candidate corresponds to a $\{0, 1\}$ vector s solving (13.3.4) with integrality condition. Thus, that problem represents the min cut problem, and we are justified in calling (13.3.4), without the integrality requirement, the fractional min cut problem for M.

Necessity of Excluded Minors Condition

We are ready to show the "only if" part of Theorem (13.3.1), which says that a connected max-flow min-cut matroid cannot have U_4^2 minors, or F_7^* minors with l. The proof is based on two reductions. The first one is accomplished by the following lemma and corollary. We omit the proof of the lemma, since it is just a particular version of the so-called duality theorem of linear programming.

(13.3.5) Lemma. *For any feasible vectors r and s of (13.3.3) and (13.3.4), respectively, we have $\underline{1}^t \cdot r \leq h \cdot s$, with equality holding if and only if both vectors r and s are optimal.*

(13.3.6) Corollary. *If every optimal solution vector r of (13.3.3) is nonintegral, then the matroid defining that problem is not a max-flow min-cut matroid.*

Proof. Let r and s be optimal solutions for (13.3.3) and (13.3.4), respectively. Because of the assumptions and Lemma (13.3.5), we must have $\max F < \underline{1}^t \cdot r = h^t \cdot s \leq h(D)$. Thus, $\max F \neq \min h(D)$, and the matroid defining (13.3.3) cannot be a max-flow min-cut matroid.   □

For the second reduction, we rewrite Lemma (5.4.3) to get the following result.

(13.3.7) Lemma. *Let M be a connected matroid with an element l. If M has U_4^2 minors, then M has a U_4^2 minor with l.*

By the preceding reductions, we may prove the necessity of the excluded minors condition of Theorem (13.3.1) by producing, for any connected matroid with a U_4^2 minor with l or with an F_7^* minor with l, a nonnegative integral capacity vector h such that any optimal solution vector for (13.3.3) is nonintegral. The next lemma says that such a vector h can always be found.

(13.3.8) Lemma. *Let a connected matroid M with an element l have a U_4^2 minor with l or an F_7^* minor with l. Then there is a nonnegative integral vector h such that all optimal solution vectors r for (13.3.3) with that h are nonintegral.*

Proof. Let N be a minor of M with l and isomorphic to U_4^2 or F_7^*. By the discussion of Section 3.4, we may assume that $N = M/U \backslash W$, where U does not contain any cycle of M, and W does not contain any cocycle. Let M have n elements. Define the vector h by $h_e = 1$ for each element of N except l, $h_e = n$ for each $e \in U$, and $h_e = 0$ for each $e \in W$. Routine calculations show that any optimal solution vector r for (13.3.3) with this h is nonintegral. We leave the verification to the reader. $\square$

Sufficiency of Excluded Minors Condition

We prove the "if" part of Theorem (13.3.1), which says that any connected matroid M with a special element l, without U_4^2 minors, and without F_7^* minors with l is a max-flow min-cut matroid.

We invoke a result of polyhedral combinatorics to simplify the proof. That result concerns a certain integrality property of linear programs called *total dual integrality* and is due to Edmonds and Giles.

(13.3.9) Theorem. *Suppose the matrix A and the vector c of the linear program*

$$\begin{aligned} \max \quad & c^t \cdot f \\ \text{s. t.} \quad & A \cdot f \leq b \\ & f \geq 0 \end{aligned}$$

(13.3.10)

are integral and permit a feasible solution for the dual linear program of (13.3.10), which is

$$\begin{aligned} \min \quad & b^t \cdot g \\ \text{s. t.} \quad & A^t \cdot g \geq c \\ & g \geq 0 \end{aligned}$$

(13.3.11)

Furthermore, assume that the linear program (13.3.10) has an integral optimal solution for every integral vector b for which it has a feasible solution. Then all extreme point solutions for the dual linear program (13.3.11) are integral.

We apply Theorem (13.3.9) as follows. We view the fractional max flow problem (13.3.3) as an instance of (13.3.10). Then the linear programming dual of (13.3.3), which is the fractional min cut problem (13.3.4), is the problem (13.3.11). Note that (13.3.3) has a feasible solution if and only if $h \geq 0$. Suppose for a given M with l, and for any integral $h \geq 0$, we can show that the fractional max flow problem has an optimal solution vector that is integral. By Theorem (13.3.9), the fractional min cut problem then has, for any integral $h \geq 0$, an optimal solution that is integral. By Lemma (13.3.5), the optimal objective function values of the two problems agree, and thus we have $\max F = \min h(D)$. We conclude that M is a max-flow min-cut matroid.

By the above arguments, we may establish the max-flow min-cut property for M by showing that the fractional max flow problem (13.3.3) has, for any integral $h \geq 0$, an optimal solution vector that is integral. We divide the latter task into two parts. In the first one, we assume M to be regular. In the second one, M is assumed to be nonregular.

Regular Matroid Case

We begin with the first part. So assume that a connected matroid M on a set E and with special element l is regular. Let h be any nonnegative integral capacity vector. By Theorem (9.2.9), M has a real totally unimodular representation matrix B. Adjoin an identity matrix I to B, getting a real matrix $A = [I \mid B]$ whose columns are indexed by E. Consider the following linear program, where the solution vector f is a real column vector indexed by E.

$$
\begin{aligned}
\max \quad & f_l \\
\text{s. t.} \quad & A \cdot f = 0 \\
& f \geq -h \\
& f \leq h
\end{aligned}
$$

(13.3.12)

The linear program is clearly bounded and has the zero vector as solution. Thus, there is an optimal solution vector, say $\tilde{f}$. By linear programming results, $\tilde{f}$ may be assumed to be the solution to a system of equations of the form $\tilde{A} \cdot f = \tilde{h}$, where $\tilde{A}$ is a square nonsingular matrix consisting of some rows of A plus some unit vector rows, and where $\tilde{h}$ contains zeros and some entries of h and $-h$. Evidently, $\tilde{A}$ is totally unimodular. By Lemma (9.2.2), $\tilde{A}^{-1}$ is totally unimodular as well, and thus $\tilde{f} = \tilde{A}^{-1} \cdot \tilde{h}$ is integral. We may suppose $\tilde{f} \geq 0$ since any negative entry $\tilde{f}_e$ of $\tilde{f}$ can be transformed to a positive one by a clearly permissible scaling of column e of A by -1. Thus, $\tilde{f}$ is nonnegative and integral.

Until stated otherwise, we assume that $\tilde{f}_l > 0$. From $\tilde{f}$, we derive $\tilde{f}_l$ circuits that we later show to correspond to an optimal solution of the fractional max flow problem (13.3.3). We obtain these circuits by repeatedly solving a certain linear inequality system. In the first iteration, that system is

$$A \cdot g = 0;$$

(13.3.13)
$$g_e \geq 0; \text{ if } e \neq l \text{ and } \tilde{f}_e > 0$$
$$g_e = 0; \text{ if } e \neq l \text{ and } \tilde{f}_e = 0$$
$$g_l = 1$$

The vector $g = \tilde{f}/\tilde{f}_l$ is feasible for (13.3.13). Arguing analogously to the case involving (13.3.12), we are thus assured of an integral nonnegative solution $\tilde{g}$ for (13.3.13). Indeed, this time the vector playing the role of the earlier $\tilde{h}$ is a unit vector, and $\tilde{g}$ may be taken to be the characteristic vector of a circuit of M with l.

We now replace $\tilde{f}$ in (13.3.13) by $\tilde{f}' = \tilde{f} - \tilde{g}$ and deduce from the so-modified system (13.3.13) another circuit of M with l. We repeat this iterative derivation of circuits until $\tilde{f}_l$ has been reduced to 0. At that time, we have $\tilde{f}_l$ circuits, each containing the element l. By the derivation, any element e of M occurs in at most h_e of these circuits.

Now assume that $\tilde{f}_l = 0$. Then we do not select any circuit at all. The discussion to follow applies to this case as well as to the one where $\tilde{f}_l > 0$ and where we do select circuits.

Define a vector $\tilde{r}$ from the circuits of M with l that we just have selected by setting $\tilde{r}_C$ equal to the number of times the circuit C occurs in the collection. Thus, $\underline{1}^t \cdot \tilde{r} = \tilde{f}_l$. We claim that $\tilde{r}$ solves the fractional max flow problem (13.3.3). By the construction, $\tilde{r}$ is feasible for (13.3.3). We prove optimality as follows. Take any feasible solution r for (13.3.3). From each nonzero entry r_C of r, we derive a nonzero solution f^C for the equation $A \cdot f = 0$ as follows. First, we obtain a nontrivial $\{0, \pm 1\}$ solution for $A \cdot f = 0$ such that the support vector of that solution is the characteristic vector of C. This can clearly be done, since A is totally unimodular. Next, we scale that $\{0, \pm 1\}$ solution by r_C to get a vector f^C. We define f to be the sum of the vectors f^C if $f_l \neq 0$, and to be the zero vector otherwise. By the derivation, f is a solution for (13.3.12), with objective function value $f_l = \underline{1}^t \cdot r$. Recall that the vector $\tilde{f}$ is optimal for (13.3.12). Thus, $\underline{1}^t \cdot \tilde{r} = \tilde{f}_l \geq f_l = \underline{1}^t \cdot r$, which proves $\tilde{r}$ to be optimal for (13.3.3). We conclude that M is a max-flow min-cut matroid.

Nonregular Matroid Case

We turn to the second part, where M is not regular. We divide the proof into three subcases. First, we assume M to be F_7 with l; second, to have at least eight elements and to be 2-separable; and third, to have at least eight elements and to be 3-connected.

Fano Matroid Subcase

Routine calculations prove that F_7 with l is a max-flow min-cut matroid. To assist the reader with the checking, we list the matrix H of (13.3.3) for F_7 with l below, but otherwise omit all details. We have partitioned H to exhibit its structure.

(13.3.14)

$$
\begin{array}{ccc|ccc|c}
1 & 0 & 0 & 1 & 0 & 0 & 1 \\
0 & 1 & 0 & 0 & 1 & 0 & 1 \\
0 & 0 & 1 & 0 & 0 & 1 & 1 \\
\hline
1 & 0 & 0 & 0 & 1 & 1 & 0 \\
0 & 1 & 0 & 1 & 0 & 1 & 0 \\
0 & 0 & 1 & 1 & 1 & 0 & 0 \\
\end{array}
$$

Matrix H of (13.3.3) for F_7 with l

For both the second and third subcase of the proof, we introduce the following terminology in connection with the fractional max flow problem (13.3.3). For any matroid under consideration, we define a *collection of weighted circuits* to be a collection of the circuits with l where a real nonnegative weight has been assigned to each circuit. We say that a collection of weighted circuits is *feasible* or *optimal* if the vector r with the weights of the circuits as entries is feasible or optimal for (13.3.3). The collection has *flow value* α if the corresponding objective function value of (13.3.3) is α. Finally, we say that a collection of weighted circuits *uses* an edge e α times if the weights of the circuits containing the element e sum to α.

The arguments of the second and third subcase use induction. The smallest instance involves the already treated F_7. Thus, for some $n \geq 7$, we assume the desired conclusion for matroids with at most n elements and take M to have $n+1$ elements. By the excluded minors condition, we know that M is in $\mathcal{M}$. Thus, the decomposition results of Section 13.2 apply.

2-Sum Decomposition Subcase

We use the decomposition Theorem (13.2.1), which we repeat below.

(13.3.15) Theorem. *Any 2-separable matroid $M \in \mathcal{M}$ on a set E has a 2-sum decomposition where both components M_1 and M_2 are connected minors of M, contain l, and thus are in $\mathcal{M}$. In addition, M_2 has an element $y \notin l$ so that any set $C \subseteq E$ is a circuit of M with l if and only if (i) or (ii) below holds.*

  (i) *C is a circuit of M_2 with l but not y.*
  (ii) *$C = (C_1 - \{l\}) \cup (C_2 - \{y\})$ where C_1 is a circuit of M_1 with l, and where C_2 is a circuit of M_2 with both l and y.*

The notation below is that of Theorem (13.3.15). In particular, M_1 and M_2 are the two components of a 2-sum decomposition of M.

First we find a collection of weighted circuits $\mathcal{C}$ that solves the fractional max flow problem for M for an arbitrary nonnegative integral vector h. Using parts (i) and (ii) of Theorem (13.3.15), we derive from $\mathcal{C}$ two collections $\mathcal{C}_1$ and $\mathcal{C}_2$ of weighted circuits for M_1 and M_2, as follows. Let C be a circuit of $\mathcal{C}$ of M with positive weight. By Theorem (13.3.15), C is a circuit of M_2 with l but not y, or $C = (C_1 - \{l\}) \cup (C_2 - \{y\})$ where C_1 is a circuit of M_1 with l, and where C_2 is a circuit of M_2 with both l and y. In the first case, we assign the weight of C to that circuit of M_2. In the second case, we assign the weight of C to both the circuit C_1 of M_1 and the circuit C_2 of M_2. By this construction, $\mathcal{C}_2$ has the same flow value as $\mathcal{C}$. Furthermore, $\mathcal{C}_1$ uses the element l of M_1 just as often as $\mathcal{C}_2$ of M_2 uses the element y, say α times.

Round up α to the next integer, getting, say, α'. Declare α' to be the capacity of the element y of M_2. Except for the element l of M_1 and for the elements l and y of M_2, assign the entries of the capacity vector h as capacities to the elements of M_1 and M_2.

Solve the fractional max flow problem for M_1. By induction and the earlier proof for regular matroids, we may assume that a collection $\mathcal{C}_1'$ of circuits with integral weights is found. Since $\mathcal{C}_1$ produces a feasible solution with flow value α for that problem, the new collection $\mathcal{C}_1'$ must have a flow value of at least α'.

Solve the fractional max flow problem for M_2. Once more, we may suppose that a collection $\mathcal{C}_2'$ of circuits with integral weights is found. The flow value of $\mathcal{C}_2'$ must be at least as large as that of $\mathcal{C}_2$, which in turn we know to be equal to that of $\mathcal{C}$.

Suppose $\mathcal{C}_2'$ uses the element y α'' times, for some $\alpha'' \leq \alpha'$. Derive a collection $\mathcal{C}_1''$ from $\mathcal{C}_1'$ by arbitrarily reducing weights of some circuits until the element l is used exactly α'' times. We combine $\mathcal{C}_1''$ with $\mathcal{C}_2'$ via Theorem (13.3.15) to a collection of circuits for M with integral weights and with flow value at least as large as that of $\mathcal{C}$, as desired.

Delta-Sum Decomposition Subcase

We turn to the final subcase, where the nonregular M has at least eight elements and is 3-connected. We use Theorem (13.2.4), which we list again below.

(13.3.16) Theorem. *Any 3-connected nonregular matroid $M \in \mathcal{M}$ on a set E is isomorphic to the Fano matroid F_7, or has a Δ-sum decomposition where the components M_1 and M_2 are connected minors of M, contain l, and thus are in $\mathcal{M}$. In the Δ-sum decomposition, both connecting triangles of M_1 and M_2 contain l. Let the remaining elements of the connecting* triangle of M_1 (resp. M_2) be a and b (resp. v and w). Then any set $C \subseteq E$ is a circuit of M with l if and only if (i), (ii), (iii), or (iv) below holds.

(i) C is a circuit C_1 of M_1 with l but without a and b.

(ii) C is a circuit C_2 of M_2 with l but without v and w.

(iii) $C = (C_a - \{a\}) \cup (C_v - \{v\})$ where C_a is a circuit of M_1 with l and a but without b, and where C_v is a circuit of M_2 with l and v but without w.

(iv) $C = (C_b - \{b\}) \cup (C_w - \{w\})$ where C_b is a circuit of M_1 with l and b but without a, and where C_w is a circuit of M_2 with l and w but without v.

The proof to come is similar to that for the 2-sum case. But there are subtle differences, as we shall see. We start again with a collection $\mathcal{C}$ of weighted circuits for M that solves the fractional max flow problem for an arbitrary nonnegative integral vector h. By Theorem (13.3.16), M has a Δ-sum decomposition into two matroids M_1 and M_2. We derive from $\mathcal{C}$ two collections $\mathcal{C}_1$ and $\mathcal{C}_2$ of weighted circuits for M_1 and M_2 by processing each circuit C of $\mathcal{C}$ as follows.

Suppose C is a circuit of case (i) or (ii) of Theorem (13.3.3). Then C occurs in M_1 or M_2, and we assign the same weight to that circuit in the corresponding collection $\mathcal{C}_1$ or $\mathcal{C}_2$. Suppose C is a circuit of case (iii) of Theorem (13.3.3). Thus, $C = (C_a - \{a\}) \cup (C_v - \{v\})$, where C_a is a circuit of M_1 with l and a but without b, and C_v is a circuit of M_2 with l and v but without w. Then we assign to both C_a and C_v the weight of C. Finally, case (iv) of Theorem (13.3.3) is handled analogously to case (iii).

Suppose the collection $\mathcal{C}_1$ uses the element a (resp. b) α times (resp. β times). By the above derivation, $\mathcal{C}_2$ uses the element v (resp. element w) also α times (resp. β times). If there are several choices for $\mathcal{C}$, we prefer one that minimizes $\alpha + \beta$.

We claim that $\alpha = 0$ or $\beta = 0$. Suppose both α and β are positive. Thus, there are, in the notation of parts (iii) and (iv) of Theorem (13.3.16), four circuits C_a, C_b, C_v, and C_w with positive weights, where C_a and C_b occur in M_1, and C_v and C_w in M_2. Let γ be the minimum of these four weights. Recall that $\{a, b, l\}$ is a triangle of M_1. Lemma (3.3.8) says that the symmetric difference of two disjoint unions of circuits of a binary matroid is a disjoint union of circuits. By two applications of this result, we see that the elements of $C_a \cup C_b \cup \{a, b, l\}$ not contained in exactly two of the circuits C_a, C_b, and $\{a, b, l\}$ of M_1 form a disjoint union of circuits that contains l but not a or b. Let C' be the circuit of that disjoint union containing l.

Derive two collections $\mathcal{C}'_1$ and $\mathcal{C}'_2$ from $\mathcal{C}_1$ and $\mathcal{C}_2$ as follows. First, add γ to the weight of the circuit C' in $\mathcal{C}_1$. Then reduce the weights of the four circuits C_a, C_b, C_v, and C_w by γ. Derive a collection $\mathcal{C}'$ for M from $\mathcal{C}'_1$ and $\mathcal{C}'_2$ of M_1 and M_2 in the by now obvious way. By the construction, $\mathcal{C}$ and

C' have the same flow value. Define α' and β' for C' analogously to α and β of C. By the construction, $\alpha' + \beta' < \alpha + \beta$, a contradiction.

We thus may assume without loss of generality that $\beta = 0$. In the notation of Theorem (13.3.16), C produces circuits of type C_1 and C_a for C_1, and of type C_2 and C_v for C_2. Round up α to the next integer, getting α'. Assign α' as capacity to the element a of M_1 (resp. element v of M_2), and declare the element b of M_1 (resp. element w of M_2) to have capacity zero. To all other elements of M_1 and M_2, assign the capacity according to the vector h for M.

Solve the fractional max flow problem for M_1. By induction and the earlier proof for regular matroids, we may assume that a collection C_1' of circuits with integral weights is found. Suppose that collection does not use the element a exactly α' times. Then, analogously to the earlier situation where both $\alpha, \beta > 0$, one easily shows that C does not minimize $\alpha + \beta$.

Now solve the fractional max flow problem for M_2. We may assume that a collection C_2' of circuits with integral weights is found that uses the element v exactly α' times. It is a simple matter to combine C_1' and C_2' of M_1 and M_2 to a collection C' of circuits with integral weights for M. The latter collection is readily seen to have a flow value that is at least as large as that of C. Thus, we have completed the last step in the proof of the excluded minors characterization of the connected max-flow min-cut matroids of Theorem (13.3.1).

In the next section, we utilize a part of the preceding proof of Theorem (13.3.1) to validate a construction of the connected max-flow min-cut matroids. We also describe polynomial algorithms for several problems involving these matroids.

13.4 Construction of Max-Flow Min-Cut Matroids and Polynomial Algorithms

In this section, we establish a construction of the connected max-flow min-cut matroids. We utilize the 2-sum and Δ-sum of Section 13.2, and rely on the construction of the regular matroids given by Theorem (11.3.16). From the proof of the construction, we derive a polynomial algorithm for deciding whether a binary matroid has the max-flow min-cut property. Finally, we describe polynomial algorithms that, for any connected max-flow min-cut matroid, solve the max flow problem, the min cut problem, and a certain shortest circuit problem. We begin with the construction of the connected max-flow min-cut matroids.

Construction of Max-Flow Min-Cut Matroids

We use the same terminology as for the construction of the regular matroids described in Theorem (11.3.16). In particular, the two initial matroids of any construction sequence, as well as all matroids that recursively are composed with the matroid already on hand, are the *building blocks*. Among the building blocks is the by now familiar matroid R_{10}. The two representation matrices for that matroid are given by (10.2.8) and (10.4.5).

We split the description of the construction into two cases, depending on whether the resulting max-flow min-cut matroid is to be regular. We begin with the case where this is so.

(13.4.1) Theorem.   *Any connected regular max-flow min-cut matroid with special element l is graphic, cographic, or isomorphic to R_{10}, or may be constructed recursively by 2-sums and Δ-sums using as building blocks graphic matroids, cographic matroids, or matroids isomorphic to R_{10}. In the case of the recursive construction, one of the two initial building blocks contains the special element l, and no other building block contains that element.*

Proof. The theorem is nothing but Theorem (11.3.16) except for the occurrence of the special element l. It is a trivial matter to adapt the proof of Theorem (11.3.16) to account for that element. We leave it to the reader to fill in the details. □

Next we deal with the nonregular case.

(13.4.2) Theorem.   *Any connected nonregular max-flow min-cut matroid with special element l is isomorphic to F_7, or may be recursively constructed by 2-sums given by (13.2.2), (13.2.3), and by Δ-sums given by (13.2.19), (13.2.21). In the case of the recursive construction, one of the two initial building blocks is isomorphic to F_7 and contains l. Furthermore, each additional building block also contains l, and is isomorphic to F_7 or is a connected regular max-flow min-cut matroid.*

A key result for the proof of Theorem (13.4.2) is the following composition theorem.

(13.4.3) Theorem.   *Let M_1 and M_2 be two connected max-flow min-cut matroids that are represented by the matrices B^1 and B^2 of (13.2.3) or (13.2.21). Then the 2-sum or Δ-sum M of M_1 and M_2 as represented by the matrix B of (13.2.2) or (13.2.19) is a max-flow min-cut matroid.*

Proof. The part of the proof of Theorem (13.3.1) concerned with 2-sum and Δ-sum decompositions proves the result. □

Proof of Theorem (13.4.2).   By Theorem (13.4.3), the compositions specified in Theorem (13.4.2) maintain the max-flow min-cut property if it is present in the components. Except for this observation, the remaining arguments are analogous to those proving Theorem (11.3.16). Indeed, the Δ-sum situation is easier to handle than the Δ-sum case of Theorem (11.3.16) since the element l straddles any 3-separation induced by an N_8 minor. Thus, the reader should have no difficulties in filling in the details. $\qquad\qquad\qquad\qquad\qquad\qquad\qquad\qquad\qquad\qquad\qquad\qquad$ □

Polynomial Test for Max-Flow Min-Cut Property

The proofs of Theorems (13.4.1) and (13.4.2) just discussed are easily translated to polynomial algorithms that find the constructions claimed by these theorems. Any such method may be used to test for the max-flow min-cut property in binary matroids. The next theorem and corollary record these facts.

(13.4.4) Theorem. *There is a polynomial algorithm that for any connected max-flow min-cut matroid finds an applicable construction as described by Theorems (13.4.1) and (13.4.2).*

(13.4.5) Corollary. *There is a polynomial algorithm for determining whether an arbitrary binary connected matroid with a special element l has the max-flow min-cut property.*

Next we devise polynomial algorithms for the max flow problem, the min cut problem, and a certain shortest circuit problem. In each case, we assume the matroid to have the max-flow min-cut property. In general, these problems seem to be difficult for arbitrary binary matroids. For example, the min cut problem and the as-yet-unspecified shortest circuit problem can be shown to be $\mathcal{NP}$-hard.

Polynomial Algorithm for Shortest Circuit Problem

The *shortest circuit* problem is as follows. For each element e of a connected binary matroid M with special element l, a nonnegative rational distance d_e is given. Then one must find a circuit C of M containing l such that the sum of the distances d_e, $e \in C$, called the *length* of C, is minimal. This problem is well solved for the max-flow min-cut matroids, as follows.

(13.4.6) Theorem. *There is a polynomial algorithm for the shortest circuit problem of the max-flow min-cut matroids.*

Proof. Let M be the given connected max-flow min-cut matroid, with given nonnegative distance vector d. We first identify a construction as specified by Theorems (13.4.1) and (13.4.2). If M is isomorphic to F_7, then enumeration solves the problem. If M is regular, then we represent M by a real totally unimodular matrix B, define $A = [I \mid B]$, and solve the following linear program using any one of several known polynomial algorithms for linear programs.

$$
\begin{aligned}
\min \quad & d^t \cdot f + d^t \cdot g \\
\text{s. t.} \quad & A \cdot f - A \cdot g = 0 \\
& f \geq 0 \\
& g \geq 0 \\
& f_l = 1 \\
& g_l = 0
\end{aligned}
$$

(13.4.7)

Arguing as in Section 13.3, the linear program (13.4.7) must have $\{0, \pm 1\}$ solution vectors f and g. Indeed, the set $\{e \mid f_e = \pm 1 \text{ or } g_e = \pm 1\}$ may be assumed to be the desired shortest circuit of M with l.

In the remaining case, M is not regular. Then by Theorem (13.4.2), M is a 2-sum or Δ-sum where one component is regular or isomorphic to F_7. We discuss the Δ-sum case, and leave the easier 2-sum situation for the reader. Let M_1 and M_2 be the components of the Δ-sum decomposition of M. In agreement with the discussion of Section 13.2, let $\{a, b, l\}$ and $\{v, w, l\}$ be the triangles of M_1 and M_2 created by the decomposition.

We may assume that M_1 is isomorphic to F_7 or is regular. Except for a, b, and l, we assign the appropriate distance values of the vector d to the elements of M_1. To the former elements, we assign 0 as distance. Temporarily, we delete the element b from M_1, and solve the shortest circuit problem for that minor of M_1. Let C_a be the circuit so found, with length m_a. Similarly, by deletion of a from M_1, we obtain C_b and m_b. Finally, we delete both a and b, and get C_l and m_l. Note that $m_l \leq m_a + m_b$, as is readily proved by comparing m_l with the sum of the distances of the symmetric difference of C_a, C_b, and $\{a, b, l\}$.

Until stated otherwise, we assume that both $a \in C_a$ and $b \in C_b$. By Theorem (13.2.4), we conclude the following. M has C_l as shortest circuit, or M has a shortest circuit that intersects M_1 in C_a or C_b, or M has a shortest circuit that is contained in M_2. To decide which case applies, we assign m_a and m_b as distances to the elements v and w, respectively, of M_2 and recursively solve the shortest circuit problem in that matroid using the construction we already have on hand for it.

Let C be the circuit so found for M_2, with length m_2. If $m_2 \geq m_l$, then C_l is the desired shortest circuit of M. Otherwise, C contains just v, or just w, or none of v and w, or both of v and w. In the first three cases, the respective shortest circuit for M is $(C_a - \{a\}) \cup (C - \{v\})$, or $(C_b - \{b\}) \cup (C - \{w\})$, or C. In the fourth case, the previously derived inequality $m_l \leq m_a + m_b$ implies that $m_2 \geq m_l$, a situation already treated.

We now discuss the case for M_1 where $a \notin C_a$ or $b \notin C_b$. Thus, the shortest circuit of M_1 using a or b is at least as long as C_l, and hence we are justified to assign a distance larger than m_a or m_b to v or w of M_2 to get the correct conclusion by the preceding algorithm. It is easy to see that this method is polynomial.

Polynomial Algorithm for Max Flow and Min Cut Problems

We turn to the max flow problem and the min cut problem. The reader has surely recognized that the min cut problem is the shortest circuit problem for the dual matroid of the given max-flow min-cut matroid. Thus, we could adapt the preceding proof procedure to directly solve the min cut problem. We will not do so. Instead, we invoke Theorem (13.3.9) and a result of linear programming to achieve a brief presentation. The latter result is concerned with the efficient solution of linear programs and is due to Grötschel, Lovász, and Schrijver.

(13.4.8) Theorem.   *Suppose there is a polynomial algorithm with the following features. For any linear program of a given class and any rational vector, the algorithm decides whether the vector is feasible for the linear program. If the answer is negative, the algorithm also obtains a violated constraint. Then there is a polynomial algorithm for the linear programs of the class and their duals. If any such linear program or its dual has extreme points, then an extreme point solution will be produced for that linear program.*

The next theorem contains the result that the max flow problem and the min cut problem can be solved in polynomial time.

(13.4.9) Theorem.   *There is a polynomial algorithm that for any connected max-flow min-cut matroid and for any nonnegative integral capacity vector solves the max flow problem and the min cut problem.*

Proof. Let M be a connected max-flow min-cut matroid, and let h be a given vector of nonnegative integral capacities for the max flow problem and the min cut problem.

For the solution of the max flow problem, we first find a construction as described in Theorem (13.4.4). Note that the decompositions of the construction always involve at least one component that is isomorphic to F_7 or regular. Thus, we can carry out the proof procedure of Theorem (13.3.1), knowing that we only need to solve max flow subproblems for matroids that are isomorphic to F_7 or regular. The F_7 case is straightforward. For regular components, linear programs of the form (13.3.12) and inequality systems given by (13.3.13) must be solved. These tasks are handled efficiently by any polynomial algorithm for linear programs. Thus, the entire proof procedure for Theorem (13.3.1) can be converted to a polynomial algorithm for the max flow problem.

We turn to the min cut case. By the definition of the max-flow min-cut property, any max flow solution also solves the fractional max flow problem (13.3.3). Thus, by Theorem (13.3.9), the fractional min cut problem (13.3.4), which is the dual of the fractional max flow problem (13.3.3), has only integral extreme solutions. Thus, by Theorem (13.4.8), a polynomial solution algorithm for the min cut problem exists if we have a polynomial algorithm that decides whether a given rational vector s is feasible for (13.3.4). The latter algorithm must also identify a violated constraint of (13.3.4) if s is not feasible.

The feasibility test for s is trivial if that vector has any negative entries. So assume s to be nonnegative. View the entries of s as distances for the elements of M different from l, and assign 0 as distance to the latter element. Then deciding whether s is feasible for (13.3.4) is clearly equivalent to deciding whether the length of a shortest circuit of M with l is greater than or equal to 1. To answer the latter question, we compute a shortest circuit C of M using the polynomial algorithm of Theorem (13.4.6). If the length of C is at most 1, then the vector s is feasible for (13.3.4). Otherwise, the characteristic vector of C defines an inequality of (13.3.4) that is violated by s. ☐

In the next section, we meet an important graph application of the max-flow min-cut matroids.

13.5   Graphs without Odd-K_4 Minors

In this section, we transform a graph problem involving certain signed graphs into a matroid problem involving max-flow min-cut matroids. As a result, we can apply the construction and polynomial testing algorithm for max-flow min-cut matroids to obtain a construction and polynomial testing algorithm for the signed graphs. We begin with an application that gives rise to the signed graphs. Let P be a bounded polyhedron of the form $\{x \in \mathbb{R}^n \mid A \cdot x \leq a; \ x \geq 0\}$. We are to determine an inequality system $D \cdot x \leq d$ such that the polyhedron $Q = \{x \mid A \cdot x \leq a; \ D \cdot x \leq d; \ x \geq 0\}$ has only integral vertices and contains all integral points of P. The polyhedron Q is usually called the *integer hull* of P. The arrays D and d can have complex structure even for small arrays A and a. Thus, determination of the structure of D and d for special cases of A and a, either by some combinatorial description or in terms of a constructive scheme, has become one of the basic problems of polyhedral combinatorics.

An equally important converse problem is as follows. One postulates some construction for D and d and characterizes the cases of A and a for which the construction does work. For the description of one such construction, we need a rounding operation for rational vectors. The operation rounds down each entry of a given vector b to the next integer. We denote the resulting vector by $\lfloor b \rfloor$. The construction is as follows. Each row D_i of D is obtained from some linear combination of the rows of A by rounding down the entries. Thus, for some rational row vector $\lambda^i \geq 0$, we have $D_i = \lfloor \lambda^i \cdot A \rfloor$. The corresponding value d_i of the vector d is $\lfloor \lambda^i \cdot b \rfloor$. We call this construction *simple rounding*.

Just a few classes of combinatorially interesting polyhedra are known for which simple rounding does produce the desired D and d. The most important case is due to Edmonds and has as A any $\{0, \pm 1\}$ matrix with two ± 1s in each column. The vector a may contain any integers. The polyhedron $P = \{x \in \mathbb{R}^n \mid A \cdot x \leq a; \ x \geq 0\}$ for this case is known as the *fractional matching polyhedron*. On the other hand, it is well known that simple rounding generally does not work for the polyhedra produced by $\{0, \pm 1\}$ matrices A and integral vectors a where A has two ± 1s in each row instead of each column. Polyhedra of the latter variety are very important, since they arise from a number of combinatorial problems. An example is the vertex cover problem, where one must cover the edges of an undirected graph by vertices. The matrix A is then the real and negated version of the transpose of the node/edge incidence matrix of the graph, and the vector a contains only -1s.

There are special cases, though, where simple rounding does work for the latter polyhedra. To describe one such class, we derive a special graph $G(A)$ from any $\{0, \pm 1\}$ matrix A with two ± 1s in each row. Each column of A corresponds to a node of $G(A)$, and each row to an edge. Specifically, if a row of A has two entries of opposite (resp. same) sign, say in column j and k, then an edge connects the nodes j and k of $G(A)$ and is declared to be *even* (resp. *odd*). We call $G(A)$ a *signed* graph.

We define a cycle C of a signed graph to be *odd* if it contains an odd number of odd edges. If we scale a column j of A by -1, getting, say, A', then the graph $G(A')$ may be obtained from $G(A)$ by declaring each even (resp. odd) edge incident at node j to be odd (resp. even). We also say that $G(A')$ is obtained from $G(A)$ by *scaling*. Note that any cycle of $G(A)$ is even or odd if and only if this is so for the corresponding cycle of $G(A')$. Any graph obtained from $G(A)$ by repeated column scaling we call *equivalent* to $G(A)$. As expected, "is equivalent to" is an equivalence relation.

We derive minors from a signed graph by a sequence of deletions and contractions interspersed with scaling steps, with the restriction that any contraction may involve even edges only. It is easily checked that two minors produced by the same deletions and contractions, in any order, are equivalent, provided the reductions obey the cycle/cocycle condition of Section 2.2.

Consider the signed graph G obtained from K_4, the complete graph on four vertices, by declaring the edges of one triangle to be odd and the remaining three edges to even. Evidently, the triangles of G are precisely its odd circuits. Declare any signed graph that is equivalent to the just-specified one to be an *odd* K_4.

The signed graphs $G(A)$ without odd-K_4 minors have very pleasant properties. Indeed, the preceding simple rounding construction does produce the desired integer hull for a variant of the previously specified polyhedron involving the matrix A. The result, due to Gerards and Schrijver, is as follows.

(13.5.1) Theorem. *Let A be a $\{0, \pm 1\}$ matrix with two ± 1s in each row, and let P be the polyhedron $\{x \in \mathbb{R}^n \mid a^1 \le A \cdot x \le a^2;\ c^1 \le x \le c^2\}$. Then simple rounding can derive, for all integer vectors a^1, a^2, c^1, and c^2, a description of the integer hull of P if and only if $G(A)$ has no odd-K_4 minors.*

At this point, we have a nice theorem of polyhedral combinatorics involving the graphs without odd-K_4 minors, but actually have no clue what these graphs look like. Nor do we know how to decide efficiently whether a signed graph has odd-K_4 minors.

To gain the desired insight, we move from signed graphs G to certain binary matroids M. Let G be given. Define the edge set of G plus an additional element l to be the groundset of M. Next we produce a representation matrix for M. We begin with the usual binary node/edge incidence matrix F for G, and record whether edges are even or odd by appending to F an additional row having 0s (resp. 1s) in the columns corresponding to the even (resp. odd) edges of G. We adjoin a column unit vector with index l, having its 1 in the new row. Let the resulting matrix be F'. Finally, we declare the circuits of M to be the index sets of the GF(2)-mindependent column subsets of F'. Since scaling in G does not affect evenness or oddness of cycles, all graphs equivalent to G produce the same matroid M.

From F', we obtain a matrix $[I \mid B]$ by elementary row operations (in GF(2)). Thus B is a representation matrix for M. The subsequent analysis is simplified if we let l index a row of B, or equivalently, if we accept the unit vector of F' indexed by l as part of the identity I of $[I \mid B]$. Indeed, in that case we only need to choose a tree T of G and scale G so that all tree edges become even, and finally use the fundamental cycles of the new G to write down the columns of B. By the derivation, any such fundamental cycle is even (resp. odd) if the out-of-tree edge producing that cycle is even (resp. odd). It is easy to see that M is connected if G is 2-connected and has at least one odd cycle. Also, every 2-connected minor of G with at least one odd cycle corresponds to a minor of M with l, as expected. But what property does M have when G has no odd-K_4 minors? The next lemma, also due to Gerards and Schrijver, gives the surprising answer.

(13.5.2) Lemma. *Let G be a 2-connected signed graph, and let M*

be the corresponding binary connected matroid. Then G has no odd-K_4 minors if and only if M has no F_7^* minors with l, and thus, if and only if M is a max-flow min-cut matroid.

Proof. Assume that G has an odd-K_4 minor, say $\overline{G}$. Because of scaling, we may presume that a triangle of that minor has all edges odd, and that the remaining edges, which form a 3-star, are even. Derive a representation matrix $\overline{B}$ for the corresponding minor $\overline{M}$ of M as described above, using the 3-star as tree. Then one readily sees that $\overline{M}$ is an F_7^* minor of M with l. By reversing the arguments, we see that any F_7^* minor of M with l corresponds to an odd-K_4 minor of G. □

By Lemma (13.5.2) and Corollary (13.4.5), we already have a polynomial algorithm for deciding whether a signed graph has no odd-K_4 minors. We also have, indirectly, a complete analysis of the graphs without odd-K_4 minors, in the form of the decomposition Theorems (13.2.1) and (13.2.4) and the construction Theorems (13.4.1) and (13.4.2) for the max-flow mincut matroids. To understand the structure of the graphs without odd-K_4 minors, we only need to translate these results into graph language. We carry out this task next.

Construction of Graphs without Odd-K_4 Minors

Let M be the connected max-flow min-cut matroid corresponding to a 2-connected signed graph G without odd-K_4 minors. We assume that G has at least one odd circuit. As before, l is the special element of M. Define G' to be the unsigned version of G. Since M/l is the graphic matroid of the 2-connected G', that matroid is connected. We will work out a description of G that involves scaling. That way, a particularly simple description results. We divide the discussion into five cases, depending on whether G or M has a 2-separation, and whether M is graphic, cographic, regular nongraphic and noncographic, or nonregular.

2-Separation Cases

Suppose $(E_1 \cup \{l\}, E_2)$ is a 2-separation of M. Suppose that $|E_1| = 1$, say $E_1 = \{e\}$. Recall that M/l is connected and graphic. Thus, e and l must be two series elements of M. Accordingly, G can be scaled so that e is the only odd edge. Let G be that scaled graph. We call a signed graph with all odd edges incident at one vertex, and thus G, a graph with one *partially odd* vertex. Later, we will see another instance of such a graph.

Suppose $|E_1| \geq 2$. Then (E_1, E_2) is a 2-separation of G. Select a base $X_1 \cup X_2$ of M where $X_1 \subseteq E_1$ and $l \in X_1$, and where X_2 is a maximal independent subset of E_2. In the corresponding representation matrix for

M, the row l has only 0s in the columns of $E_2 - X_2$. Correspondingly, G has, up to scaling, odd edges only in E_1. Assume such scaling. We derive from G the 2-sum components G_1 and G_2 as depicted in (8.2.8), where the explicitly shown connecting edges are declared to be even. Note that G_2 has only even edges. Later, we will encounter another 2-sum decomposition, so we define the present one to be a 2-*sum of type 1*.

As a result of the preceding discussion, we may assume from now on that M is 3-connected. This implies that any parallel class of edges of G contains at most two edges, one even and one odd.

Suppose the deletion of parallel edges from G creates a 2-separable graph. Then G has a 2-separation (E_1, E_2) where neither E_1 nor E_2 is just a set of parallel edges. Assume $|E_1| = 2$. Since E_1 is not just a set of parallel edges, the two edges of E_1 must be incident at a degree 2 node. But by scaling, both edges can be made even, so $(E_1, E_2 \cup \{l\})$ is a 2-separation of M contrary to the assumption that M is 3-connected. Hence, $|E_1|, |E_2| \geq 3$. Clearly, $(E_1 \cup \{l\}, E_2)$ and $(E_1, E_2 \cup \{l\})$ are 3-separations of M. In the terminology of Section 13.2, the element l straddles the 3-separation $(E_1, E_2 \cup \{l\})$ of M. Furthermore, that 3-separation induces a Δ-sum decomposition of M into M_1 and M_2, as depicted by the matrices B, B^1, and B^2 of (13.2.19) and (13.2.21). Since l straddles the 3-separation of M, the column l of the submatrix A^2 of B of (13.2.19) is spanned by the columns of the submatrix D of B. So if we GF(2)-pivot in column l of B and subsequently delete the pivot row, then the matrix D of B is reduced to a matrix with GF(2)-rank 1. We rely on this fact next when we interpret the Δ-sum decomposition of M in terms of a decomposition of G.

Suppose in each one of the matrices B, B^1, and B^2 of (13.2.19) and (13.2.21) we perform a GF(2)-pivot on the 1 in row a and column l. Let $\tilde{B}$, $\tilde{B}^1$, and $\tilde{B}^2$ result. Delete row l from $\tilde{B}$, $\tilde{B}^1$, and $\tilde{B}^2$, getting B', B'^1, and B'^2. Clearly, the matrix B' corresponds to G', the unsigned version of G. Since l straddles the 3-separation of M and B, the matrices B'^1 and B'^2 are readily seen to correspond to a somewhat unusual 2-sum decomposition of G. Indeed, the component graphs are as depicted in (8.2.8) once we replace the connecting edge x of G_1 in (8.2.8) by two parallel edges a and b, and the connecting edge y of G_2 in (8.2.8) by two parallel edges v and w. Assume such a replacement has been done, and denote by G'_1 and G'_2 the graphs corresponding to B'^1 and B'^2. Because of scaling, we may suppose that the signature of G agrees with that given by row l of $\tilde{B}$. Then according to row l of $\tilde{B}^1$ and $\tilde{B}^2$, we may sign G'_1 and G'_2 to obtain graphs for the matroids M_1 and M_2 of the Δ-sum decomposition, as follows. In G'_1 (resp. G'_2), we declare the edge a (resp. w) to be odd, and the edge b (resp. v) to be even. The remaining edges of G'_1 and G'_2 are signed in agreement with the signature of the edges of G. Let G_1 and G_2 be the graphs so obtained for M_1 and M_2. We call this special 2-sum decomposition a 2-*sum of type 2*.

Theorem (11.2.10), Corollary (11.2.12), and Theorem (13.4.3), which cover the composition of regular matroids and of max-flow min-cut matroids, imply that the above 2-sum decompositions of type 1 and 2 are reversible. That is, if both components G_1 and G_2 correspond to regular matroids (resp. max-flow min-cut matroids), then the 2-sum G of type 1 or 2 also corresponds to a regular matroid (resp. max-flow min-cut matroid).

We have completed the 2-separation cases. From now on, we assume that M is 3-connected, and that G is 3-connected up to parallel edges. In fact, any parallel class of edges of G consists of at most two edges, one even and one odd.

Graphic Matroid Case

We assume M to be graphic. Thus, there is a graph H so that M is the graphic matroid $M(H)$ of H. Recall that G' is the unsigned version of G. Since M/l is the graphic matroid of H/l as well as of G', the graphs H/l and G' must be 2-isomorphic. We know that G is 3-connected up to parallel edges, so by Theorem (3.2.36), $H/l = G'$. The graph H may thus be derived from G' by splitting some vertex v into two nodes, which are then connected by the edge l. Pick a spanning tree X' of G' that has the node v as tip node. Then $X = X' \cup \{l\}$ is a tree of H. The representation matrix of M corresponding to the base X has l as row index. The 1s in that row correspond to a subset $\overline{E}$ of the edges of G incident at v. Thus, up to scaling, the odd edges of G are precisely the edges of $\overline{E}$. Assuming such scaling, G is therefore a graph with one partially odd vertex. By reversing the construction, we see that any signed graph G with one partially odd vertex produces a graphic M.

Cographic Matroid Case

We assume M to be cographic but not graphic. Hence, M^* is graphic. As before, G' represents M/l, so both M/l and $(M/l)^* = M^* \setminus l$ are planar. We conclude that G' and G may be taken to be 3-connected plane graphs plus parallel edges, and that there is a graph H such that $M(H) = M^*$ and $G' = (H \setminus l)^*$. In H, let v_1 and v_2 be the endpoints of the edge l. Pick a tree of H that does not contain the edge l. In the corresponding representation matrix for M^*, the 1s of column l represent a path P from v_1 to v_2. Thus, in $H \setminus l$, each vertex other than v_1 and v_2 has an even number of edges of P incident. By scaling in the plane graph G, the edges of P become precisely the odd edges. Thus, up to scaling, exactly the two faces of G corresponding to v_1 and v_2 of $H \setminus l$ have an odd number of odd edges incident. The latter property is invariant under scaling, so the same property holds for G. We call a planar graph with this property, and thus G, a graph with *two odd faces*.

Regular Nongraphic and Noncographic Matroid Case

We turn to the case where the 3-connected matroid M is regular but not graphic and not cographic. By Theorem (13.4.1), M is isomorphic to R_{10}, or is a Δ-sum with components M_1 and M_2 where one component does not contain l. The R_{10} case is not possible, since M/l is graphic, and since the contraction of any element of R_{10} produces an $M(K_{3,3})^*$ minor, as shown in the proof of Lemma (10.4.4). Thus, we only need to examine the case of a Δ-sum. Let $(E_1 \cup \{l\}, E_2)$ be the corresponding 3-separation of M. By the derivation of Theorem (13.4.1) via Theorem (11.3.16), we may assume that $|E_1 \cup \{l\}|, |E_2| \geq 6$. The pair (E_1, E_2) cannot be a 2-separation of G since otherwise $|E_1|, |E_2| \geq 5$ implies that the earlier discussed 2-separation case applies.

Suppose $(E_1, E_2 \cup \{l\})$ is also a 3-separation of M. In the terminology of Section 13.2, the element l straddles the 3-separation $(E_1 \cup \{l\}, E_2)$. Let B of (13.2.19) be a representation matrix for M exhibiting the 3-separation $(E_1, E_2 \cup \{l\})$, except that we ignore in B the explicitly shown 0s and 1s in the submatrix indexed by $\{a, b, d\}$ and $\{l, v, w\}$. Since l straddles the 3-separation, the submatrix D of B spans the column l of A^2. Then a GF(2)-pivot on any 1 in column l of A^2, followed by the deletion of the pivot row and pivot column, reduces D, which has GF(2)-rank $D = 2$, to a matrix with GF(2)-rank 1. We conclude that (E_1, E_2) is a 2-separation of M/l, and hence of G, contrary to assumption. Hence, l does not straddle the 3-separation. Thus, analogously to the case of a 2-sum of type 1, there is a base $X_1 \cup X_2$ for M where $X_1 \subseteq E_1$ and $l \in X_1$, and where X_2 is a maximal independent subset of E_2. Furthermore, up to scaling, all edges of G in E_2 are even. For the remainder of this case, we suppose G to be of this form.

Assume that (E_1, E_2) is a graph 3-separation of G. Arguing as in the 2-sum case, G has a Δ-sum decomposition where we declare all edges of the connecting triangles to be even, and where the component defined from E_2 has even edges only. Without chance of confusion, we call this process a Δ-*sum* decomposition of G. By the composition Corollary (11.2.12) for regular matroids, the Δ-sum decomposition of G is reversible.

So suppose (E_1, E_2) is not a graph 3-separation of G. This is possible, by Theorem (3.2.25). Recall that each parallel class of G has at most two edges, one even and one odd. Since the edges of E_2 are even, all parallel edges must occur in E_1. Define G'' to be G minus all odd parallel edges. Correspondingly, define E_1'' and E_2'' from E_1 and E_2, respectively. Since $|E_1|, |E_2| \geq 5$, we have $|E_1''|, |E_2''| \geq 3$. Let M'' be the minor of M corresponding to G''. Then (E_1'', E_2'') is a 3-separation of M'' that induces the 3-separation $(E_1 \cup \{l\}, E_2)$ of M. If (E_1'', E_2'') is a graph 3-separation of G'', then (E_1, E_2) is a graph 3-separation of G contrary to assumption. Thus, by Theorem (3.2.25), E_1'' is a cutset of cardinality 3, and removal of that cutset from G produces two connected graphs H_{21} and H_{22} whose edge sets form a partition of E_2. Furthermore, each one of the latter graphs contains a cycle.

If H_{21} or H_{22} has at least four edges, then a graph 3-separation of G exists where one side contains only even edges, and the earlier case applies. Thus, each one of H_{21} and H_{22} is just a triangle, and G consists of a pair of triangles whose nodes are pairwise joined by three edges, say e, f, and g, plus edges parallel to the latter edges. Then G corresponds to a cographic matroid case unless G is of the form

(13.5.3)

Exceptional signed graph case

where e, f, and g are odd, and where all other edges are even. The matroid M for this graph is regular, as is easily checked.

At this point, we want to summarize the above analysis for the situation where M is regular. For that summary, in Theorem (13.5.4) below, we need a characterization of the regularity of M in terms of signed graphs. To this end, we define an *odd double triangle* to be the following signed graph. We start with a triangle, all of whose edges are even. Then we add to each edge one parallel odd edge. Using a representation matrix for F_7, for example the matrix of (10.2.4), one readily confirms that M has an F_7 minor with l if and only if G has an odd double triangle minor. It is easy to check that M is regular if and only if M has no F_7 minors with l and no F_7^* minors with l. Indeed, this claim follows almost immediately from the discussion of Section 13.2. We thus have the following theorem for the case where M is regular.

(13.5.4) Theorem. *Let G be a 2-connected signed graph without odd-K_4 minors and without odd double triangle minors, but with an odd circuit. Then up to scaling, G has exactly one partially odd vertex, or is planar and has exactly two odd faces, or is the graph of (13.5.3), or may be constructed recursively by 2-sums of type 1 or 2, and by Δ-sums. Up to scaling, the building blocks are the cited graphs and graphs having only even edges.*

Proof. The above analysis and a simple inductive argument prove the result.   □

We turn to the final case, where M is nonregular.

Nonregular Matroid Case

Suppose G produces a nonregular max-flow min-cut matroid M. The translation of the construction Theorem (13.4.2) for M gives the following result.

(13.5.5) Theorem. *Let G be a 2-connected signed graph without odd-K_4 minors, but with an odd double triangle minor. Then up to scaling, G is an odd double triangle, or may be recursively constructed by 2-sums of type 1 or 2. In the case of the recursive construction, one of the initial building blocks is an odd double triangle. Furthermore, each additional building block is also an odd double triangle, or corresponds to a signed graph having no odd-K_4 minors and no double triangle minors.*

Proof. As shown above, matroid 2-sums and matroid Δ-sums with l as straddling element correspond to graph 2-sums of type 1 or 2. Thus, the theorem follows from Theorem (13.4.2). □

In the next section, we include applications, extensions, and references.

13.6 Applications, Extensions, and References

The first comprehensive treatment of the max flow problem for graphs is contained in Ford and Fulkerson (1962). The definition and characterization of max-flow min-cut matroids by excluded minors is due to Seymour (1977a). Related is a characterization of matroid ports in Seymour (1976), (1977b). The decomposition of the max-flow min-cut matroids is described in Tseng and Truemper (1986). That reference proves a decomposition result slightly stronger than Theorem (13.2.13): Any N_8 minor of a max-flow min-cut matroid M has a 3-separation that up to indices other than l is given by (13.2.5) and that induces a 3-separation of M. A short proof of Theorem (13.2.13) is given in Bixby and Rajan (1989).

Basic linear programming results may be found in any standard text, for example in Dantzig (1963), Chvátal (1983), or Schrijver (1986). The concept of total dual integrality and the related Theorem (13.3.9) are due to Edmonds and Giles (1977).

The existence of polynomial algorithms for the various problems of Section 13.4 is established in Truemper (1987a). That reference proves the min cut problem and the shortest circuit problem to be $\mathcal{NP}$-hard, and also shows that the max flow problem is solved by an integer extreme point solution of the linear program (13.3.3), not just an integer solution as proved here. Theorem (13.4.8) is due to Grötschel, Lovász, and Schrijver (1988). Actually, simpler machinery suffices to prove the results of

Section 13.4, as shown in Truemper (1987a). The recognition problem of max-flow min-cut path matrices is treated in Hartvigsen and Wagner (1988).

The simple rounding scheme is analyzed in Chvátal (1973). The most important case treatable by simple rounding is the matching problem. Its solution is due to Edmonds (1965c), (1965d); see Lovász and Plummer (1986) for an excellent exposition of this result and of many related ones. Theorem (13.5.1) and Lemma (13.5.2) are from Gerards and Schrijver (1986). General results for packing and covering problems involving matroid circuits are described in Seymour (1980c). Additional decomposition theorems and other results for signed graphs are given in Gerards (1988), (1989b). Gerards, Lovász, Schrijver, Seymour, and Truemper (1991) summarize various decomposition results for signed graphs.

In Section 13.4, it is shown that the fractional min cut problem (13.3.4) has only integer extreme point solutions, provided max-flow min-cut matroids are involved. The latter condition is sufficient, but not necessary. Thus, one may want to characterize when precisely (13.3.4) has only integer extreme point solutions. Much progress has been made on this difficult problem (see Lehman (1981), Cornuéjols and Novick (1990), and Seymour (1990)), but finding a complete characterization seems to be very difficult.

References

Agrawal, A., and Satyanarayana, A. (1984), An $O(|E|)$ time algorithm for computing the reliability of a class of directed networks, *Operations Research* 32 (1984) 493–515.

Agrawal, A., and Satyanarayana, A. (1985), Network reliability analysis using 2-connected digraph reductions, *Networks* 15 (1985) 239–256.

Ahuja, R. K., Magnanti, T. L., and Orlin, J. B. (1989), Network Flows, in: *Handbooks in Operations Research and Management Science*, Vol. 1: *Optimization* (G. L. Nemhauser, A. H. Rinnooy Kan, and M. J. Todd, eds.), North-Holland, Amsterdam, 1989, pp. 211–369.

Aigner, M. (1979), *Combinatorial Theory*, Springer Verlag, Berlin, 1979.

Akers, S. B., Jr. (1960), The use of wye-delta transformations in network simplification, *Operations Research* 8 (1960) 311–323.

Anstee, R. P., and Farber, M. (1984), Characterizations of totally balanced matrices, *Journal of Algorithms* 5 (1984) 215–230.

Appel, K., and Haken, W. (1977), Every planar map is four colorable. Part I: Discharging, *Illinois Journal of Mathematics* 21 (1977) 429–490.

Appel, K., Haken, W., and Koch, J. (1977), Every planar map is four colorable. Part II: Reducibility, *Illinois Journal of Mathematics* 21 (1977) 491–567.

Arnborg, S., and Proskurowski, A. (1986), Characterization and recognition of partial 3-trees, *SIAM Journal on Algebraic and Discrete Methods* 7 (1986) 305–314.

Asano, T., Nishizeki, T., and Seymour, P. D. (1984), A note on nongraphic matroids, *Journal of Combinatorial Theory (B)* 37 (1984) 290–293.

Auslander, L., and Trent, H. M. (1959), Incidence matrices and linear graphs, *Journal of Mathematics and Mechanics* 8 (1959) 827–835.

Auslander, L., and Trent, H. M. (1961), On the realization of a linear graph given its algebraic specification, *Journal of the Acoustical Society of America* 33 (1961) 1183–1192.

Bachem, A., and Kern, W. (1992), *Oriented Matroids*, in preparation.

Barahona, F. (1983), The max-cut problem on graphs not contractible to K_5, *Operations Research Letters* 2 (1983) 107–111.

Barahona, F., and Grötschel, M. (1986), On the cycle polytope of a binary matroid, *Journal of Combinatorial Theory (B)* 40 (1986) 40–62.

Barahona, F., Grötschel, M., Jünger, M., and Reinelt, G. (1988), An application of combinatorial optimization to statistical physics and circuit layout design, *Operations Research* 36 (1988) 493–513.

Barahona, F., and Mahjoub, A. R. (1986), On the cut polytope, *Mathematical Programming* 36 (1986) 157–173.

Barahona, F., and Mahjoub, A. R. (1989a), Compositions of graphs and polyhedra. I. Balanced induced subgraphs and acyclic subgraphs, working paper, University of Waterloo, 1989.

Barahona, F., and Mahjoub, A. R. (1989b), Compositions of graphs and polyhedra. II. Stable sets, working paper, University of Waterloo, 1989.

Barahona, F., and Mahjoub, A. R. (1990a), Compositions of graphs and polyhedra. III. Graphs with no W_4 minor, working paper, University of Waterloo, 1990.

Barahona, F., and Mahjoub, A. R. (1990b), Compositions of graphs and polyhedra. IV. Acyclic spanning subgraphs, working paper, University of Waterloo, 1990.

Barnette, D. W., and Grünbaum, B. (1969), On Steinitz's theorem concerning convex 3-polytopes and on some properties of planar graphs, in: *The Many Facets of Graph Theory, Springer Lecture Notes* 110 (1969) 27–40, Springer Verlag, Berlin.

Baum, S., and Trotter, L. E., Jr. (1978), Integer rounding and polyhedral decomposition for totally unimodular systems, in: *Optimization and Operations Research*, Proceedings Bonn 1977 (R. Henn, B. Korte, and W. Oettli, eds.), *Lecture Notes in Economics and Mathematical Systems* 157 (1978) 15–23, Springer Verlag, Berlin.

Berge, C. (1972), Balanced matrices, *Mathematical Programming* 2 (1972) 19–31.

Berge, C. (1973), *Graphs and Hypergraphs*, North-Holland, Amsterdam, 1973.

Berge, C. (1980), Balanced matrices and property (G), *Mathematical Programming Study* 12 (1980) 163–175.

Bern, M. W., Lawler, E. L., and Wong, A. L. (1987), Linear-time computation of optimal subgraphs of decomposable graphs, *Journal of Algorithms* 8 (1987) 216–235.

Bixby, R. E. (1972), *Composition and Decomposition of Matroids and Related Topics*, thesis, Cornell University, 1972.

Bixby, R. E. (1974), ℓ-matrices and a characterization of binary matroids, *Discrete Mathematics* 8 (1974) 139–145.

Bixby, R. E. (1975), A composition for matroids, *Journal of Combinatorial Theory (B)* 18 (1975) 59–72.

Bixby, R. E. (1976), A strengthened form of Tutte's characterization of regular matroids, *Journal of Combinatorial Theory (B)* 20 (1976) 216–221.

Bixby, R. E. (1977), Kuratowski's and Wagner's theorems for matroids, *Journal of Combinatorial Theory (B)* 22 (1977) 31–53.

Bixby, R. E. (1979), On Reid's characterization of the ternary matroids, *Journal of Combinatorial Theory (B)* 26 (1979) 174–204.

Bixby, R. E. (1982a), Matroids and operations research, in: *Advanced Techniques in the Practice of Operations Research* (6 tutorials presented at the semi-annual joint ORSA/TIMS meeting, Colorado Springs, 1980; H. J. Greenberg, F. H. Murphy, and S. H. Shaw, eds.), North-Holland, New York, 1982, pp. 333–458.

Bixby, R. E. (1982b), A simple theorem on 3-connectivity, *Linear Algebra and Its Applications* 45 (1982) 123–126.

Bixby, R. E. (1984a), Recent algorithms for two versions of graph realization and remarks on applications to linear programming, in: *Progress in Combinatorial Optimization* (W. R. Pulleyblank, ed.), Academic Press, Toronto, 1984, pp. 39–68.

Bixby, R. E. (1984b), A composition for perfect graphs, in: *Topics on Perfect Graphs* (C. Berge and V. Chvátal, eds.), *Annals of Discrete Mathematics* 21 (1984) 221–224.

Bixby, R. E., and Coullard, C. R. (1984), On chains of 3-connected matroids, *Discrete Applied Mathematics* 15 (1986) 155–166.

Bixby, R. E., and Coullard, C. R. (1987), Finding a small 3-connected minor maintaining a fixed minor and a fixed element, *Combinatorica* 7 (1987) 231–242.

Bixby, R. E., and Cunningham, W. H. (1979), Matroids, graphs, and 3-connectivity, in: *Graph Theory and Related Topics* (J. A. Bondy and U. S. R. Murty, eds.), Academic Press, New York, 1979, pp. 91–103.

Bixby, R. E., and Cunningham, W. H. (1980), Converting linear programs to network problems, *Mathematics of Operations Research* 5 (1980) 321–357.

Bixby, R. E., and Cunningham, W. H. (1987), Short cocircuits in binary matroids, *European Journal of Combinatorics* 8 (1987) 213–225.

Bixby, R. E., Cunningham, W. H., and Rajan, R. (1986), A decomposition algorithm for matroids, working paper, Rice University, 1986.

Bixby, R. E., and Rajan, R. (1989), A short proof of the Truemper-Tseng theorem on max-flow min-cut matroids, *Linear Algebra and Its Applications* 114/115 (1989) 277–292.

Bixby, R. E., and Wagner, D. K. (1988), An almost linear-time algorithm for graph realization, *Mathematics of Operations Research* 13 (1988) 99–123.

Björner, A., Las Vergnas, M., Sturmfels, B., White, N., and Ziegler, G. M. (1992), *Oriented Matroids*, Cambridge University Press, Cambridge, 1992.

Bland, R. G., and Edmonds, J. (1978), unpublished.

Bland, R. G., and Las Vergnas, M. (1978), Orientability of Matroids, *Journal of Combinatorial Theory (B)* 24 (1978) 94–123.

Bock, F. C. (1971), An algorithm to construct a minimum directed spanning tree in a directed network, in: *Developments in Operations Research*, Vol. 1 (B. Avi-Itzhak, ed.), Gordon and Breach, New York, 1971, pp. 29–44.

Bondy, J. A., and Murty, U. S. R. (1976), *Graph Theory with Applications*, Macmillan, London, 1976.

Boulala, M., and Uhry, J. P. (1979), Polytope des indépendants dans un graphe série-parallèle, *Discrete Mathematics* 27 (1979) 225–243.

Brown, D. P. (1976), Circuits and unimodular matrices, *SIAM Journal on Applied Mathematics* 31 (1976) 468–473.

Brown, D. P. (1977), Compound and unimodular matrices, *Discrete Mathematics* 19 (1977) 1–5.

Bruno, J., and Weinberg, L. (1971), The principal minors of a matroid, *Linear Algebra and Its Applications* 4 (1971) 17–54.

Brylawski, T. H. (1971), A combinatorial model for series-parallel networks, *Transactions of the American Mathematical Society* 154 (1971) 1–22.

Brylawski, T. H. (1972), A decomposition for combinatorial geometries, *Transactions of the American Mathematical Society* 171 (1972) 235–282.

Brylawski, T. H. (1975), Modular constructions for combinatorial geometries, *Transactions of the American Mathematical Society* 203 (1975) 1–44.

Brylawski, T. H., and Lucas, D. (1973), Uniquely representable combinatorial geometries, in: *Teorie Combinatorie* (Proceedings of 1973 International Colloquium, Accademia Nazionale dei Lincei) (B. Segre, ed.), Rome, Italy, 1976, pp. 83–104.

Burlet, M., and Fonlupt, J. (1984), Polynomial algorithm to recognize a Meyniel graph, in: *Topics on Perfect Graphs* (C. Berge and V. Chvátal, eds.), *Annals of Discrete Mathematics* 21 (1984) 225–252. Also in: *Progress in Combinatorial Optimization* (W. R. Pulleyblank, ed.), Academic Press, Toronto, 1984, pp. 69–99.

Burlet, M., and Uhry, J. P. (1982), Parity graphs, in: *Bonn Workshop on Combinatorial Optimization* (A. Bachem, M. Grötschel, and B. Korte, eds.), *Annals of Discrete Mathematics* 16 (1982) 1–26.

Camion, P. (1963a), Caractérisation des matrices unimodulaires, *Cahiers du Centre d'Études de Recherche Opérationelle* 5 (1963) 181–190.

Camion, P. (1963b), *Matrices Totalement Unimodulaire et Problèmes Combinatoires*, thesis, Université Libre de Bruxelles, Brussels, 1963.

Camion, P. (1965), Characterization of totally unimodular matrices, *Proceedings of the American Mathematical Society* 16 (1965) 1068–1073.

Camion, P. (1968), Modules unimodulaires, *Journal of Combinatorial Theory* 4 (1968) 301–362.

Cederbaum, I. (1957), Matrices all of whose elements and subdeterminants are 1, −1 or 0, *Journal of Mathematics and Physics* 36 (1957) 351–361.

Chandrasekaran, R. (1969), Total unimodularity of matrices, *SIAM Journal on Applied Mathematics* 17 (1969) 1032–1034.

Chandrasekaran, R., and Shirali, S. (1984), Total weak unimodularity: testing and applications, *Discrete Mathematics* 51 (1984) 137–145.

Chandru, V., Coullard, C. R., and Wagner, D. K. (1985), On the complexity of recognizing a class of generalized networks, *Operations Research Letters* 4 (1985) 75–78.

Chartrand, G., and Harary, F. (1967), Planar permutation graphs, *Annales de l'Institut Henri Poincaré Section B* 3 (1967) 433–438.

Chu, Y.-J., and Liu, T.-H. (1965), On the shortest arborescence of a directed graph, *Scientia Sinica (Peking)* 4 (1965) 1396–1400.

Chvátal, V. (1973), Edmonds polytopes and a hierarchy of combinatorial problems, *Discrete Mathematics* 4 (1973) 305–337.

Chvátal, V. (1983), *Linear Programming*, Freeman, New York, 1983.

Chvátal, V. (1985), Star-cutsets and perfect graphs, *Journal of Combinatorial Theory (B)* 39 (1985) 189–199.

Chvátal, V. (1986), private communication.

Chvátal, V. (1987), On the P_4-structure of perfect graphs. III. Partner decompositions, *Journal of Combinatorial Theory (B)* 43 (1987) 349–353.

Chvátal, V., and Hoang, C. T. (1985), On the P_4-structure of perfect graphs. I. Even decompositions, *Journal of Combinatorial Theory (B)* 39 (1985) 209–219.

Chvátal, V., Lenhart, W. J., and Sbihi, N. (1990), Two-colourings that decompose perfect graphs, *Journal of Combinatorial Theory (B)* 49 (1990) 1–9.

Commoner, F. G. (1973), A sufficient condition for a matrix to be totally unimodular, *Networks* 3 (1973) 351–365.

Colbourn, C. J. (1987), *The Combinatorics of Network Reliability*, Oxford University Press, New York, 1987.

Conforti, M., and Cornuéjols, G. (1990), A decomposition theorem for balanced matrices, in: *Proceedings of Conference on Integer Programming and Combinatorial Optimization*, University of Waterloo, 1990 (R. Kannan and W. R. Pulleyblank, eds.), University of Waterloo Press, Waterloo, Canada, 1990.

Conforti, M., Cornuéjols, G., and Rao, M. R. (1991), Balanced matrices, in preparation.

Conforti, M., and Laurent, M. (1988), On the facial structure of independence system polyhedra, *Mathematics of Operations Research* 13 (1988) 543–555.

Conforti, M., and Laurent, M. (1989), On the geometric structure of independence systems, *Mathematical Programming* 45 (1989) 255–277.

Conforti, M., and Rao, M. R. (1987), Structural properties and decomposition of restricted and strongly unimodular matrices, *Mathematical Programming* 38 (1987) 17–27.

Conforti, M., and Rao, M. R. (1989), Odd cycles and matrices with integrality properties, *Mathematical Programming* 45 (1989) 279–294.

Conforti, M., and Rao, M. R. (1991a), Structural properties and decomposition of linear balanced matrices, *Mathematical Programming*, to appear.

Conforti, M., and Rao, M. R. (1991b), Properties of balanced and perfect matrices, *Mathematical Programming*, to appear.

Conforti, M., and Rao, M. R. (1991c), Articulation sets in linear perfect matrices and hypergraphs. I. Forbidden configurations and star cutsets, *Discrete Mathematics*, to appear.

Conforti, M., and Rao, M. R. (1991d), Articulation sets in linear perfect matrices and hypergraphs. II. The wheel theorem and clique articulations, *Discrete Mathematics*, to appear.

Cornuéjols, G., and Cunningham, W. H. (1985), Compositions for perfect graphs, *Discrete Mathematics* 55 (1985) 245–254.

Cornuéjols, G., Naddef, D., and Pulleyblank, W. R. (1983), Halin graphs and the travelling salesman problem, *Mathematical Programming* 26 (1983) 287–294.

Cornuéjols, G., Naddef, D., and Pulleyblank, W. (R.) (1985), The traveling salesman problem in graphs with 3-edge cutsets, *Journal of the Association for Computing Machinery* 32 (1985) 383–410.

Cornuéjols, G., and Novick, B. (1990), Ideal 0,1 matrices or the weak max-flow min-cut property, working paper, Carnegie-Mellon University, 1990.

Coullard, C. R. (1986), Counterexamples to conjectures on 4-connected matroids, *Combinatorica* 6 (1986) 315–320.

Coullard, C. R., Gardner, L. L., and Wagner, D. K. (1991), Decomposition of 3-connected graphs, *Combinatorica*, to appear.

Coullard, C. R., and Oxley, J. G. (1991), Extensions of Tutte's wheels-and-whirls theorem, *Journal of Combinatorial Theory (B)*, to appear.

Coullard, C. R., and Pulleyblank, W. R. (1989), On cycle cones and polyhedra, *Linear Algebra and Its Applications* 114/115 (1989) 613–640.

Coullard, C. R., and Reid, T. J. (1988), Element subsets of 3-connected matroids, *Congressus Numerantium* 66 (1988) 81–92.

Crama, Y., Hammer, P. L., and Ibaraki, T. (1986), Strong unimodularity for matrices and hypergraphs, *Discrete Applied Mathematics* 15 (1986) 221–239.

Crapo, H. H., and Rota, G.-C. (1970), *On the Foundations of Combinatorial Theory: Combinatorial Geometries* (preliminary edition), M.I.T. Press, Cambridge, 1970.

Cunningham, W. H. (1973), *A Combinatorial Decomposition Theory*, thesis, University of Waterloo, 1973.

Cunningham, W. H. (1979), Binary matroid sums, *Quarterly Journal of Mathematics Oxford* 30 (1979) 271–281.

Cunningham, W. H. (1981), On matroid connectivity, *Journal of Combinatorial Theory (B)* 30 (1981) 94–99.

Cunningham, W. H. (1982a), Separating cocircuits in binary matroids, *Linear Algebra and Its Applications* 43 (1982) 69–86.

Cunningham, W. H. (1982b), Polyhedra for composed independence systems, *Annals of Discrete Mathematics* 16 (1982) 57–67.

Cunningham, W. H. (1982c), Decomposition of directed graphs, *SIAM Journal on Algebraic and Discrete Methods* 3 (1982) 214–228.

Cunningham, W. H. (1986), Improved bounds for matroid partition and intersection algorithms, *SIAM Journal on Computing* 15 (1986) 948–957.

Cunningham, W. H., and Edmonds, J. (1978), Decomposition of Linear Systems, unpublished.

Cunningham, W. H., and Edmonds, J. (1980), A combinatorial decomposition theory, *Canadian Journal of Mathematics* 32 (1980) 734–765.

Dantzig, G. B. (1963), *Linear Programming and Extensions*, Princeton University Press, Princeton, New Jersey, 1963.

Dantzig, G. B., and Fulkerson, D. R. (1956), On the max-flow min-cut theorem of networks, in: *Linear Inequalities and Related Systems* (H. W. Kuhn and A. W. Tucker, eds.), Princeton University Press, Princeton, New Jersey, 1956, pp. 215–221.

Dirac, G. A. (1952), A property of 4-chromatic graphs and some remarks on critical graphs, *Journal of the London Mathematical Society* 27 (1952) 85–92.

Duchamp, A. (1974), Une caractérisation des matroïdes binaires par une propriété de coloration, *Comptes Rendus Hebdomadaires des Séances de l'Académie des Sciences Paris (A)* 278 (1974) 1163–1166.

Duffin, R. J. (1965), Topology of series-parallel networks, *Journal of Mathematical Analysis and Applications* 10 (1965) 303–318.

Edmonds, J. (1965a), Minimum partition of a matroid into independent subsets, *Journal of Research of the National Bureau of Standards (B)* 69 (1965) 67–72.

Edmonds, J. (1965b), Lehman's switching game and a theorem of Tutte and Nash-Williams, *Journal of Research of the National Bureau of Standards (B)* 69 (1965) 73–77.

Edmonds, J. (1965c), Maximum matching and a polyhedron with {0, 1} vertices, *Journal of Research of the National Bureau of Standards (B)* 69 (1965) 125–130.

Edmonds, J. (1965d), Paths, trees, and flowers, *Canadian Journal of Mathematics* 17 (1965) 449–467.

Edmonds, J. (1967a), Optimum branchings, *Journal of Research of the National Bureau of Standards (B)* 71 (1967) 233–240.

Edmonds, J. (1967b), Systems of distinct representatives and linear algebra, *Journal of Research of the National Bureau of Standards (B)* 71 (1967) 241–245.

Edmonds, J. (1970), Submodular functions, matroids, and certain polyhedra, in: *Combinatorial Structures and Their Applications* (R. Guy, H. Hanani, N. Sauer, and J. Schönheim, eds.), Gordon and Breach, New York, 1970, pp. 69–87.

Edmonds, J. (1979), Matroid intersection, in: *Discrete Optimization I* (P. L. Hammer, E. L. Johnson, and B. H. Korte, eds.), *Annals of Discrete Mathematics* 4 (1979) 39–49.

Edmonds, J., and Fulkerson, D. R. (1965), Transversals and matroid partition, *Journal of Research of the National Bureau of Standards (B)* 69 (1965) 147–153.

Edmonds, J., and Giles, R. (1977), A min-max relation for submodular functions on graphs, in: *Studies in Integer Programming* (P. L. Hammer, E. L. Johnson, B. H. Korte, and G. L. Nemhauser, eds.), *Annals of Discrete Mathematics* 1 (1977) 185–204.

Edmonds, J., and Johnson, E. L. (1973), Matching, Euler tours and the Chinese postman, *Mathematical Programming* 5 (1973) 88–124.

Edmonds, J., Lovász, L., and Pulleyblank, W. R. (1982), Brick decompositions and the matching rank of graphs, *Combinatorica* 2 (1982) 247–274.

El-Mallah, E. S., and Colbourn, C. J. (1990), On two dual classes of planar graphs, *Discrete Mathematics* 80 (1990) 21–40.

Epifanov, G. V. (1966), Reduction of a plane graph to an edge by a star-triangle transformation, *Doklady* 166 (1966) 13–17.

Faddeev, D. K., and Faddeeva, V. N. (1963), *Computational Methods of Linear Algebra*, Freeman, San Francisco, 1963.

Farley, A. M. (1981), Networks immune to isolated failures, *Networks* 11 (1981) 255–268.

Farley, A. M., and Proskurowski, A. (1982), Networks immune to isolated line failures, *Networks* 12 (1982) 393–403.

Fellows, M. R. (1989), The Robertson–Seymour theorems: A survey of applications (Proceedings AMS–IMS–SIAM Joint Summer Conference on Graphs and Algorithms, Boulder, Colorado, 1987) (R. B. Richter, ed.), *Contemporary Mathematics* 89 (1989) 1–18.

Feo, T. A. (1985), *Efficient reduction of planar networks for solving certain combinatorial problems*, thesis, University of California, Berkeley, 1985.

Feo, T. A., and Provan, J. S. (1988), Delta-wye transformations and the efficient reduction of two-terminal planar graphs, working paper, University of North Carolina, Chapel Hill, 1988.

Folkman, J., and Lawrence, J. (1978), Oriented matroids, *Journal of Combinatorial Theory (B)* 25 (1978) 199–236.

Fonlupt, J., and Naddef, D. (1991), The traveling salesman problem in graphs with some excluded minors, *Mathematical Programming*, to appear.

Fonlupt, J., and Raco, M. (1984), Orientation of matrices, *Mathematical Programming Study* 22 (1984) 86–98.

Ford, Jr, L. R., and Fulkerson, D. R. (1962), *Flows in Networks*, Princeton University Press, Princeton, New Jersey, 1962.

Fournier, J.-C. (1981), A characterization of binary geometries by a double elimination axiom, *Journal of Combinatorial Theory (B)* 31 (1981) 249–250.

Frank, A. (1981), A weighted matroid intersection algorithm, *Journal of Algorithms* 2 (1981) 328–336.

Frank, A. (1990a), Conservative weightings and ear-decompositions of graphs, in: *Proceedings of Conference on Integer Programming and Combinatorial Optimization*, University of Waterloo, 1990, (R. Kannan and W. R. Pulleyblank, eds.), University of Waterloo Press, Waterloo, Ontario, Canada, 1990.

Frank, A. (1990b), Submodular functions in graph theory, Report 900660-OR, University of Bonn, 1990.

Fujishige, S. (1980), An efficient PQ-graph algorithm for solving the graph-realization problem, *Journal of Computer and Systems Sciences* 21 (1980) 63–86.

Fujishige, S. (1983), Canonical decompositions of symmetric submodular systems, *Discrete Applied Mathematics* 5 (1983) 175–190.

Fujishige, S. (1984), Submodular systems and related topics, *Mathematical Programming Study* 22 (1984) 113–131.

Fujishige, S. (1985), A decomposition of distributive lattices, *Discrete Mathematics* 55 (1985) 35–55.

Fujishige, S. (1989), Linear and nonlinear optimization problems with submodular constraints, in: *Mathematical Programming. Recent Developments and Applications* (M. Iri and K. Tanabe, eds.), KTK Scientific Publishers, Tokyo, and Kluwer Academic Publishers, Dordrecht, 1989, pp. 203–225.

Fujishige, S. (1991), *Submodular functions and optimization* (Annals of Discrete Mathematics, Vol. 47), North-Holland, Amsterdam, 1991.

Fulkerson, D. R., Hoffman, A. J., and Oppenheim, R. (1974), On balanced matrices, *Mathematical Programming Study* 1 (1974) 120–132.

Gabor, C. P., Supowit, K. J., and Hsu, W.-L. (1989), Recognizing circle graphs in polynomial time, *Journal of the Association for Computing Machinery* 36 (1989) 435–473.

Garey, M. R., and Johnson, D. S. (1979), *Computers and Intractability: a Guide to the Theory of NP-Completeness*, Freeman, San Francisco, 1979.

Gerards, A. M. H. (1988), *Graphs and Polyhedra*, thesis, Catholic University of Brabant, 1988.

Gerards, A. M. H. (1989a), A short proof of Tutte's characterization of totally unimodular matrices, *Linear Algebra and Its Applications* 114/115 (1989) 207–212.

Gerards, A. M. H. (1989b), A min-max relation for stable sets in graphs with no odd-K_4, *Journal of Combinatorial Theory (B)* 47 (1989) 330–348.

Gerards, A. M. H. (1990), On Tutte's characterization of graphic matroids – a graphic proof, working paper, Centre for Mathematics and Computer Science, Amsterdam, 1990.

Gerards, A. M. H., Lovász, L., Schrijver, A., Seymour, P. D., and Truemper, K. (1991), Regular matroids from graphs, manuscript, 1991.

Gerards, A. M. H., and Schrijver, A. (1986), Matrices with the Edmonds-Johnson property, *Combinatorica* 6 (1986) 365–379.

Ghouila-Houri, A. (1962), Caractérisation des matrices totalement unimodulaires, *Comptes Rendus Hebdomadaires des Séances de l'Académie des Sciences Paris* 254 (1962) 1192–1194.

Gitler, I. (1991), *Delta-Wye-Delta Transformations – Algorithms and Applications*, thesis, University of Waterloo, 1991.

Gondran, M. (1973), Matrices totalement unimodulaires, Électricité de France, Bulletin de la Direction des Études et Recherches, Série C (Mathématiques, Informatique) No. 1 (1973) 55–73.

Gould, R. (1958), Graphs and vector spaces, *Journal of Mathematics and Physics* 37 (1958) 193–214.

Grötschel, M., Lovász, L., and Schrijver, A. (1988), *Geometric Algorithms and Combinatorial Optimization*, Springer Verlag, Heidelberg, 1988.

Grötschel, M., and Truemper, K. (1989a), Master polytopes for cycles in binary matroids, *Linear Algebra and Its Applications* 114/115 (1989) 523–540.

Grötschel, M., and Truemper, K. (1989b), Decomposition and optimization over cycles in binary matroids, *Journal of Combinatorial Theory (B)* 46 (1989) 306–337.

Grünbaum, B. (1967), *Convex Polytopes*, Interscience-Wiley, London, 1967.

Hadlock, F. (1975), Finding a maximum cut of a planar graph in polynomial time, *SIAM Journal on Computing* 4 (1975) 221–225.

Hadwiger, H. (1943), Über eine Klassifikation der Streckenkomplexe, *Vierteljahrsschrift der naturforschenden Gesellschaft in Zürich* 88 (1943) 133–142.

Halin, R. (1964), Über einen Satz von K. Wagner zum Vierfarbenproblem, *Mathematische Annalen* 153 (1964) 47–62.

Halin, R. (1967), Zur Klassifikation der endlichen Graphen nach H. Hadwiger und K. Wagner, *Mathematische Annalen* 172 (1967) 46–78.

Halin, R. (1969), Zur Theorie der n-fach zusammenhängenden Graphen, *Abhandlungen aus dem Mathematischen Seminar der Universität Hamburg* 33 (1969) 133–164.

Halin, R. (1981), *Graphentheorie II*, Wissenschaftliche Buchgesellschaft, Darmstadt, 1981.

Harary, F. (1969), *Graph Theory*, Addison-Wesley, Reading, Massachussetts, 1969.

Hartvigsen, D. B., and Wagner, D. K. (1988), Recognizing max-flow mincut path matrices, *Operations Research Letters* 7 (1988) 37–42.

Hassin, R. (1982), Minimum cost flow with set-constraints, *Networks* 12 (1982) 1–21.

Hassin, R. (1988), Solution bases of multiterminal cut problems, *Mathematics of Operations Research* 13 (1988) 535–542.

Hassin, R. (1990a), An algorithm for computing maximum solution bases, *Operations Research Letters* 9 (1990) 315–318.

Hassin, R. (1990b), Simultaneous solution of families of problems, (Proceedings SIGAL International Symposium on Algorithms, Tokyo, 1990), *Algorithms, Lecture Notes in Computer Science* 450, (T. Asano, T. Ibaraki, H. Imai, T. Nishizeki, eds.), Springer Verlag, Berlin, 1990, pp. 288–299.

Hassin, R. (1991), Multiterminal Xcut problems, *Annals of Operations Research*, to appear.

Heller, I. (1957), On linear systems with integral valued solutions, *Pacific Journal of Mathematics* 7 (1957) 1351–1364.

Heller, I. (1963), On unimodular sets of vectors, in: *Recent Advances in Mathematical Programming* (R. L. Graves and P. Wolfe, eds.), McGraw-Hill, New York, 1963, pp. 39–53.

Heller, I., and Hoffman, A. J. (1962), On unimodular matrices, *Pacific Journal of Mathematics* 12 (1962) 1321–1327.

Heller, I., and Tompkins, C. B. (1956), An extension of a theorem of Dantzig's, in: *Linear Inequalities and Related Systems* (H. W. Kuhn and A. W. Tucker, eds.), Princeton University Press, Princeton, New Jersey, 1956, pp. 247–254.

Hempel, L., Herrmann, U., Hölzer, W., and Wetzel, R. (1989), A necessary and sufficient condition for the total unimodularity of a matrix in terms of graph theory, *Optimization* 20 (1989) 901–914.

Hoang, C. T. (1985), On the P_4-structure of perfect graphs. II. Odd decompositions, *Journal of Combinatorial Theory (B)* 39 (1985) 220–232.

Hoffman, A. J. (1960), Some recent applications of the theory of linear inequalities to extremal combinatorial analysis, in: *Combinatorial Analysis* (Proceedings of Symposia in Applied Mathematics, Vol. X) (R. Bellman and M. Hall, Jr., eds.), American Mathematical Society, Providence, Rhode Island, 1960, 113–127.

Hoffman, A. J. (1974), A generalization of max flow-min cut, *Mathematical Programming* 6 (1974) 352–359.

Hoffman, A. J. (1976), Total unimodularity and combinatorial theorems, *Linear Algebra and Its Applications* 13 (1976) 103–108.

Hoffman, A. J. (1979), The role of unimodularity in applying linear inequalities to combinatorial theorems, in: *Discrete Optimization I* (P. L. Hammer, E. L. Johnson, and B. H. Korte, eds.), *Annals of Discrete Mathematics* 4 (1979) 73–84.

Hoffman, A. J., and Kruskal, J. B. (1956), Integral boundary points of convex polyhedra, in: *Linear Inequalities and Related Systems* (H. W. Kuhn and A. W. Tucker, eds.), Princeton University Press, Princeton, New Jersey, 1956, pp. 223–246.

Hoffman, A. J., and Kuhn, H. W. (1956), Systems of distinct representatives and linear programming, *The American Mathematical Monthly* 63 (1956) 455–460.

Hoffman, A. J., and Oppenheim, R. (1978), Local unimodularity in the matching polytope, in: *Algorithmic Aspects of Combinatorics* (B. Alspach, P. Hell, and D. J. Miller, eds.), *Annals of Discrete Mathematics* 2 (1978) 201–209.

Hopcroft, J. E., and Tarjan, R. E. (1973), Dividing a graph into triconnected components, *SIAM Journal on Computing* 2 (1973) 135–158.

Hsu, W.-L. (1986), Coloring planar perfect graphs by decomposition, *Combinatorica* 6 (1986) 381–385.

Hsu, W.-L. (1987a), Recognizing planar perfect graphs, *Journal of the Association for Computing Machinery* 34 (1987) 255–288.

Hsu, W.-L. (1987b), Decomposition of perfect graphs, *Journal of Combinatorial Theory (B)* 43 (1987) 70–94.

Hsu, W.-L. (1988), The coloring and maximum independent set problems on planar perfect graphs, *Journal of the Association for Computing Machinery* 35 (1988) 535–563.

Ingleton, A. W. (1959), A note on independence functions and rank, *Journal of the London Mathematical Society* 34 (1959) 49–56.

Ingleton, A. W. (1971), Representation of matroids, in: *Combinatorial Mathematics and Its Applications* (proceedings of a conference held at the Mathematical Institute, Oxford) (D. J. A. Welsh, ed.), Academic Press, London, 1971, pp. 149–167.

Inukai, T., and Weinberg, L. (1979), Graph-realizability of matroids, *Annals of New York Academy of Sciences* 319 (1979) 289–305.

Inukai, T., and Weinberg, L. (1981), Whitney connectivity of matroids, *SIAM Journal on Algebraic and Discrete Methods* 2 (1981) 108–120.

Iri, M. (1968), On the synthesis of loop and cutset matrices and the related problems, *RAAG Memoirs* A 4 (1968) 4–38.

Iri, M. (1969), The maximum-rank minimum-term-rank for the pivotal transforms of a matrix, *Linear Algebra and Its Applications* 2 (1969) 427–446.

Iri, M. (1979), A review of recent work in Japan on principal partitions of matroids and their applications, *Annals of the New York Academy of Sciences* 319 (1979) 306–319.

Iri, M. (1983), Applications of matroid theory, in: *Mathematical Programming: The State of the Art* (A. Bachem, M. Grötschel, and B. Korte, eds.), Springer Verlag, Berlin, 1983, pp. 158–201.

Iri, M., and Fujishige, S. (1981), Use of matroid theory in operations research, circuits, and systems theory, *International Journal of Systems Science* 12 (1981) 27–54.

Jensen, P. M., and Korte, B. (1982), Complexity of matroid property algorithms, *SIAM Journal on Computing* 11 (1982) 184–190.

Kahn, J. (1982), Characteristic sets of matroids, *Journal of the London Mathematical Society* 26 (1982) 207–217.

Kahn, J. (1984), A geometric approach to forbidden minors for GF(3), *Journal of Combinatorial Theory (A)* 37 (1984) 1–12.

Kahn, J. (1985), A problem of P. Seymour on nonbinary matroids, *Combinatorica* 5 (1985) 319–323.

Kahn, J. (1988), On the uniqueness of matroid representations over GF(4), *Bulletin of the London Mathematical Society* 20 (1988) 5–10.

Kahn, J., and Kung, J. P. S. (1982), Varieties of combinatorial geometries, *Transactions of the American Mathematical Society* 271 (1982) 485–499.

Kahn, J., and Seymour, P. D. (1988), On forbidden minors for GF(3), *Proceedings of the American Mathematical Society* 102 (1988) 437–440.

Karp, R. M. (1971), A simple derivation of Edmonds' algorithm for optimum branchings, *Networks* 1 (1971) 265–272.

Kelmans, A. K. (1987), A short proof and a strengthening of the Whitney 2-isomorphism theorem on graphs, *Discrete Mathematics* 64 (1987) 13–25.

Kishi, G., and Kajitani, Y. (1967), Maximally distant trees in a linear graph (in Japanese), *Transactions of the Institute of Electronics and Communication Engineers of Japan* 51A (1968) 196–203.

König, D. (1936), *Theorie der endlichen und unendlichen Graphen*, Akademische Verlagsgesellschaft, Leipzig, 1936 (reprinted: Chelsea, New York, 1950, and Teubner, Leipzig, 1986).

Kress, M., and Tamir, A. (1980), The use of Jacobi's lemma in unimodularity theory, *Mathematical Programming* 18 (1980) 344–348.

Krogdahl, S. (1977), The dependence graph for bases in a matroid, *Discrete Mathematics* 19 (1977) 47–59.

Kung, J. P. S. (1986a), Numerically regular hereditary classes of combinatorial geometries, *Geometriae Dedicata* 21 (1986) 85–105.

Kung, J. P. S. (1986b), Growth rates and critical exponents of classes of binary combinatorial geometries, *Transactions of the American Mathematical Society* 293 (1986) 837–859.

Kung, J. P. S. (1986c), *A Source Book in Matroid Theory*, Birkhäuser, Boston, 1986.

Kung, J. P. S. (1987), Excluding the cycle geometries of the Kuratowski graphs from binary geometries, *Proceedings of the London Mathematical Society* 55 (1987) 209–242.

Kung, J. P. S. (1988), The long-line graph of a combinatorial geometry. I. Excluding $M(K_4)$ and the $(q + 2)$-point line as minors, *Quarterly Journal of Mathematics Oxford* 39 (1988) 223–234.

Kung, J. P. S. (1990a), The long-line graph of a combinatorial geometry. II. Geometries representable over two fields of different characteristics, *Journal of Combinatorial Theory (B)* 50 (1990) 41–53.

Kung, J. P. S. (1990b), Combinatorial geometries representable over GF(3) and GF(q). I. The number of points, *Discrete and Computational Geometry* 5 (1990) 83–95.

Kung, J. P. S., and Oxley, J. G. (1988), Combinatorial geometries representable over GF(3) and GF(q). II. Dowling geometries, *Graphs and Combinatorics* 4 (1988) 323–332.

Kuratowski, K. (1930), Sur le problème des courbes gauches en topologie, *Fundamenta Mathematicae* 15 (1930) 271–283.

Lancaster, P., and Tismenetsky, M. (1985), *The Theory of Matrices with Applications*, Academic Press, Orlando, Florida, 1985.

Lawler, E. L. (1975), Matroid intersection algorithms, *Mathematical Programming* 9 (1975) 31–56.

Lawler, E. L. (1976), *Combinatorial Optimization: Networks and Matroids*, Holt, Rinehart and Winston, New York, 1976.

Lawler, E. L., and Martel, C. U. (1982a), Computing maximal "polymatroidal" network flows, *Mathematics of Operations Research* 7 (1982) 334–347.

Lawler, E. L., and Martel, C. U. (1982b), Flow network formulations of polymatroid optimization problems, in: *Bonn Workshop on Combinatorial Optimization* (A. Bachem, M. Grötschel, and B. Korte, eds.), *Annals of Discrete Mathematics* 16 (1982) 189–200.

Lazarson, T. (1958), The representation problem for independence functions, *Journal of the London Mathematical Society* 33 (1958) 21–25.

Lehman, A. (1963), Wye-delta transformations in probabilistic networks, *Journal of the Society of Industrial and Applied Mathematics (SIAM)* 11 (1963) 773–805.

Lehman, A. (1964), A solution of the Shannon switching game, *Journal of the Society of Industrial and Applied Mathematics (SIAM)* 12 (1964) 687–725.

Lehman, A. (1981), The width-length inequality and degenerative projective planes, unpublished manuscript, 1981.

Lemos, M. (1991), K-elimination property for circuits of matroids, *Journal of Combinatorial Theory (B)* 51 (1991) 211–226.

Löfgren, L. (1959), Irredundant and redundant Boolean branch-networks, *IRE Transactions on Circuit Theory* CT-6 (Special Supplement) (1959) 158–175.

Lovász, L. (1980), Matroid matching and some applications, *Journal of Combinatorial Theory (B)* 28 (1980) 208–236.

Lovász, L. (1983), Ear-decompositions of matching-covered graphs, *Combinatorica* 3 (1983) 105–117.

Lovász, L. (1987), Matching structure and the matching lattice, *Journal of Combinatorial Theory (B)* 43 (1987) 187–222.

Lovász, L., and Plummer, M. D. (1975), On bicritical graphs, in: *Infinite and Finite Sets* (Colloquia Mathematica Societatis János Bolyai 10) (A. Hajnal, R. Rado, and V. T. Sós, eds.), North-Holland, Amsterdam, 1975, pp. 1051–1079.

Lovász, L., and Plummer, M. D. (1986), *Matching Theory*, Akadémiai Kiadó, Budapest, 1986.

MacLane, S. (1936), Some interpretations of abstract linear dependence in terms of projective geometry, *American Journal of Mathematics* 58 (1936) 236–240.

Mahjoub, A. R. (1988), On the stable set polytope of a series-parallel graph, *Mathematical Programming* 40 (1988) 53–57.

Maurras, J. F., Truemper, K., and Akgül, M. (1981), Polynomial algorithms for a class of linear programs, *Mathematical Programming* 21 (1981) 121–136.

Menger, K. (1927), Zur allgemeinen Kurventheorie, *Fundamenta Mathematicae* 10 (1927) 96–115.

Minty, G. J. (1966), On the axiomatic foundations of the theories of directed linear graphs, *Journal of Mathematics and Mechanics* 15 (1966) 485–520.

Monma, C. L., and Sidney, J. B. (1979), Sequencing with series-parallel precedence constraints, *Mathematics of Operations Research* 4 (1979) 215–224.

Moore, E. F., and Shannon, C. E. (1956), Reliable circuits using less reliable relays I–II, *Journal of the Franklin Institute* 262 (1956) 191–208 and 281–297.

Motzkin, T. S. (1956), The assignment problem, in: *Numerical Analysis* (Proceedings of Symposia in Applied Mathematics, Vol. VI) (H. Curtiss, ed.), McGraw-Hill, New York, 1956, pp. 109–125.

Murota, K. (1987), *Systems Analysis by Graphs and Matroids*, Springer Verlag, Berlin, 1987.

Murota, K., and Iri, M. (1985), Structural solvability of systems of equations — a mathematical formulation for distinguishing accurate and inaccurate numbers in structural analysis of systems, *Japan Journal of Applied Mathematics* 2 (1985) 247–271.

Murota, K., Iri, M., and Nakamura, M. (1987), Combinatorial canonical form of layered mixed matrices and its application to block-triangularization of systems of linear/nonlinear equations, *SIAM Journal on Algebraic and Discrete Methods* 8 (1987) 123–149.

Murty, U. S. R. (1976), Extremal matroids with forbidden restrictions and minors, in: *Proceedings of the Seventh Southeastern Conference on Combinatorics* (F. Hoffman, L. Lesniak, R. Mullin, K. B. Reid, and R. Stanton, eds.), *Congressus Numerantium* 17 (1976) 463–468.

Nakamura, M., and Iri, M. (1979), Fine structures of matroid intersections and their applications, in: *Proceedings of the 1979 International Symposium on Circuits and Systems*, Tokyo, 1979, pp. 996–999.

Narayanan, H., and Vartak, N. (1981), An elementary approach to the principal partition of a matroid, *Transactions of the Institute of Electronics and Communication Engineers of Japan* E64 (1981) 227–234.

Nash-Williams, C. St. J. A. (1961), Edge-disjoint spanning trees of finite graphs, *Journal of the London Mathematical Society* 36 (1961) 445–450.

Nash-Williams, C. St. J. A. (1964), Decomposition of finite graphs into forests, *Journal of the London Mathematical Society* 39 (1964) 12.

Nash-Williams, C. St. J. A. (1966), An application of matroids to graph theory, *Theory of Graphs International Symposium (Rome)*, Dunod, Paris, 1966, pp. 263–265.

Negami, S. (1982), A characterization of 3-connected graph containing a given graph, *Journal of Combinatorial Theory (B)* 32 (1982) 69–74.

Nemhauser, G., and Wolsey, L. (1988), *Integer and Combinatorial Optimization*, Wiley, New York, 1988.

Nishizeki, T., and Saito, N. (1975), Necessary and sufficient conditions for a graph to be three-terminal series-parallel, *IEEE Transactions on Circuits and Systems* CAS-22 (1975) 648–653.

Nishizeki, T., and Saito, N. (1978), Necessary and sufficient conditions for a graph to be three-terminal series-parallel-cascade, *Journal of Combinatorial Theory (B)* 24 (1978) 344–361.

Ore, O. (1962), *Theory of Graphs* (American Mathematical Society Colloquium Publications, Vol. 38), American Mathematical Society, Providence, Rhode Island, 1962.

Ore, O. (1967), *The Four-Color Problem* (Pure and Applied Mathematics, Vol. 27), Academic Press, New York, 1967.

Orlova, G. I., and Dorfman, Y. G. (1972), Finding the maximum cut in a graph (in Russian), Izvestija Akademii Nauk SSSR, Tehničeskaja Kibernetika (1972) (3) 155–159, (English translation: *Engineering Cybernetics* 10 (1972) 502–506).

Oxley, J. G. (1981a), On a matroid generalization of graph connectivity, *Mathematical Proceedings of the Cambridge Philosophical Society* 90 (1981) 207–214.

Oxley, J. G. (1981b), On 3-connected matroids, *Canadian Journal of Mathematics* 33 (1981) 20–27.

Oxley, J. G. (1984), On singleton 1-rounded sets of matroids, *Journal of Combinatorial Theory (B)* 37 (1984) 189–197.

Oxley, J. G. (1986), On the matroids representable over GF(4), *Journal of Combinatorial Theory (B)* 41 (1986) 250–252.

Oxley, J. G. (1987a), On nonbinary 3-connected matroids, *Transactions of the American Mathematical Society* 300 (1987) 663–679.

Oxley, J. G. (1987b), The binary matroids with no 4-wheel minor, *Transactions of the American Mathematical Society* 301 (1987) 63–75.

Oxley, J. G. (1987c), A characterization of the ternary matroids with no $M(K_4)$-minor, *Journal of Combinatorial Theory (B)* 42 (1987) 212–249.

Oxley, J. G. (1989a), A characterization of certain excluded-minor classes of matroids, *European Journal of Combinatorics* 10 (1989) 275–279.

Oxley, J. G. (1989b), The regular matroids with no 5-wheel minor, *Journal of Combinatorial Theory (B)* 46 (1989) 292–305.

Oxley, J. G. (1990a), On an excluded-minor class of matroids, *Discrete Mathematics* 82 (1990) 35–52.

Oxley, J. G. (1990b), A characterization of a class of non-binary matroids, *Journal of Combinatorial Theory (B)* 49 (1990) 181–189.

Oxley, J. G. (1991), *Matroid Theory*, Oxford University Press, Oxford, to appear.

Oxley, J. G., and Reid, T. J. (1990), The smallest rounded set of binary matroids, *European Journal of Combinatorics* 11 (1990) 47–56.

Oxley, J. G., and Row, D. (1989), On fixing elements in matroid minors, *Combinatorica* 9 (1989) 69–74.

Ozawa, T. (1971), On the minimal graphs of a certain class of graphs, *IEEE Transactions on Circuit Theory* CT-18 (1971) 387.

Padberg, M. W. (1974), Perfect zero-one matrices, *Mathematical Programming* 6 (1974) 180–196.

Padberg, M. W. (1975), Characterisations of totally unimodular, balanced and perfect matrices, in: *Combinatorial Programming: Methods and Applications* (B. Roy, ed.), Reidel, Dordrecht (The Netherlands), 1975, pp. 275–284.

Padberg, M. W. (1976), A note on the total unimodularity of matrices, *Discrete Mathematics* 14 (1976) 273–278.

Padberg, M. W. (1988), Total unimodularity and the Euler-subgraph problem, *Operations Research Letters* 7 (1988) 173–179.

Politof, T. (1988a), $\Delta - Y$ reducible graphs, working paper WPS 88-12-52, Concordia University, Montreal, 1988.

Politof, T. (1988b), $Y - \Delta$ reducible graphs, working paper WPS 88-12-51, Concordia University, Montreal, 1988.

Politof, T., and Satyanarayana, A. (1986), Network reliability and inner-four-cycle-free graphs, *Mathematics of Operations Research* 11 (1986) 484–505. A correction given in: Resende, L. I. P. (1988), New expressions for the extended $\Delta - Y$ reductions, working paper, University of California, Berkeley, 1988.

Politof, T., and Satyanarayana, A. (1990), A linear time algorithm to compute the reliability of planar cube-free networks, *IEEE Transactions on Reliability* 39 (1990) 557–563.

Rado, R. (1957), Note on independence functions, *Proceedings of the London Mathematical Society* 7 (1957) 300–320.

Rajan, A. (1986), *Algorithmic Applications of Connectivity and Related Topics in Matroid Theory*, thesis, Northwestern University, 1986.

Ratliff, H. D., and Rosenthal, A. S. (1983), Order picking in a rectangular warehouse: A solvable case of the travelling salesman problem, *Operations Research* 31 (1983) 507–521.

Rebman, K. R. (1974), Total unimodularity and the transportation problem: a generalization, *Linear Algebra and Its Applications* 8 (1974) 11–24.

Recski, A. (1989), *Matroid Theory and Its Applications in Electrical Networks and Statics*, Springer Verlag, Heidelberg, 1989.

Reid, T. J. (1990), Fixing elements in minors of binary matroids, *Congressus Numerantium* 73 (1990) 215–222.

Reid, T. J. (1991a), Triangles in 3-connected matroids, *Discrete Mathematics*, to appear.

Reid, T. J. (1991b), A note on roundedness in 4-connected matroids, *Discrete Mathematics*, to appear.

Reid, T. J. (1991c), On fixing edges in graph minors, *Graphs and Combinatorics*, to appear.

Reid, T. J. (1991d), The binary matroids having an element which is in every four-wheel minor, *Ars Combinatoria*, to appear.

Robertson, N. (1984), Minimal cyclic-4-connected graphs, *Transactions of the American Mathematical Society* 284 (1984) 665–687.

Robertson, N. (1988), private communication.

Robertson, N., and Seymour, P. D. (1983), Graph Minors. I. Excluding a forest, *Journal of Combinatorial Theory (B)* 35 (1983) 39–61.

Robertson, N., and Seymour, P. D. (1984), Graph Minors III. Planar tree-width, *Journal of Combinatorial Theory (B)* 36 (1984) 49–64.

Robertson, N., and Seymour, P. D. (1985), Graph Minors – A Survey, in: *Surveys in Combinatorics* 1985 (Proceedings, Tenth British Combinatorial Conference) (I. Anderson, ed.), *London Mathematical Society Lecture Note Series* 103 (1985) 153–171.

Robertson, N., and Seymour, P. D. (1986a), Graph minors. II. Algorithmic aspects of tree-width, *Journal of Algorithms* 7 (1986) 309–322.

Robertson, N., and Seymour, P. D. (1986b), Graph minors. V. Excluding a planar graph, *Journal of Combinatorial Theory (B)* 41 (1986) 92–114.

Robertson, N., and Seymour, P. D. (1986c), Graph minors. VI. Disjoint paths across a disc, *Journal of Combinatorial Theory (B)* 41 (1986) 115–138.

Robertson, N., and Seymour, P. D. (1988), Graph minors. VII. Disjoint paths on a surface, *Journal of Combinatorial Theory (B)* 45 (1988) 212–254.

Robertson, N., and Seymour, P. D. (1990a), Graph minors. IV. Tree-width and well-quasi-ordering, *Journal of Combinatorial Theory (B)* 48 (1990) 227–254.

Robertson, N., and Seymour, P. D. (1990b), Graph minors. VIII. A Kuratowski theorem for general surfaces, *Journal of Combinatorial Theory (B)* 48 (1990) 255–288.

Robertson, N., and Seymour, P. D. (1990c), Graph minors. IX. Disjoint crossed paths, *Journal of Combinatorial Theory (B)* 49 (1990) 40–77.

Robertson, N., and Seymour, P. D. (1991a), Graph minors. X. Obstructions to tree-decomposition, *Journal of Combinatorial Theory (B)* 52 (1991) 153–190.

Robertson, N., and Seymour, P. D. (1991b), Graph minors. XI. Tree-decompositions and well-quasi-ordering, working paper, Bell Communications Research, Morristown, New Jersey, 1991.

Robertson, N., and Seymour, P. D. (1991c), Graph minors. XII. Circuits on a surface, working paper, Bell Communications Research, Morristown, New Jersey, 1991.

Robertson, N., and Seymour, P. D. (1991d), Graph minors. XIII. The disjoint paths problem, working paper, Bell Communications Research, Morristown, New Jersey, 1991.

Robertson, N., and Seymour, P. D. (1991e), Graph minors. XIV. Distance on a surface, working paper, Bell Communications Research, Morristown, New Jersey, 1991.

Robertson, N., and Seymour, P. D. (1991f), Graph minors. XV. Extending an embedding, working paper, Bell Communications Research, Morristown, New Jersey, 1991.

Robertson, N., and Seymour, P. D. (1991g), Graph minors. XVI. Giant steps, working paper, Bell Communications Research, Morristown, New Jersey, 1991.

Robertson, N., and Seymour, P. D. (1991h), Graph minors. XVII. Excluding a non-planar graph, working paper, Bell Communications Research, Morristown, New Jersey, 1991.

Robertson, N., and Seymour, P. D. (1991i), Graph minors. XVIII. Taming a vortex, in preparation.

Robertson, N., and Seymour, P. D. (1991j), Graph minors. XIX. Well-quasi-ordering on a surface, in preparation.

Robertson, N., and Seymour, P. D. (1991k), Graph minors. XX. Wagner's conjecture, in preparation.

Robinson, G. C., and Welsh, D. J. A. (1980), The computational complexity of matroid properties, *Mathematical Proceedings of the Cambridge Philosophical Society* 87 (1980) 29–45.

Rosenthal, A., and Frisque, D. (1977), Transformations for simplifying network reliability calculations, *Networks* 7 (1977) 97–111.

Rota, G.-C. (1970), Combinatorial theory, old and new, *Actes du Congrès International des Mathématiciens* (Nice), Vol. 3, Gauthier–Villars, Paris, 1970, pp. 229–233.

Roudneff, J.-P., and Wagowski, M. (1989), Characterizations of ternary matroids in terms of circuit signatures, *Journal of Combinatorial Theory (B)* 47 (1989) 93–106.

Saaty, T. L., and Kainen, P. C. (1977), *The Four Color Problem*, McGraw–Hill, New York (reprinted: Dover Publications, New York, 1986).

Satyanarayana, A., and Wood, R. K. (1985), A linear-time algorithm for computing K-terminal reliability in series-parallel networks, *SIAM Journal on Computing* 14 (1985) 818–832.

Schrijver, A. (1983), Min-max results in combinatorial optimization, in: *Mathematical Programming: The State of the Art* (A. Bachem, M. Grötschel, and B. Korte, eds.), Springer Verlag, Berlin, 1983, pp. 439–500.

Schrijver, A. (1984), Total dual integrality from directed graphs, crossing families, and sub- and supermodular functions, in: *Progress in Combinatorial Optimization* (W. R. Pulleyblank, ed.), Academic Press, Toronto, 1984, 315–361.

Schrijver, A. (1986), *Theory of Linear and Integer Programming*, Wiley, Chichester, 1986.

Seymour, P. D. (1976), A forbidden minor characterization of matroid ports, *Quarterly Journal of Mathematics Oxford* 27 (1976) 407–413.

Seymour, P. D. (1977a), The matroids with the max-flow min-cut property, *Journal of Combinatorial Theory (B)* 23 (1977) 189–222.

Seymour, P. D. (1977b), A note on the production of matroid minors, *Journal of Combinatorial Theory (B)* 22 (1977) 289–295.

Seymour, P. D. (1979a), Matroid representation over GF(3), *Journal of Combinatorial Theory (B)* 26 (1979) 159–173.

Seymour, P. D. (1979b), Sums of circuits, in: *Graph Theory and Related Topics* (J. A. Bondy and U. S. R. Murty, eds.), Academic Press, New York, 1979, pp. 341–355.

Seymour, P. D. (1980a), On Tutte's characterization of graphic matroids, in: *Combinatorics* 79 (Part 1) (M. Deza and I. G. Rosenberg, eds.), *Annals of Discrete Mathematics* 8 (1980) 83–90.

Seymour, P. D. (1980b), Decomposition of regular matroids, *Journal of Combinatorial Theory (B)* 28 (1980) 305–359.

Seymour, P. D. (1980c), Packing and covering with matroid circuits, *Journal of Combinatorial Theory (B)* 28 (1980) 237–242.

Seymour, P. D. (1981a), Matroids and multicommodity flows, *European Journal of Combinatorics* 2 (1981) 257–290.

Seymour, P. D. (1981b), private communication.

Seymour, P. D. (1981c), Recognizing graphic matroids, *Combinatorica* 1 (1981) 75–78.

Seymour, P. D. (1981d), On Tutte's extension of the four-colour problem, *Journal of Combinatorial Theory (B)* 31 (1981) 82–94.

Seymour, P. D. (1981e), On minors of non-binary matroids, *Combinatorica* 1 (1981) 387–394.

Seymour, P. D. (1981f), Some applications of matroid decomposition, in: *Algebraic Methods in Graph Theory* (Colloquia Mathematica Societatis János Bolyai 25) (L. Lovász and V. T. Sós, eds.), North-Holland, Amsterdam, 1981, pp. 713–726.

Seymour, P. D. (1985a), Applications of the regular matroid decomposition, in: *Matroid Theory* (Colloquia Mathematica Societatis János Bolyai 40) (L. Lovász and A. Recski, eds.), North-Holland, Amsterdam, 1985, pp. 345–357.

Seymour, P. D. (1985b), Minors of 3-connected matroids, *European Journal of Combinatorics* 6 (1985) 375–382.

Seymour, P. D. (1986a), Triples in matroid circuits, *European Journal of Combinatorics* 7 (1986) 177–185.

Seymour, P. D. (1986b), Adjacency in binary matroids, *European Journal of Combinatorics* 7 (1986) 171–176.

Seymour, P. D. (1988), On the connectivity function of a matroid, *Journal of Combinatorial Theory (B)* 45 (1988) 25–30.

Seymour, P. D. (1990), On Lehman's width-length characterization, in: *Polyhedral Combinatorics* (DIMACS Series in Discrete Mathematics and Theoretical Computer Science, Vol. 1) (W. Cook and P. D. Seymour, eds.), American Mathematical Society, 1990.

Seymour, P. D., and Walton, P. N. (1981), Detecting matroid minors, *Journal of the London Mathematical Society* 23 (1981) 193–203.

Soun, Y., and Truemper, K. (1980), Single Commodity Representation of Multicommodity Networks, *SIAM Journal on Algebraic and Discrete Methods* 1 (1980) 348–358.

Strang, G. (1980), *Linear Algebra and Its Applications* (2nd edition), Academic Press, New York, 1980.

Swaminathan, R., and Wagner, D. K. (1990), Vertex 2-isomorphism, Technical Report CC-90-11, Institute for Interdisciplinary Engineering Studies, Purdue University, 1990.

Takamizawa, K., Nishizeki, T., and Saito, N. (1982), Linear time computability of combinatorial problems on series-parallel graphs, *Journal of the Association for Computing Machinery* 29 (1982) 623–641.

Tamir, A. (1976), On totally unimodular matrices, *Networks* 6 (1976) 373–382.

Tamir, A. (1987), Totally balanced and totally unimodular matrices defined by center location problems, *Discrete Applied Mathematics* 16 (1987) 245–263.

Tan, J. J.-M. (1981), *Matroid 3-connectivity*, thesis, Carleton University, 1981.

Thomassen, C. (1980), Planarity and duality of finite and infinite graphs, *Journal of Combinatorial Theory (B)* 29 (1980) 244–271.

Thomassen, C. (1989), Whitney's 2-switching theorem, cycle spaces, and arc mappings of directed graphs, *Journal of Combinatorial Theory (B)* 46 (1989) 257–291.

Tomizawa, N. (1976a), An $O(m^3)$ algorithm for solving the realization problem on graphs — on combinatorial characterizations of graphic $(0, 1)$-matrices (in Japanese), in: *Papers of the Technical Group on Circuit and System Theory of the Institute of Electronics and Communication Engineers of Japan* CST (1976) 75–106.

Tomizawa, N. (1976b), Strongly irreducible matroids and principal partition of a matroid into strongly irreducible minors (in Japanese), *Transactions of the Institute of Electronics and Communication Engineers of Japan* J59-A (1976) 83–91.

Tomizawa, N., and Fujishige, S. (1982), Historical survey of extensions of the concept of principal partition and their unifying generalization to hypermatroids, in: *Proceedings, International Symposium on Circuits and Systems of IEEE Circuits and Systems Society*, Rome, 1982, pp. 142–145.

Truemper, K. (1977), Unimodular matrices of flow problems with additional constraints, *Networks* 7 (1977) 343–358.

Truemper, K. (1978), Algebraic characterizations of unimodular matrices, *SIAM Journal on Applied Mathematics* 35 (1978) 328–332.

Truemper, K. (1980a), On Whitney's 2-isomorphism theorem for graphs, *Journal of Graph Theory* 4 (1980) 43–49.

Truemper, K. (1980b), Complement total unimodularity, *Linear Algebra and Its Applications* 30 (1980) 77–92.

Truemper, K. (1982a), On the efficiency of representability tests for matroids, *European Journal of Combinatorics* 3 (1982) 275–291.

Truemper, K. (1982b), Alpha-balanced graphs and matrices and GF(3)-representability of matroids, *Journal of Combinatorial Theory (B)* 32 (1982) 112–139.

Truemper, K. (1984), Partial matroid representations, *European Journal of Combinatorics* 5 (1984) 377–394.

Truemper, K. (1985a), A decomposition theory for matroids. I. General results, *Journal of Combinatorial Theory (B)* 39 (1985) 43–76.

Truemper, K. (1985b), A decomposition theory for matroids. II. Minimal violation matroids, *Journal of Combinatorial Theory (B)* 39 (1985) 282–297.

Truemper, K. (1986), A decomposition theory for matroids. III. Decomposition conditions, *Journal of Combinatorial Theory (B)* 41 (1986) 275–305.

Truemper, K. (1987a), Max-flow min-cut matroids: Polynomial testing and polynomial algorithms for maximum flow and shortest routes, *Mathematics of Operations Research* 12 (1987) 72–96.

Truemper, K. (1987b), On matroid separations of graphs, *Journal of Graph Theory* 11 (1987) 531–536.

Truemper, K. (1988), A decomposition theory for matroids. IV. Decomposition of graphs, *Journal of Combinatorial Theory (B)* 45 (1988) 259–292.

Truemper, K. (1989a), On the delta-wye reduction for planar graphs, *Journal of Graph Theory* 13 (1989) 141–148.

Truemper, K. (1990), A decomposition theory for matroids. V. Testing of matrix total unimodularity, *Journal of Combinatorial Theory (B)* 49 (1990) 241-281.

Truemper, K. (1991), A decomposition theory for matroids. VI. Almost regular matroids, *Journal of Combinatorial Theory (B)* (1991), to appear.

Truemper, K. (1992), A decomposition theory for matroids. VII. Analysis of minimal violation matrices, *Journal of Combinatorial Theory (B)* (1992),to appear.

Truemper, K., and Chandrasekaran, R. (1978), Local unimodularity of matrix-vector pairs, *Linear Algebra and Its Applications* 22 (1978) 65–78.

Truemper, K., and Soun, Y. (1979), Minimal forbidden subgraphs of unimodular multicommodity networks, *Mathematics of Operations Research* 4 (1979) 379–389.

Tseng, F. T., and Truemper, K. (1986), A decomposition of the matroids with the max-flow min-cut property, *Discrete Applied Mathematics* 15 (1986) 329–364. Also: Addendum, *Discrete Applied Mathematics* 20 (1988) 87–88.

Tsuchiya, T., Ohtsuki, T., Ishizaki, Y., Watanabe, H., Kajitani, Y., and Kishi, G. (1967), Topological degrees of freedom of electrical networks, *Fifth Annual Allerton Conference on Circuit and System Theory* (1967) 644–653.

Tutte, W. T. (1958), A homotopy theorem for matroids I, II, *Transactions of the American Mathematical Society* 88 (1958) 144–160, 161–174.

Tutte, W. T. (1959), Matroids and graphs, *Transactions of the American Mathematical Society* 90 (1959) 527–552.

Tutte, W. T. (1960), An algorithm for determining whether a given binary matroid is graphic, *Proceedings of the American Mathematical Society* 11 (1960) 905–917.

Tutte, W. T. (1961), On the problem of decomposing a graph into n connected factors, *Journal of the London Mathematical Society* 36 (1961) 221–230.

Tutte, W. T. (1964), From matrices to graphs, *Canadian Journal of Mathematics* 16 (1964) 108–127.

Tutte, W. T. (1965), Lectures on matroids, *Journal of Research of the National Bureau of Standards (B)* 69 (1965) 1–47.

Tutte, W. T. (1966a), *Connectivity in Graphs*, University of Toronto Press, Toronto, 1966.

Tutte, W. T. (1966b), Connectivity in matroids, *Canadian Journal of Mathematics* 18 (1966) 1301–1324.

Tutte, W. T. (1971), *Introduction to the Theory of Matroids*, American Elsevier, New York, 1971.

Uhry, B. (1979), Polytope des indépendants d'un graph série-parallèle, *Discrete Mathematics* 27 (1979) 225–243.

Vamos, P. (1968), On the representation of independence structures, unpublished.

Vamos, P. (1978), The missing axiom of matroid theory is lost forever, *Journal of the London Mathematical Society* 18 (1978) 403–408.

Veinott, Jr., A. F., and Dantzig, G. B. (1968), Integral extreme points, *SIAM Review* 10 (1968) 371–372.

Wagner, D. K. (1983), *An Almost Linear-Time Graph Realization Algorithm*, thesis, Northwestern University, 1983.

Wagner, D. K. (1985a), On theorems of Whitney and Tutte, *Discrete Mathematics* 57 (1985) 147–154.

Wagner, D. K. (1985b), Connectivity in bicircular matroids, *Journal of Combinatorial Theory (B)* 39 (1985) 308–324.

Wagner, D. K. (1987), Forbidden subgraphs and graph decomposition, *Networks* 17 (1987) 105–110.

Wagner, D. K. (1988), Equivalent factor matroids of a graph, *Combinatorica* 8 (1988) 373–377.

Wagner, K. (1937a), Über eine Eigenschaft der ebenen Komplexe, *Mathematische Annalen* 114 (1937) 570–590.

Wagner, K. (1937b), Über eine Erweiterung eines Satzes von Kuratowski, *Deutsche Mathematik* 2 (1937) 280–285.

Wagner, K. (1960), Bemerkungen zu Hadwiger's Vermutung, *Mathematische Annalen* 141 (1960) 433–451.

Wagner, K. (1970), *Graphentheorie*, Bibliographisches Institut, Mannheim, 1970.

Wald, J. A., and Colbourn, C. J. (1983a), Steiner trees, partial 2-trees, and minimum IFI networks, *Networks* 13 (1983) 159–167.

Wald, J. A., and Colbourn, C. J. (1983b), Steiner trees in probabilistic networks, *Microelectronics and Reliability* 23 (1983) 837–840.

Welsh, D. J. A. (1976), *Matroid Theory*, Academic Press, London, 1976.

de Werra, D. (1981), On some characterisations of totally unimodular matrices, *Mathematical Programming* 20 (1981) 14–21.

White, N. (1986), *Theory of Matroids*, Cambridge University Press, Cambridge, 1986.

White, N. (1987), *Combinatorial Geometries*, Cambridge University Press, Cambridge, 1987.

White, N. (1991), *Matroid Applications*, Cambridge University Press, Cambridge, 1991.

Whitesides, S. H. (1984), A method for solving certain graph recognition and optimization problems, with applications to perfect graphs, in: *Topics on Perfect Graphs* (C. Berge and V. Chvátal, eds.), *Annals of Discrete Mathematics* 21 (1984) 281–297.

Whitney, H. (1932), Non-separable and planar graphs, *Transactions of the American Mathematical Society* 34 (1932) 339–362.

Whitney, H. (1933a), 2-isomorphic graphs, *American Journal of Mathematics* 55 (1933) 245–254.

Whitney, H. (1933b), Planar graphs, *Fundamenta Mathematicae* 21 (1933) 73–84.

Whitney, H. (1935), On the abstract properties of linear dependence, *American Journal of Mathematics* 57 (1935) 509–533.

Wilson, R. J. (1972), *Introduction to Graph Theory*, Longman Group Limited, London, 1972.

Yannakakis, M. (1985), On a class of totally unimodular matrices, *Mathematics of Operations Research* 10 (1985) 280–304.

Young, H. P. (1971), A quick proof of Wagner's equivalence theorem, *Journal of the London Mathematical Society* 3 (1971) 661–664.

Zaslavsky, T. (1981a), The geometry of root systems and signed graphs, *American Mathematical Monthly* 88 (1981) 88–105.

Zaslavsky, T. (1981b), Characterizations of signed graphs, *Journal of Graph Theory* 5 (1981) 401–406.

Zaslavsky, T. (1982), Signed graphs, *Discrete Applied Mathematics* 4 (1982) 47-74.

Zaslavsky, T. (1987), Balanced decompositions of a signed graph, *Journal of Combinatorial Theory (B)* 43 (1987) 1–13.

Zaslavsky, T. (1989), Biased graphs. I. Bias, balance, and gains, *Journal of Combinatorial Theory (B)* 47 (1989) 32–52.

Zaslavsky, T. (1991), Biased graphs. II. The three matroids, *Journal of Combinatorial Theory (B)* 51 (1991) 46–72.

Ziegler, G. M. (1990), Matroid representations and free arrangements, *Transactions of the American Mathematical Society* 320 (1990) 525–541.

Ziegler, G. M. (1991), Binary supersolvable matroids and modular constructions, *Proceedings of the American Mathematical Society*, to appear.

Author Index

Subject Index